WITHDRAWN
UTSA LIBRARIES

RENEWALS 458-4574

Advanced Materials for Membran Separations

ACS SYMPOSIUM SERIES **876**

Advanced Materials
for Membrane Separations

Ingo Pinnau, Editor
Membrane Technology and Research, Inc.

Benny D. Freeman, Editor
University of Texas at Austin

**Sponsored by the
ACS Division of Polymeric Materials: Science and
Engineering, Inc.**

American Chemical Society, Washington, DC

Library of Congress Cataloging-in-Publication Data

Advanced materials for membrane separations / Ingo Pinnau, editor, Benny D. Freeman, editor.

    p. cm.—(ACS symposium series ; 876)

    "Sponsored by the ACS Division of Polymeric Materials: Science and Engineering, Inc."

    Includes bibliographical references and index.

    ISBN 0–8412–3833–2

    1. Membrane separation—Congresses.

    I. Pinnau, I. (Ingo). II. Freeman, B. D. (Benny D.). III. American Chemical Society. Meeting (222nd : 2001 : Chicago, Ill.) IV. Series.

TP248.25.M46A36   2004
660′.28424—dc22   2003063837

The paper used in this publication meets the minimum requirements of American National Standard for Information Sciences—Permanence of Paper for Printed Library Materials, ANSI Z39.48–1984.

Copyright © 2004 American Chemical Society

Distributed by Oxford University Press

The cover art is courtesy of Jim Rockoff. © Copyright Hydranautics. Used with permission.)

All Rights Reserved. Reprographic copying beyond that permitted by Sections 107 or 108 of the U.S. Copyright Act is allowed for internal use only, provided that a per-chapter fee of $27.25 plus $0.75 per page is paid to the Copyright Clearance Center, Inc., 222 Rosewood Drive, Danvers, MA 01923, USA. Republication or reproduction for sale of pages in this book is permitted only under license from ACS. Direct these and other permission requests to ACS Copyright Office, Publications Division, 1155 16th St., N.W., Washington, DC 20036.

The citation of trade names and/or names of manufacturers in this publication is not to be construed as an endorsement or as approval by ACS of the commercial products or services referenced herein; nor should the mere reference herein to any drawing, specification, chemical process, or other data be regarded as a license or as a conveyance of any right or permission to the holder, reader, or any other person or corporation, to manufacture, reproduce, use, or sell any patented invention or copyrighted work that may in any way be related thereto. Registered names, trademarks, etc., used in this publication, even without specific indication thereof, are not to be considered unprotected by law.

PRINTED IN THE UNITED STATES OF AMERICA

Library
University of Texas
at San Antonio

Foreword

The ACS Symposium Series was first published in 1974 to provide a mechanism for publishing symposia quickly in book form. The purpose of the series is to publish timely, comprehensive books developed from ACS sponsored symposia based on current scientific research. Occasionally, books are developed from symposia sponsored by other organizations when the topic is of keen interest to the chemistry audience.

Before agreeing to publish a book, the proposed table of contents is reviewed for appropriate and comprehensive coverage and for interest to the audience. Some papers may be excluded to better focus the book; others may be added to provide comprehensiveness. When appropriate, overview or introductory chapters are added. Drafts of chapters are peer-reviewed prior to final acceptance or rejection, and manuscripts are prepared in camera-ready format.

As a rule, only original research papers and original review papers are included in the volumes. Verbatim reproductions of previously published papers are not accepted.

ACS Books Department

Contents

Preface...xi

1. **Gas and Liquid Separations Using Membranes: An Overview**.............1
   Benny D. Freeman and Ingo Pinnau

2. **Molecular Simulation of Gas Sorption and Diffusion
   in Silicon-Based Polymers**...24
   J. R. Fried and Bo Li

3. **New Perspectives of Gas Sorption in Solution-Diffusion
   Membranes**...39
   N. F. A. van der Vegt

4. **Solubility of Gases and Vapors in Glassy Polymer Blends**...................55
   F. Grassia, M. Giacinti Baschetti, F. Doghieri, and G. C. Sarti

5. **Predicting Gas Solubility in Glassy Polymers through
   Nonequilibrium EOS**...74
   F. Doghieri, M. Quinzi, D. G. Rethwisch, and G. C. Sarti

6. **Nanostructure of Free Volume in Glassy Polymers as Studied
   by Probe Methods and Computer Simulation**...................................91
   Yu. Yampolskii, V. Shantarovich, D. Hofmann, and M. Heuchel

7. **Fluoropolymer–Hydrocarbon Polymer Composite
   Membranes for Natural Gas Separation**...106
   Rajeev S. Prabhakar and Benny D. Freeman

8. **Structure, Gas Sorption, and Gas Diffusion of a Low-Density
   Liquid Crystalline Polyester**..129
   M. Tsukahara, Y. Tsujita, H. Yoshimizu, and T. Kinoshita

9. **Gas Permeation Properties of Silane-Cross-Linked
   Poly(propylene oxide)**...139
   Kazukiyo Nagai and Tsutomu Nakagawa

10. **Gas Transport Properties of Tetramethyl Polysulfones Containing Trimethylsilyl Group Side Substituents**............154
Michael D. Guiver, Ying Dai, Gilles P. Robertson, Kwi Jong Lee, Jae Young Jho, and Yong Soo Kang

11. **Pure- and Mixed-Gas Permeation Properties of Poly(p-tert-butyl diphenylacetylene)**............167
Ingo Pinnau, Zhenjie He, Toshio Masuda, and Toshikazu Sakaguchi

12. **Membrane Separations Using Functionalized Polyphosphazene Materials**............177
Frederick F. Stewart, Christopher J. Orme, Mason K. Harrup, Robert P. Lash, Don H. Weinkauf, and John D. McCoy

13. **Novel Carbon–Silica Membranes for Improved Gas Separation**......190
Young Moo Lee and Ho Bum Park

14. **Gas Separation and Pervaporation through Microporous Carbon Membranes Derived from Phenolic Resin**............203
Hidetoshi Kita, Koji Nanbu, Hiroshi Maeda, and Ken-ichi Okamoto

15. **Novel Nanostructured Polymer–Inorganic Hybrid Membranes for Vapor–Gas Separation**............218
Zhenjie He, Ingo Pinnau, and Atsushi Morisato

16. **Molecular Chemistry Control of Hybrid Morphology and Transport Properties in Polyimide–Silica Membranes**............234
Chris J. Cornelius, Eva Marand, Pavla Meakin, and Anita J. Hill

17. **Proton Conductivity and Vapor Permeation Properties of Polyimides Containing Sulfonic Acid Groups**............253
Tatsuya Watari, Jianhua Fang, Xiaoxia Guo, Kazuhiro Tanaka, Hidetoshi Kita, and Ken-ichi Okamoto

18. **Formation of Composite Membranes with Ultrathin Skins Using New Methods of Organic Film Formation: Gas-Selective Membranes**............269
Merlin L. Bruening

19. **Surface Modification of Polymeric Membranes by UV Grafting**......281
Doo Hyun Lee, Hyun-Il Kim, and Sung Soo Kim

20. Effect of Microwave Irradiation on CO_2 Permeability in Cellulose Acetate Membranes...................300
Y. Tsujita, Y. Nakai, H. Yoshimizu, and M. Yamauchi

21. Improving the Permselectivity of Commercial Cation-Exchange Membranes for Electrodialysis Applications...................311
Sophie Tan, Alexis Laforgue, and Daniel Bélanger

22. Control of Transport Modes of Ions Using a Temperature-Responsive Charged Membrane...................324
Mitsuru Higa and Tomoko Yamakawa

23. Ion-Exchange Membranes from Blends of Sulfonated Polyphosphazene and Kynar FLEX PVDF...................335
Ryszard Wycisk, Roy Carter, Peter N. Pintauro, and Catherine Byrne

24. Behavior of Absorbed Water in and Oxygen Permeability of Hydrophilic Membranes with Bulky Hydrophobic Side Chain Groups...................352
Shuichi Takahashi, Akiko Saito, and Tsutomu Nakagawa

25. Chemically Modified Polysulfone Hollow Fibers with Zwitterionic Sulfoalkylbetaine Group Having Improved Blood Compatibility...................366
Akon Higuchi, Hirokazu Hashiba, Rika Hayashi, Boo Ok Yoon, Mitsuo Hattori, and Mariko Hara

26. Selective Protein Adsorption and Cell Attachment to Rubbed Fluorinated Polyimides...................383
H. Kawakami, K. Ashiba, and Y. Okuyama

27. Separation of Endocrine Disruptors from Aqueous Solutions by Pervaporation: Relationship between the Separation Factor and Solute Physical Parameters...................394
Boo Ok Yoon, Takao Asano, Kenta Nakaegawa, Mai Ishige, Mariko Hara, and Akon Higuchi

28. Improvement of Selectivities of Microphase-Separated Membranes for the Removal of Volatile Organic Compounds...................411
Tadashi Uragami, Hiroshi Yamada, Terumi Meotoiwa, and Takashi Miyata

Indexes

Author Index..429

Subject Index...431

Preface

This book grew from a symposium at the 222nd National American Chemical Society (ACS) meeting in Chicago, Illinois, August 26–30, 2001. This book was preceded by another membrane-based symposium (and two other Symposium Series Books) held at the Las Vegas, Nevada ACS meeting in 1997. The topics covered in the 2001 ACS symposium summarize significant recent advances in various research areas for development of novel materials used in membrane separations.

A key focus area of the symposium was the preparation and characterization of novel materials for membrane separations. Therefore several chapters report properties on new types of high-performance polymers. Advances in theoretical modeling and computer simulation are covered in this book, and results from these studies are becoming much more common and more consistent with experimental data. Newly discovered physical phenomena, such as the suppression of hydrocarbon sorption in fluoropolymers, are being considered to help solve outstanding separations problems involving plasticization in polymeric membranes. Novel classes of inorganic as well as hybrid inorganic–organic nanostructured membrane materials are now being studied widely, and the topics in this book reflect this trend. As ion transport in membranes plays an increasingly important role in the materials science of fuel cells and solid polymer electrolytes, it is not surprising that several contributions in this area are present. Finally, the large presence and significant utility of membranes in biomedical and biotechnological applications is found in multiple chapters. In summary, this book provides a broad snapshot of the field of advanced materials science research for membranes.

We are very grateful to the authors of the chapters in this book. Without their time, energy, and commitment to excellence, this book would not have been possible. The session organizers at the ACS meeting were also instrumental in the success of the symposium and, ultimately, in the quality of the work appearing in this book. Finally, we gratefully acknowledge the critically important contributions made by our support staff, Sande Storey and Magnus Eunson.

The editors dedicate this book to Professor Tsutomu Nakagawa of Meiji University in Kawasaki City, Japan. His outstanding commitment to membrane science and education is evident in each and every generation of his students and colleagues.

Ingo Pinnau
Membrane Technology and Research, Inc.
1360 Willow Road, Suite 103
Menlo Park, CA 94025
650–328–2228 x 113 (telephone)
650-328–6580 (fax)
ipin@mtrinc.com (email)

Benny D. Freeman
Center for Energy and Environmental Resources
University of Texas at Austin
10100 Burnet Road, Building 133
Austin, TX 78758
512–232–2803 (telephone)
512–232–2807 (fax)
freeman@che.utexas.edu (email)

Advanced Materials for Membrane Separations

Chapter 1

Gas and Liquid Separations Using Membranes: An Overview

Benny D. Freeman[1] and Ingo Pinnau[2]

[1]Department of Chemical Engineering, The University of Texas at Austin, Austin, TX 78712
[2]Membrane Technology and Research, Inc., 1360 Willow Road, Suite 103, Menlo Park, CA 94025

An overview of several aspects of materials science associated with gas and liquid separation membranes is presented. Gas separation materials are divided into three categories depending on whether solubility selectivity, diffusivity selectivity or possibly either dominates the overall selectivity. Examples of separations in each of these categories and potential opportunities for future materials science research in these areas are discussed. Recent developments for more-fouling-resistant membranes for liquid separations are presented. Nonporous thin-film composite ultrafiltration membranes and surface-modified interfacial composite reverse osmosis membranes exhibit significantly improved anti-fouling properties compared to conventional porous UF and RO membranes.

© 2004 American Chemical Society

Polymeric membranes are used widely for both gas and liquid phase separations in industries ranging from the chemical and petrochemical sectors to the food processing, water purification, and medical fields. Important considerations contributing to the widespread use of membranes in these diverse areas include the inherently low energy required by membranes to affect a separation (since no phase change is required in many membrane applications), the simplicity and reliability of membrane processes, and the highly space-efficient nature of membrane systems. This book focuses on recent research advances in the materials science aspects of membranes. Because polymeric membranes provide incomplete separation and because they often have chemical or thermal stability properties that are not sufficient for desired applications, the focus of much of the research described in this book and much of the active research in the field is related to the development of materials that (i) can achieve higher selectivity as well as permeability and (ii) are more resistant to end-use process environments. Additionally, in the liquid separation arena, fouling of membranes is a ubiquitous phenomenon inhibiting the performance and limiting the lifetime of membranes. This overview chapter provides a short summary of the fundamentals of gas and liquids separations using membranes, key scientific and technical issues facing the field, and opportunities for future research.

Gas Separations

Polymer membranes are used industrially for a variety of gas separations, including (i) oxygen and water removal from air to produce high purity, dry nitrogen, (ii) hydrogen removal from mixtures with nitrogen, carbon monoxide or light hydrocarbons, acid gas removal from natural gas, and (iii) separation of organic vapors (*e.g.*, ethylene or propylene) from mixtures with light gases (*e.g.*, nitrogen or air) *(1,2)*. All polymer membranes used in these applications operate according to the so-called solution-diffusion mechanism *(3)*. Within this framework, advanced by Sir Thomas Graham more than a century ago *(4)*, transport of gas molecules from the upstream (i.e., high pressure or high chemical potential) side of a membrane to the downstream (i.e., low pressure or low chemical potential) face of a membrane occurs via a 3-step process: (i) sorption of gas molecules into the upstream (or high pressure) side of the membrane, (ii) diffusion down the concentration gradient, and (iii) desorption from the downstream (or low pressure) face of the membrane. The second step, diffusion through the membrane, is the rate-limiting step, and the rate-limiting step in diffusion is the thermally stimulated molecular motion of the polymer segments, which provides transient gaps in the polymer matrix that permit penetrant diffusion steps to occur. Gas flux is typically expressed in terms of the permeability coefficient of gas A, P_A:

$$P_A \equiv \frac{N_A l}{p_1 - p_2} \qquad (1)$$

where N_A is the steady state flux of species A, l is the membrane thickness, and p_i is the pressure (pure gas) or partial pressure (gas mixture) of component A at the upstream (i=1) or downstream (i=2) face of the membrane. Within the context of the solution-diffusion model and Fick's law and in the limit when the downstream pressure is much less than the upstream pressure, the permeability may be expressed as the product of the gas solubility in the upstream face of the membrane and the average effective gas diffusion coefficient in the membrane: $P_A = S_A x D_A$. The ability of a polymer membrane to separate two gases (A and B) is characterized by the ideal selectivity, α_{AB}:

$$\alpha_{AB} = \frac{P_A}{P_B} = \frac{S_A}{S_B} \times \frac{D_A}{D_B} \qquad (2)$$

where S_A/S_B, the solubility selectivity, is the ratio of the solubilities of the two components, and D_A/D_B, the diffusivity selectivity, is the ratio of the diffusion coefficients of the two gases in the polymer membrane. Component A is the more permeable species.

Because permeability depends on the product of solubility and diffusivity, gas A may be more permeable than gas B in a particular polymer because: (i) A is more soluble in the polymer than B, (ii) A has a higher diffusion coefficient than B, or (iii) both. Similarly, Eq. 2 suggests two routes by which polymer materials design may enhance selectivity: (i) alter solubility selectivity and/or (ii) alter diffusivity selectivity. Broadly speaking, gas separation membrane materials have been selected to separate molecules primarily based on differences in gas molecule diffusivity in a polymer (so-called strongly size-sieving or high diffusivity-selective materials) or on differences in gas solubility in polymers (so-called solubility-selective or weakly size-sieving materials).

In the absence of specific polymer/gas interactions, gas solubility often increases rather regularly with some measure of the tendency of the gas molecule to condense, such as critical temperature, normal boiling point, enthalpy of condensation, etc. Larger, more condensable molecules (*e.g.*, *n*-butane) are more soluble than smaller, less condensable species (*e.g.*, H_2, N_2, etc.), and the increase in solubility with an increase in gas condensability is essentially independent of polymer type. Specific examples of this behavior are given in the chapter by Prabhakar and Freeman. On the other hand, gas diffusion coefficients in polymers decrease rather regularly with an increase in gas molecule size. Moreover, the membrane material can have a very strong influence on the sensitivity of diffusion coefficients to gas molecule size. As shown in Figure 1, rigid, low free volume, glassy polymers such as polysulfone, which is used for air separation and a variety of hydrogen separation applications, have a much stronger dependence of gas diffusion coefficients on gas molecule size than flexible rubbery polymers *(5,6)*. Polysulfone would be an example of a strongly size-sieving material, whereas poly(dimethyl siloxane) is a weakly size-sieving material. The difference in permeability coefficients from one penetrant to another in a given polymer reflects the often conflicting trends between solubility and diffusivity. For this reason, as shown in Figure 2, polysulfone permeability decreases rather monotonically with increasing

4

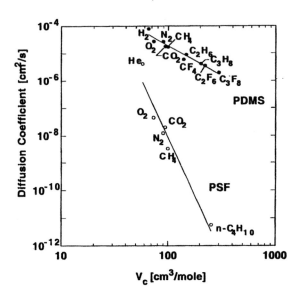

Figure 1. Diffusion coefficients of gases in glassy, rigid polysulfone (PSF) and flexible, rubbery poly(dimethyl siloxane) (PDMS).
(Reproduced from reference 5. Copyright 1999 American Chemical Society.)

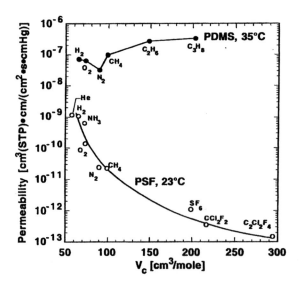

Figure 2. Permeability coefficients of gases in glassy, rigid polysulfone (PSF) and flexible, rubbery poly(dimethyl siloxane) (PDMS).
(Reproduced from reference 14. Copyright 1999 American Chemical Society.)

penetrant size, whereas poly(dimethyl siloxane) is actually more permeable to larger molecules than to smaller molecules *(5)*.

Table I provides a map of various gas separations of interest and the types of membrane materials that might be useful. In each case, membrane materials that are more permeable to A than to B are desired, and the penetrant pairs are listed as A/B. Each of these cases will be discussed in turn.

Case I

Case I involves the selective removal of small molecules (*e.g.*, H_2) from mixtures with larger, but more soluble gases (*e.g.*, N_2, CH_4, or CO). In these situations, solubility selectivity works against high overall selectivity (because light gases such as H_2 are much less soluble than N_2, CH_4, etc.). Hence, membrane materials for these separations must achieve high selectivity based on high diffusivity selectivity. Rigid, glassy amorphous polymers such as polysulfone and aromatic polyamides *(7)*, which have very high diffusivity selectivity, have been commercialized for such separations. The ready availability of such materials and the importance of these separations to the chemical and petrochemical industries made some of these hydrogen separations among the first commercial applications of gas separation membrane technology *(8)*.

Table I. Materials selection for gas separation membranes

Cases	Examples	Membrane material
I: $S_B > S_A$, $D_A \gg D_B$	H_2/N_2, H_2/CO, H_2/CH_4, H_2/PFC, N_2/PFC	Strongly size-sieving
II: $S_A > S_B$, $D_A > D_B$	O_2/N_2, H_2O/Air, CO_2/CH_4, H_2S/CH_4, CO_2/N_2, olefin/paraffin[a]	Potentially either strongly size sieving or strongly solubility-selective
III: $S_A \gg S_B$, $D_B > D_A$	CO_2/H_2, H_2S/H_2, VOC/Air, VOC/N_2, HHC/H_2, HHC/CH_4	Weakly size-sieving (i.e., strongly solubility-selective)

NOTE: PFC=perfluorocarbons (*e.g.*, C_2F_6, C_3F_8, etc.), VOC=volatile organic compounds (*e.g.*, ethylene, propylene, vinyl chloride monomer, etc.), and HHC=higher hydrocarbons (*e.g.*, ethane, propane, butane, etc.). In each case, the objective is to have a membrane be much more permeable to gas A than to gas B. The gas pairs are given as A/B, where A is the more permeable component. [a]Olefin and paraffin (*e.g.*, ethylene and ethane) solubility are quite similar in many polymers. When this separation is discussed, examples will be given where solubility selectivity in favor of the olefin is used as the basis for separation.

The case of H_2 removal from mixtures with light gases, such as N_2, illustrates membrane materials research needs for Case I. Due to the large difference in size between H_2 and N_2, which have kinetic diameters of 2.89 Å and 3.64 Å, respectively, and the availability of very strongly size-sieving glassy polymers, many available membrane materials have selectivities of several hundred for this gas pair (and for the other gas pairs listed in Case I). However, due to process condition constraints (primarily resulting from low pressure ratios being used to recover the H_2 product in the permeate at pressures as high as possible (9)), lower selectivity materials can often be used. This possibility is attractive since materials with lower selectivity often have higher permeability, as shown in Figure 3, and higher permeability would translate to reduced membrane area and, in turn, lower capital costs for a membrane system. However, these processes may operate at pressures in excess of 100 bar feed pressure, and the mechanical properties of any material considered for this application must be sufficient to withstand large transmembrane pressure differences during the lifetime of the membrane (7). Unfortunately, materials with higher diffusivity (and, therefore, permeability) often have less favorable mechanical properties (10), so compromises between pressure resistance and permeability can be necessary. Recent work by Nunes et al. has focused on using silica nanoparticles grown in-situ in a polymer matrix as a route to reinforce the polymer and make the resulting polymer/inorganic hybrid material more resistant to pressure than the polymer alone (11).

Figure 3 illustrates a ubiquitous trait of membrane materials that achieve high selectivity based on high diffusivity selectivity; more permeable materials tend to be less selective, and less permeable materials tend to be more selective. This trend has been widely observed (12) and two theoretical models have been proposed which predict such tradeoff relations for polymer membranes (13,14). In the 1980's and the early part of the 1990's, many research programs focused on synthesizing new classes of polymers with the hope of developing materials with permeability and selectivity values higher than those given by so-called "upper bound" lines such as the one shown in Figure 3. However, after a substantial research effort, it has become clear, from both a practical as well as theoretical viewpoint, that preparing polymers endowed with separation property combinations much better than those predicted by the current upper bound lines will be very challenging.

Much of the new materials research has focused on nonpolymeric materials, such as microporous molecular sieving carbon (15), zeolites (16), ceramics, etc. (17) that obey fundamentally different limitations in separations performance than polymer membranes because the transport mechanism is different from that in polymeric membranes (18). However, such membrane materials are often much more difficult than polymers to process into the thin, high surface area membrane modules required for commercial use, and they are often quite brittle, making them mechanically fragile. Another area of current active research has focused on imbedding highly selective microporous particles into tough polymer matrices as a route to preparing hybrid materials having separation

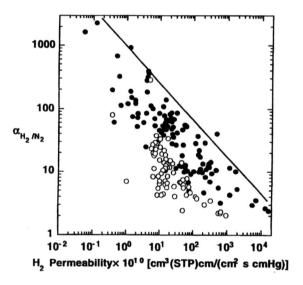

Figure 3. Tradeoff between hydrogen permeability and H_2/N_2 selectivity in glassy (●) and rubbery (o) polymers. The experimental data are from Robeson (12), and the theoretical model line of the upper bound is from Freeman. (Reproduced from reference 13. Copyright 1999 American Chemical Society.)

characteristics of microporous materials and processing/mechanical characteristics of polymers *(19,20)*. Related efforts have focused on growing fine nonporous particles, such as TiO_2, in polymer matrices and using them as crosslink sites to restrict polymer chain mobility and, in turn, improve separation properties *(21)*. To date, none of these materials have been successfully reduced to practice, due in part to significant problems related to polymer/particle interface strength and continuity and difficulties in dispersing fine particles in low free volume polymer matrices.

Case II

Case II in Table I corresponds, in principle, to the most favorable situation from a separations viewpoint. Component A would be both more soluble and more mobile than B in the polymer. Therefore, both solubility selectivity and diffusivity selectivity favor high selectivity of A over B. In principle, either strongly size-sieving or strongly solubility-selective materials could be used for such separations. For oxygen and water removal from air, typically strongly size sieving membrane materials, such as polysulfone and polyimides, are used. For carbon dioxide removal from natural gas, membranes which achieve high overall selectivity based largely on high diffusivity selectivity (*e.g.*, polyimides *(22,23)*) and based on high solubility selectivity (*e.g.*, cellulose acetate *(24)*) have been deployed commercially.

In cases where gas molecules such as O_2, N_2, etc. sorb to very low levels in the hydrophobic, glassy polymers that are commonly used, the separation properties of the polymers may change little with increasing feed pressure or, equivalently, with increasing dissolved gas concentration in the polymer. The situation is, however, quite different when the gas molecules can sorb extensively into the polymers, as in the case of CO_2 removal from natural gas. In this situation, high concentrations of gas molecules in the polymer increase the polymer free volume and molecular mobility; both of these factors decrease diffusivity selectivity and, in turn, overall selectivity. This phenomenon, called plasticization, and its deleterious effects on separation performance is described in more detail in the chapter by Prabhakar and Freeman. If a membrane material derives high overall selectivity from high diffusivity selectivity, precipitous decreases in selectivity may be observed when the polymer sorbs large amounts of gas, which may render the material unsuitable as a separation membrane *(23)*.

The issue of controlling plasticization and its impact on membrane selectivity is perhaps the most significant materials science challenge facing the use of membranes for removal of acid gases (*e.g.*, CO_2 and H_2S) from natural gas and for organic/organic (*e.g.*, propylene/propane) separations. Several approaches are being explored, and Table II provides a snapshot of materials science approaches to preparing membranes for olefin/paraffin separation. This separation is not currently practiced using membranes but, due to the importance of olefin/paraffin separation and the extremely energy intensive distillation technology currently used for this separation *(25)*, there has been significant interest in exploring alternative separation technologies, such as membranes.

9

Table II. Materials science approaches to olefin/paraffin membranes

Approach	Example Material	C_3H_6 Flux (GPU)	C_3H_6/C_3H_8 Selectivity	Comments
Strongly size-sieving polymers	6FDA/6FpDA Polyimide (26)	7[a]	7	Susceptible to plasticization and loss of selectivity
Molecular sieving carbon	Carbonized polyimide (27)	19[b]	13	Fragile, fouled by condensable contaminants
Reduce sorption of hydrocarbons to control plasticization	Hyflon AD 60 (28)	19[c]	3.3	Stable but has low selectivity
Facilitated transport (i.e. high solubility selectivity)	Poly(ethylene oxide) blended with 80 wt.% $AgBF_4$ (29)	49[d]	45	Easily poisoned by H_2S, can form silver acetylide, which is explosive (30)

NOTE: 1 GPU = 10^{-6} cm^3(STP)/(cm^2·s·cm Hg).

[a] These data were obtained using a mixed gas feed of propylene and propane in a ratio of 50:50. The temperature was 35°C, the feed pressure was 4.04 atm, and the permeate was under vacuum. The permeability data in the reference were converted to flux by assuming an effective thickness of 1,000 Å, which is typical of modern gas separation membranes.

[b] The data were obtained at 35°C using a 50:50 mixture of propylene and propane at a total feed pressure of 1 atm, and the permeate pressure was vacuum.

[c] The data were obtained using a 60:40 propylene:propane mixture at 30°C. The feed pressure was 11.2 atm, and the permeate pressure was 1 atm. Hyflon AD 60 is a copolymer prepared from 40% tetrafluoroethylene and 60% 2,2,4-trifluoro-5-trifluoromethoxy-1,3-dioxole.

[d] The data were obtained at 24°C using a 50:50 mixture of propylene and propane at a total feed pressure of 4.4 atm and a permeate pressure of 1 atm.

The difficulty of this separation derives from the fact that (i) olefins and their paraffin analogs have very similar boiling points, which makes these components have very similar solubilities in most polymers, and (ii) olefins and paraffins are similar in size, which limits diffusivity selectivity.

Very rigid, extremely size-sieving polymers have been explored for olefin/paraffin separation. The objective of such materials is to perform the separation based on the fact that olefins, such as propylene, are slightly smaller than their paraffin analogs. For example, propylene has a Lennard-Jones diameter of 4.68 Å while that of propane is 5.06 Å (31). One example of this approach is presented in Table II. Generally, in such strongly size-sieving materials, permeability coefficients (and, therefore, flux), particularly of larger organic species, are low and selectivity values are modest. Moreover, such materials are often very sensitive to plasticization when exposed to high-pressure mixtures of olefin and paraffin gases, which can result in a sharp decrease in selectivity.

Molecular-sieving carbon membranes, obtained by pyrolysis of polymer membranes, are also being explored for olefin/paraffin separation (27). These materials are not susceptible to plasticization like polymer membranes, because they do not swell by sorption of gases, but they can be susceptible to plugging or fouling by larger, more condensable contaminant molecules. Additionally, such membranes are typically very brittle and can be quite fragile and difficult to prepare in high surface area membrane modules. Nevertheless, higher selectivity than can be obtained even with strongly size-sieving polymers may be possible using carbon membranes.

Recently, fluoropolymers have been shown to exhibit significantly reduced solubility to hydrocarbon gases than expected based on the molecular characteristics (i.e., condensability and size) of the hydrocarbon gases (28,32,33). This feature of fluorinated polymer matrices is being explored as a way to suppress hydrocarbon-induced plasticization of membrane materials. Table II presents an example of this approach. The fluoropolymer showed good resistance to plasticization, but selectivity was low. However, the materials whose properties have been reported have not been optimized for gas separations. Systematic structure/property studies could lead to much better property profiles for these materials.

Olefins have double bonds, whereas paraffins are saturated hydrocarbons, and there is a long history of preparing membranes based on the concept of using the interaction between, for example, the double bond in olefin molecules and Group I-B metal cations, such as silver ions, to enhance membrane separation properties (34,35). Such facilitated transport membranes represent an extreme example of enhancing solubility selectivity as a means to improve overall selectivity. Often, supported liquid membranes (35,36), highly hydrated cation exchange membranes (37), or water-swollen polymer membranes (38) have been studied for this application. However, these membranes must be continually hydrated in order to function; if the membrane becomes desiccated, the silver cannot dissociate from its anion sufficiently to permit significant interactions with olefin molecules, and the facilitated transport effect is not observed. One recent approach to overcoming the hydration problem has been to dissolve the silver salt in a polymer, such as poly(ethylene oxide), which can form a solid solution of the silver salt in the polymer (*29*). This approach is similar to that used for solid polymer electrolytes in polymer-based batteries (*39*). As indicated in Table II, materials based on such concepts can exhibit good flux and high selectivity. However, for solid polymer electrolyte membranes based on silver salts, there are drawbacks. Silver salts may be poisoned by exposure to H_2S, and some olefin/paraffin streams contain trace levels of H_2S. Also, these streams may contain low levels of triple bonded contaminants (acetylenes), and silver salts can react with acetylene compounds to form silver acetylide, which is explosive (*30,40*). A research need in this area is the development of new olefin carriers not based on silver salts or other metals that do not exibit good long term stability.

Case III

The third case in Table I represents separations in which it is desirable for the more permeable component to be the larger component in the mixture. Diffusivity selectivity will work against such separations, so they are only feasible if membrane materials with sufficiently high solubility selectivity are used. These separations can be divided crudely into two categories: (i) those for which polymers undergo specific favorable interactions with component A (i.e., the more permeable component) but not with B in order to increase the solubility selectivity in favor of A, and (ii) those separations (*e.g.*, HHC/methane) where one cannot readily engineer interactions into the polymer membrane to favor sorption of the desired component. In this case, one can modify the polymer membrane to minimize diffusivity selectivity, thereby allowing the overall selectivity to approach the solubility selectivity.

The selective removal of CO_2 from mixtures with H_2 serves as an example of a separation where a membrane polymer can be modified so that it exhibits a favorable interaction with quadrupolar CO_2 but not with H_2 in order to increase the CO_2/H_2 solubility selectivity and overall selectivity. This separation is of potential interest in the purification of hydrogen produced from hydrocarbon reforming reactions (*41*). Table III presents a comparison of properties of various weakly size-sieving materials for this separation.

PDMS is included in Table III not because it enjoys special interactions with CO_2 but because it is the most widely used commercial membrane material for solubility-selective separations (*e.g.*, ethylene or propylene removal from nitrogen (*42*) or VOC removal from air gases (*2,43*)), and as such, it provides a useful reference point. Despite being a very weakly size-sieving rubber, it has low selectivity for this gas pair. If a rubbery polymer containing polar ether units is considered (*i.e.*, PEO), then selectivity increases by almost a factor of two. This material is not very permeable because it is highly crystalline, and crystalline order reduces permeability. The crystallinity in PEO can be disrupted by preparing heterophase block copolymers of PEO with, for example, polyamides. This material is an order of magnitude more permeable than PEO and has somewhat higher selectivity. The interaction of the membrane matrix can be increased further by considering materials such as the polyelectrolyte PVBTAF. In the presence of water, the F$^-$ anion in this polymer undergoes a reversible chemical reaction with CO_2 and water to form HCO_3^-, which acts as a facilitated transport carrier for CO_2 (44). This approach gives very high selectivity. Similarly, hydrophilic polymers, such as poly(vinyl alcohol) can be blended with amino acid salts and, in the presence of water, these materials behave as facilitated transport membranes for CO_2. In both of the facilitated transport examples, water is required on both the feed and permeate side of the membrane for high selectivity to be observed, probably as a consequence of the active participation of water in the chemical reaction that enhances CO_2 solubility. This need to maintain the membrane in a humidified state could be a drawback in the use of such materials, and it would be a good step forward in this field if stable carriers could be identified for CO_2 that exhibited less dependence on feed and permeate stream humidity levels. Additionally, if such membranes were to be used for syngas purification, it could be useful to have membranes that could operate at high temperature. With polymeric membranes, this appears to be a strong challenge for materials design. Many polymers are not stable at temperatures above 200°C, and polymeric materials tested to date lose most or all of their ability to separate CO_2 from H_2 at temperatures in excess of 150°C (41).

Table III. Properties of membrane materials for CO_2/H_2 separations

Material	CO_2 Permeability (Barrer)	CO_2/H_2 selectivity
Poly(dimethyl siloxane) PDMS (45)	3,160	3.6
Poly(ethylene oxide) (PEO) (46)	12	7
PEO-polyamide block copolymer (47)	120	10
Poly(vinylbenzyltrimethyl ammonium fluoride) (PVBTAF) (48)	3-6 GPU[a]	40-90
Poly(vinyl alcohol) blended with 50 wt.% glycine ethylene diamine salt (49)	211	30

NOTE: 1 Barrer=10^{-10} cm^3(STP)cm/(cm^2 s cm Hg)

[a] These data were obtained at 23°C using composite membranes, so the permeation data are reported in pressure-normalized flux, GPU, rather than in Barrer. For comparison, a typical commercial gas separation membrane might have an effective thickness of the order of 1,000 Å; hence, a membrane with a permeability of 1 Barrer would exhibit a pressure-normalized flux of 10 GPU. A range of CO_2 flux and CO_2/H_2 selectivity values are cited because the CO_2 flux increases and CO_2/H_2 selectivity increases as CO_2 feed pressure is reduced from 1.9 atm to 0.4 atm.

For separations such as the removal of higher hydrocarbons (so-called natural gas liquids) from natural gas for dew point and heating value control *(2)*, it is not obvious that one can alter the polymer structure to improve the solubility of a higher hydrocarbon (*e.g.*, *n*-butane) relative to that of a lighter hydrocarbon (*e.g.*, methane). In these cases, solubility selectivity strongly favors the higher hydrocarbon over the light hydrocarbon, whereas diffusivity selectivity favors the light hydrocarbon. One strategy to improve overall selectivity is to alter the polymer structure to make the diffusion coefficients of both the large and small gas molecules as similar as possible in order to bring the diffusivity selectivity as close to one as possible while still retaining the favorable solubility-selectivity characteristics. In effect, it is highly favorable to weaken the size-sieving ability of the membrane material. One path towards this end is to use flexible rubbery polymers, which have an inherently weak size-sieving ability due to their facile segmental motion. In this regard, the polymer used commercially in these applications, PDMS, has the lowest glass transition temperature of all known polymers and very low barriers to rotation about the main chain bonds. From this point of view, it is not obvious that one can prepare a rubbery polymer membrane that is significantly more flexible and, therefore, less size sieving, than PDMS.

However, there are other routes to highly solubility-selective materials that are much more permeable to, for example, *n*-butane than to methane. The most permeable polymers known are actually glassy polymers endowed with extremely high levels of free volume. The best-known of these materials is poly(1-trimethylsilyl-1-propyne) (PTMSP) *(50-52)*. However, many substituted acetylene polymers are more permeable to larger, condensable molecules than to smaller, less soluble gases *(50)*. These materials apparently derive much of their very interesting transport properties from their high free volume. The effect of free volume on diffusion coefficients is often modeled using the Cohen-Turnbull equation *(53)*:

$$D_A = D_o \exp\left(-Bv^*/V_f\right) \tag{3}$$

where D_o and B are empirical constants, v^* is proportional to gas molecule size, and V_f is the average size of a free volume element in the membrane material. Eq. 3 predicts that gas diffusion coefficients will decrease rapidly as penetrant size increases but will increase as free volume in the membrane material increases. Eq. 3 also shows that the sensitivity of gas diffusion coefficients to free volume should decrease as free volume increases. This is indeed found to be the case. In substituted acetylene polymers, such as poly(4-methyl-2-pentyne) (PMP), the addition of nonporous, nanoscale fumed silica particles disrupts polymer chain packing and increases free volume in the polymer matrix which, in turn, increases diffusion coefficients and permeability *(54-56)*. The increase in free volume with increasing concentration of fumed silica particles was confirmed by positron annihilation lifetime spectroscopy *(56, 57)* and by ^{129}Xe NMR measurements *(58)*. The enhancement in diffusion coefficients with increasing free volume is also accompanied by a weakening of the size-sieving ability of the polymer/silica composite materials, which results in a simultaneous increase in both permeability and n-butane/methane selectivity, as shown in Figure 4. Data for PDMS are shown for comparison, and it is clear that these composite materials, made from rigid polymers blended with nanoscale silica particles, exhibit markedly better transport characteristics than PDMS.

The shape of the tradeoff curve in Figure 4 is typical for separations in which one seeks to selectively permeate the larger, more soluble component (6). In such cases, permeability generally increases as a result of increasing diffusion coefficients and, often, as diffusion coefficients increase, the size-sieving ability of the membrane material decreases (consistent with Eq. 3), so overall selectivity also increases. In the area of solubility-selective polymers, much work remains to be done to optimize polymer or polymer-based composite structures to enhance separation characteristics. As in the other cases considered, the ability of membrane materials to perform reliably over long time periods in complex, chemically and/or thermally challenging environments is an excellent area for future research.

Liquid Separations

Membrane-based processes for liquid separations include microfiltration (MF), ultrafiltration (UF), nanofiltration (NF), and reverse osmosis (RO). Membranes used for MF and UF applications are typically porous and separation is achieved by a size-sieving mechanism, whereas NF and RO membranes have an essentially non-porous separating layer and separation is governed by a solution/diffusion mechanism. Most current research for development of improved membranes for liquid separations involves surface modifications to provide more fouling-resistant membranes. For example, membrane fouling is the principal problem inhibiting the widespread use of UF, NF, and RO membranes for industrial wastewater treatment. The two types of fouling, surface and internal, that occur in conventional porous membranes are illustrated in Figure 5.

Surface fouling is caused by deposition of a precipitated layer of particulates on the membrane surface. The equilibrium thickness of this layer is generally controlled by the turbulence of the fluid solution flowing across the membrane. Several studies have shown a close relationship between the level of fouling, and hence the water flux, and the turbulence of the feed solution (59,60). Of the several other parameters that affect fouling of UF, NF, and RO membranes, membrane material and the surface structure play important roles (61-63). Adhesion of fouling matter to the membrane surface is determined, in part, by the membrane material and surface roughness. Hydrophobic materials, such as polysulfone, foul very easily, whereas hydrophilic materials, such as cellulose or cellulose acetate, are much more fouling resistant.

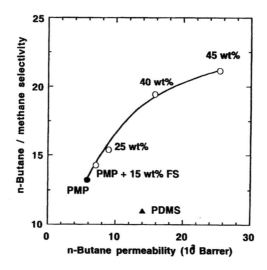

Figure 4. n-Butane permeability and n-butane/methane selectivity in poly(4-methyl-2-pentyne) (PMP) and nanocomposites of PMP and nanoscale fumed silica (FS) (54-56). Data for poly(dimethyl siloxane) (PDMS) are also included for comparison. These mixed gas data were determined using a feed mixture of 2 volume % n-butane in methane. The feed pressure was 11.2 atm, and the permeate pressure was atmospheric.

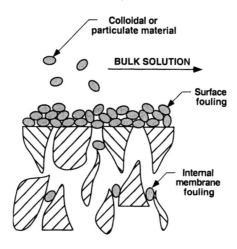

Figure 5. Schematic representation of fouling mechanisms of porous ultrafiltration membranes. Surface fouling is the deposition of solid material on the membrane surface. This fouling layer can be controlled by high turbulence, regular cleaning, and using hydrophilic membrane materials to minimize adhesion to the membrane surface. Surface fouling is generally reversible. Internal fouling is caused by penetration of solid material into the membrane, which results in plugging of the pores. Internal membrane fouling is generally irreversible.

The second type of fouling illustrated in Figure 5 is internal membrane fouling. Current UF, NF, and RO membranes have a porous surface, and fouling can be caused by penetration of solutes or colloidal particles into the interior of the membrane. Because particles accumulate and eventually cannot be removed, this type of fouling is essentially irreversible even after prolonged backflushing and chemical cleaning of the membrane.

Methods to provide more fouling-resistant UF membranes include surface modifications (64-68) and non-porous surface coatings (69-71). Of these approaches, non-porous thin-film composite UF membranes appear to be the most promising for reduction of membrane fouling, because internal membrane fouling can be completely eliminated. Non-porous UF membranes consist of two layers: (a) a highly permeable, porous support and (b) a thin, non-porous selective layer. Materials for the support include, for example, porous polysulfone, polyethersulfone, poly(vinylidene fluoride) and porous inorganic membranes. The selective, fouling-resistant coating layer material should be highly hydrophilic to provide high water flux and be neutral to minimize any solute/membrane surface interactions. Highly water-sorbing materials, such as poly(vinyl alcohol) or polyethers, have been the preferred choice for non-porous thin-film composite UF membranes. Peinemann et al. developed nonporous UF membranes based on a porous poly(vinylidene fluoride) support coated with a thin (0.2-0.5 μm) polyether-polyamide block copolymer (69). These nonporous UF membranes have much lower pure water fluxes compared to conventional porous hydrophobic UF membranes. However, when tested with oil/water emulsions, the porous UF membranes foul significantly and have lower water fluxes compared to those of non-porous UF membranes.

An example of this behavior is shown in Figure 6 (72). Several long-term module tests were performed with a motor oil/water emulsion using a microporous poly(vinylidene fluoride) [PVDF] UF spiral-wound membrane module and a module containing a PVDF membrane coated with a thin, non-porous layer of a polyamide-12/poly(ethylene oxide) block copolymer. Initially, the porous PVDF had a much higher water flux (400 $L/m^2 \cdot h$ at 150 psig) than the coated membrane (50 $L/m^2 \cdot h$ at 150 psig), but over a period of 22 days this flux declined about 40-fold to about 12 $L/m^2 \cdot h$. In contrast, the water flux of the coated membrane was almost completely retained for the entire 22-day test. After the test period, both membranes were regenerated by flushing the test system with clean water; no chemical cleaning agents were used. The porous PVDF membrane only partially regained its original flux, showing that a large fraction of the flux decline was due to permanent internal membrane fouling. The flux of the coated, non-porous membrane, however, returned to its original value. When both membranes were retested with the motor oil/water emulsion, the flux of the porous PVDF membrane declined quickly, whereas the coated membrane maintained its previous high value. Non-porous UF membranes are currently being tested for oil/water separations for shipboard treatment of bilgewater.

Current membranes used in reverse osmosis applications also exhibit significant fouling problems which limits their more widespread use in challenging applications, such as wastewater treatment. RO membrane systems typically require MF and UF pretreatment to remove suspended solids, oil, organic solvents, charged surfactants, charged colloids, and micro-organisms.

The best RO membranes for desalination of brackish water and seawater are highly crosslinked aromatic polyamide or poly(piperazineamide) thin-film composite membranes made by interfacial polymerization (73). These composite membranes consist of three layers: (i) a microporous support; (ii) an ultrathin (~ 0.1-0.2 μm-thick), dense, selective polymer layer; and (iii) a thin porous surface layer.

The surface structure of a typical aromatic polyamide RO membrane (SWC-1, Hydranautics, Oceanside, CA) is shown in Figure 7. The surface of the membrane is surprisingly porous. The membrane has a "ridge-and-valley" structure with a surface pore size in the range of 0.1-0.5 μm. This membrane fouls very easily as solutes and particulates pass through the surface pores into the internal membrane structure. Because interfacially polymerized RO composite membranes typically contain either an anionic or a cationic charge, additional fouling results from the interactions of the charged membrane material with colloids and surfactants (74,75).

Recent work on the development of more fouling resistant reverse osmosis membranes has focused on surface modifications of commercial RO membranes (76-80). One example demonstrating the utility of this approach is shown in Figure 8. A commercial RO membrane (SWC-1; Hydranautics, Oceanside, CA) was coated with a thin layer of a polyamide-6/polyethylene oxide block copolymer (Pebax 1657). The coated RO membrane had lower initial water flux but provided significant protection from membrane fouling. In crossflow permeation tests using an aqueous feed containing 1,000 ppm mineral oil stabilized with a neutral silicone-based surfactant, the flux of uncoated RO membranes decreased significantly over a period of four weeks. In the case of the uncoated seawater desalination membrane, the water flux declined to almost zero. In contrast, the flux of the coated membrane was essentially constant over the entire test period. After four weeks, the membranes were flushed with clean water for 24 hours and then re-tested with the oil/surfactant/water emulsion. The fluxes of the uncoated SWC-1 membrane remained very low, indicating severe internal fouling. On the other hand, the water flux of the coated membrane was restored to its initial value.

The reason for the significantly improved fouling resistant of the coated membranes is illustrated in Figure 9. As shown by the scanning electron photomicrographs, the uncoated SWC-1 membrane has a very rough and porous surface, typical of an interfacial composite membrane made from monomeric amines (73). Such a surface provides many cavities into which fouling material can easily penetrate and lodge. The coated RO membrane, in contrast, has a smooth and completely non-porous surface. Fouling materials on this surface are easily removed by the tangential flow of the feed solution. Future research in

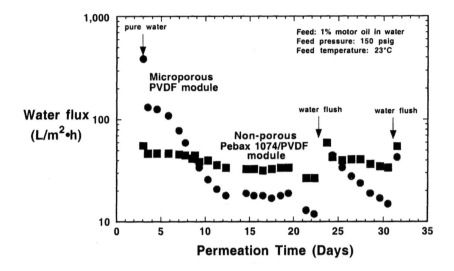

Figure 6. Water fluxes of porous PVDF and non-porous PVDF/Pebax 1074 (polyamide-12/polyethylene oxide block copolymer) spiral-wound membrane modules as a function of operating time. Feed: 1% motor oil/1,000 ppm neutral surfactant in water; feed pressure: 150 psig; T=23°C.

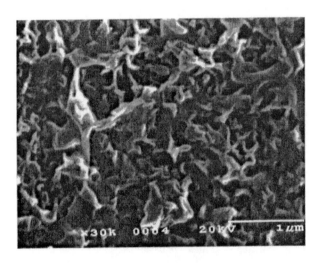

Figure 7. Scanning electron micropgraph of the surface structure of a commercial high-pressure RO membrane (SWC-1, Hydranautics, Oceanside, CA).

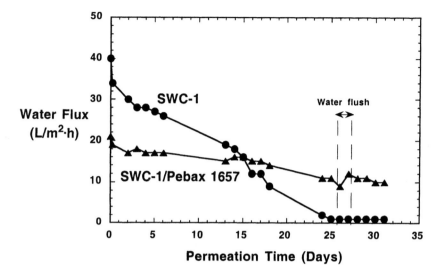

Figure 8. Water fluxes of a commercial RO membrane (SWC-1, Hydranautics, Oceanside, CA) and a SWC 1/Pebax 1657 (polyamide-6/polyethylene oxide block copolymer) composite membrane as a function of operating time. Feed: 1,000 ppm mineral oil/100 ppm neutral surfactant in water; feed pressure: 500 psig; T=23°C.

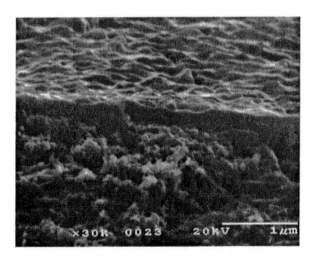

Figure 9. Scanning electron micropgraph of the cross-section and surface of a non-porous SWC-1/Pebax 1657 RO composite membrane.

20

surface-coated RO membranes should focus on optimization of the coating thickness and selection of the coating layer material.

Conclusions

Materials science plays an important role in enhancing the performance of membranes for gas and liquid separation. In gas separations, there is a great need for materials that are stable in the process environment where membranes are applied. More plasticization-resistant materials are needed in many separations where strongly size-sieving materials are being used or considered for use. More selective, stable materials are needed in separations that rely on high solubility selectivity (or weak size-sieving ability) to affect separations.

Current commercial ultrafiltration, nanofiltration and reverse osmosis membranes exhibit excellent water flux and solute rejection. However, most membrane types suffer from severe fouling when operated in challenging environments. To further broaden the use of membranes in liquid separation processes, more-fouling-resistant membranes are required.

Acknowledgments

We gratefully acknowledge partial support of this work by the Office of Naval Research.

References

1. Baker, R. W. *Membrane Technology and Applications* McGraw Hill: New York, 2000.
2. Baker, R. W. *Ind. Eng. Chem. Res.* **2002**, *41*, 1393-1411.
3. Wijmans, J. G.; Baker, R. W. *J. Membrane Sci.* **1995**, *107*, 1-21.
4. Graham, T. *Phil. Mag.* **1866**, *32*, 401-420.
5. Freeman, B. D.; Pinnau, I. In *Polymeric Membranes for Gas and Vapor Separations: Chemistry and Materials Science, ACS Symposium Series Number 733*; Freeman, B. D.; Pinnau, I., Eds.; American Chemical Society: Washington, DC, 1999; pp 1-27.
6. Freeman, B. D.; Pinnau, I. *Trends in Polymer Science* **1997**, *5*, 167-173.
7. Ekiner, O. M.; Vassilatos, G. *J. Membrane Sci.* **1990**, *53*, 259-273.
8. Lonsdale, H. K. *J. Membrane Sci.* **1982**, *10*, 81-181.
9. Coker, D. T.; Freeman, B. D.; Fleming, G. K. *AIChE J.* **1998**, *44*, 1289-1302.

10. Hirayama, Y.; Yoshinaga, T.; Nakanishi, S.; Kusuki, Y. In *Polymeric Membranes for Gas and Vapor Separations: Chemistry and Materials Science, ACS Symposium Series Number 733*; Freeman, B. D.; Pinnau, I., Eds.; American Chemical Society: Washington, DC, 1999; pp 194-214.
11. Nunes, S. P.; Peinemann, K. V.; Ohlrogge, K.; Alpers, A.; Keller, M.; Pires, A. T. N. *J. Membrane Sci.* **1999**, *157*, 219-226.
12. Robeson, L. M. *J. Membrane Sci.* **1991**, *62*, 165-185.
13. Freeman, B. D. *Macromolecules* **1999**, *32*, 375-380.
14. Alentiev, A. Y.; Yampolskii, Y. P. *J. Membrane Sci.* **2000**, *165*, 201-216.
15. Ismail, A. F.; David, L. I. B. *J. Membrane Sci.* **2001**, *193*, 1-18.
16. Caro, J.; Noack, M.; Kölsch, P.; Schäfer, R. *Microporous & Mesoporous Mat.* **2000**, *38*, 3-24.
17. Lin, Y. S.; Kumakiri, I.; Nair, B. N.; Alsyouri, H. *Sep. Purif. Methods* **2002**, *31*, 229-379.
18. Shelekhin, A. B.; Dixon, A. G.; Ma, Y. H. *AIChE J.* **1995**, *41*, 58-67.
19. Mahajan, R.; Koros, W. J. *Ind. Eng. Chem. Res.* **2000**, *39*, 2692-2696.
20. Pechar, T. W.; Tsapatsis, M.; Marand, E.; Davis, R. *Desalination* **2002**, *146*, 3-9.
21. Cornelius, C.; Hibshman, C.; Marand, E. *Sep. Purif. Tech.* **2001**, *25*, 181-193.
22. Bos, A.; Punt, I.; Strathmann, H.; Wessling, M. *AIChE J.* **2001**, *47*, 1088-1093.
23. White, L. S.; Blinka, T. A.; Kloczewski, H. A.; Wang, I. *J. Membrane Sci.* **1995**, *103*, 73-82.
24. Lee, S. Y.; Minhas, B. S.; Donohue, M. D. *AIChE Symp. Ser.* **1988**, *84*, 93-101.
25. Eldridge, B. R. *Ind. Eng. Chem. Res.* **1993**, *32*, 2208-2212.
26. Staudt-Bickel, C.; Koros, W. J. *J. Membrane Sci.* **2000**, *170*, 205-214.
27. Okamoto, K.-I.; Kawamura, S.; Yoshino, M.; Kita, H.; Hirayama, Y.; Tanihara, N.; Kusuki, Y. *Ind. Eng. Chem. Res.* **1999**, *38*, 4424-4432.
28. Pinnau, I.; He, Z.; DaCosta, A. R.; Amo, K. D.; Daniels, R. *U.S. Patent 6,361,583*, **2002**.
29. Pinnau, I.; Toy, L. G.; Casillas, C. *U.S. Patent 5,670,051*, **1997**.
30. Stettbacher, A. *Nitrocellulose* **1940**, *11*, 227-229.
31. Tanaka, K.; Taguchi, A.; Hao, J.; Kita, H.; Okamoto, K. *J. Membrane Sci.* **1996**, *121*, 197-207.
32. Merkel, T. C.; Bondar, V.; Nagai, K.; Freeman, B. D. *Macromolecules* **1999**, *32*, 370-374.
33. Prabhakar, R.; Freeman, B. *Desalination* **2002**, *144*, 79-83.
34. Hughes, R. D.; Mahoney, J. A.; Steigelmann, E. F. In *Recent Developments in Separation Science Vol. 9*, Li, N.N.; Calo, J.M. (Eds.), CRC Press: Boca Raton, FL, 1986; pp. 173-195.
35. Teramoto, M.; Matsuyama, H.; Yamshiro, T.; Katayama, Y. *J. Chem. Eng. Jpn.* **1986**, *19*, 419-424.

36. Tsou, D. T.; Blachman, W. M.; Davis, J. C. *Ind. Eng. Chem. Res.* **1994**, *33*, 3209-3216.
37. Richter, C.; Woermann, D. *J. Membrane Sci.* **1994**, *91*, 217-229.
38. Ho, W. S.; Dalrymple, D. C. *J. Membrane Sci.* **1994**, *91*, 13-25.
39. Gray, F. M. *Solid Polymer Electrolytes: Fundamentals and Technological Applications*, VCH Publishers, Inc.: New York, 1991.
40. Cotton, F. A.; Wilkinson, G.; Murillo, C. A.; Bochmann, M. *Advanced Inorganic Chemistry*, 6th Ed.; John Wiley & Sons, Inc.: New York, 1999.
41. Merkel, T. C.; Gupta, R.; Turk, B.; Freeman, B. D. *J. Membrane Sci.* **2001**, *191*, 85-94.
42. Baker, R. W.; Wijmans, J. G.; Kaschemekat, J. H. *J. Membrane Sci.* **1998**, *151*, 55-62.
43. Wang, X.; Daniels, R.; Baker, R. W. *AIChE J.* **2001**, *47*, 1094-1100.
44. Quinn, R.; Laciak, D. V.; Pez, G. P. *J. Membrane Sci.* **1997**, *131*, 61-69.
45. Merkel, T. C.; Bondar, V. I.; Nagai, K.; Freeman, B. D. *J. Polym. Sci.: Polym. Phys. Ed.* **2000**, *38*, 415-434.
46. Lin, H.; Freeman, B. D. *J. Membrane Sci.* **in press**.
47. Bondar, V. I.; Freeman, B. D.; Pinnau, I. *J. Polym. Phys., Part B, Polym. Phys.* **2000**, *38*, 2051-2062.
48. Quinn, R.; Laciak, D. V. *J. Membrane Sci.* **1997**, *131*, 49-60.
49. Ho, W. S. W. *U.S. Patent 5,611,843*, **1997**.
50. Nagai, K.; Masuda, T.; Nakagawa, T.; Freeman, B. D.; Pinnau, I. *Prog. Polym. Sci.* **2001**, *26*, 721-798.
51. Pinnau, I.; Toy, L. G. *J. Membrane Sci.* **1996**, *116*, 199-209.
52. Pinnau, I.; Casillas, C. G.; Morisato, A.; Freeman, B. D. *J. Polym. Sci.: Polym. Phys. Ed.* **1996**, *34*, 2613-2621.
53. Cohen, M. H.; Turnbull, D. *J. Chem. Phys.* **1959**, *31*, 1164-1169.
54. Pinnau, I; He, Z. *U.S. Patent 6,316,684*, **2001**.
55. Merkel, T. C.; Freeman, B. D.; He, Z.; Morisato, A.; Pinnau, I. *Proceedings of the American Chemical Society Division of Polymeric Materials: Science and Engineering* **2001**, *85*, 301-302.
56. Merkel, T. C.; Freeman, B. D.; Spontak, R. J.; He, Z.; Pinnau, I.; Meakin, P.; Hill, A. J. *Science* **2002**, *296*, 519-522.
57. Merkel, T. C.; Freeman, B. D.; Spontak, R. J.; He, Z.; Pinnau, I.; Meakin, P.; Hill, A. J. *Chem. Mater.* **2003**, *15*, 109-123.
58. Merkel, T. C.; Toy, L. G.; Andrady, A. L.; Gracz, H.; Stejskal, E. O. *Macromolecules* **2003**, *36*, 353-358.
59. Belfort, G. In *Synthetic Membrane Processes*, Belfort, G. (Ed.) Academic Press, Orlando, FL, 1984, pp. 1-19.
60. Cheryan, M. *Ultrafiltration Handbook*, Technomic Publishing Co., Lancaster, PA, 1986.
61. Fane, A.G.; Fell, C.J.D.; Waters, A.G. *J. Membrane Sci.* **1981**, *9*, 245-262.
62. Nilson, J.L. *J. Membrane Sci.* **1988**, *36*, 147-160.
63. Lipp, P.; Lee, C.H.; Fane, A.G.; Fell, C.J.D. *J. Membrane Sci.* **1988**, *36*, 161-177.

64. Hvid, K.B.; Nielsen, P.S.; Stengaard, F.F. *J. Membrane Sci.* **1990**, *53*, 189-202.
65. Nyström, M.; Järvinen, P. *J. Membrane Sci.* **1991**, *60*, 275-296.
66. Ulbricht, M.; Belfort, G. *J. Membrane Sci.* **1996**, *111*, 193-215.
67. Pieracci, J.; Crivello, J.V.; Belfort, G. *J. Membrane Sci.* **1999**, *156*, 233-240.
68. Reddy, A.V.R.; Mohan, D.J.; Bhattacharya, A.; Shah, V.J.; Ghosh, P.K. *J. Membrane Sci.* **2003**, *214*, 211-221.
69. Nunes, S.P.; Sforça, M.L.; Peinemann, K.-V. *J. Membrane Sci.* **1995**, *106*, 49-56.
70. Li, R.H.; Barbari, T.A. *J. Membrane Sci.* **1995**, *105*, 71-78.
71. Dai, W.S.; Barbari, T.A. *J. Membrane Sci.* **1999**, *156*, 67-79.
72. Pinnau, I.; Ly, J.H.; Baker, R.W.; Freeman, B.D. unpublished data, presented at Annual SERDP Review Meeting, Arlington, VA, May 17, 2001.
73. Petersen, R.J. *J. Membrane Sci.* **1993**, *83*, 81-150.
74. Zhu, X.; Elimelech, M. *Environ. Sci. Technol.* **1997**, *31*, 3654-3662.
75. Vrijenhoek, E.M.; Hong, S.; Elimelech, M. *J. Membrane Sci.* **2001**, *188*, 115-128.
76. Gilron, J.; Belfer, S. Väisänen, P.; Nyström, M. *Desalination* **2001**, *140*, 167-179.
77. Freger, V.; Gilron, J.; Belfer, S. *J. Membrane Sci.* **2002**, *209*, 283-292.
78. Gerad, R.; Hachisuka, H.; Hirose, M. *Desalination* **1998**, *119*, 47-55.
79. Hachisuka, H.; Ikeda, K. *U.S. Patent 6,177,011*, **2001**.
80. Mickols, W.E. *U.S. Patent 6,280,853*, **2001**.

Chapter 2

Molecular Simulation of Gas Sorption and Diffusion in Silicon-Based Polymers

J. R. Fried and Bo Li

Center for Computer-Aided Molecular Design, Department of Chemical and Materials Engineering, University of Cincinnati, Cincinnati, OH 45221–0171

Molecular dynamics, Grand Canonical Monte Carlo, and Transition-State Theory (TST) simulations employing the COMPASS force field have be used to explore gas diffusion, solubility, and free volume in polydimethylsiloxane (PDMS), polydimethylsilmethylene (PDMSM), and four asymmetrically-substituted polysilanes — polyphenylmethyl-silane, polycyclohexylmethylsilane, poly(n-propylmethyl-silane), and polytrifluoropropylmethylsilane. Gases studied include He, O_2, N_2, CO_2, and CH_4. Gas self-diffusion coefficients in polysilanes are higher than those in PDMSM but lower than those in PDMS; however, the polysilanes have the highest gas solubility among the silicon-based polymers. Free volume in the polysilanes fall between those of PDMS and PDMSM as do the diffusion coefficients. The highly extended main chains of all four polysilanes are rigid; however, the substituent groups (e.g., methyl, propyl, and phenyl) are mobile and may contribute to their moderate gas diffusivity. As a result of their high gas solubility, polysilanes have high permeability. Polytrifluoropropylmethylsilane may have particularly attractive CO_2 permeability and CO_2/CH_4 permselectivity.

© 2004 American Chemical Society

The extremely high gas permeability of polydimethylsiloxane (PDMS) has been widely reported (1–5). Unfortunately, experimental data for other silicon-based polymers is very limited; however, there is some published permeability data for a number of gases in the polycarbosilane — polydimethylsilmethylene (PDMSM) (2,3,4,6,7). These studies have shown that PDMSM is nearly an order of magnitude less permeable than PDMS. For example, O_2 permeability is reported to be 933 barrer for PDMS but only 101 barrer for PDMSM at 308 K (2).

Polysilanes are all-silicon backbone polymers that have high thermal stability and are resistant to hydrolysis. In addition, polysilanes exhibit some interesting properties that can be attributed to electron delocalization of the σ-conjugated backbone. These properties include strong absorption and fluorescence in the near-UV region (300 to 400 nm), nonlinear optical behavior, thermochromism, and photoconductivity. Polysilanes have many potential uses such as silicon carbide preceramics, photoresists, electroluminescent devices, and polymeric photoinitiators (8–11). Mark has suggested that the permeability of polysilanes is lower than that of polysiloxanes (12). Unfortunately, no permeability data has been reported for this interesting class of polymers.

The high permeability of PDMS is attributed to the extraordinary flexibility of the siloxane chain as a result of the long Si–O skeletal bond (1.64 Å) compared to C–C (1.53 Å). Also, the Si–O–Si bond angle of 143° is much more open than the tetrahedral bond of ~100° (12). High main-chain flexibility results in a very low T_g (150 K) and high free volume. Replacement of the methyl substituent with more bulky side chains increases T_g and decreases both chain flexibility and gas diffusivity; however, gas solubility is, generally, much less affected by the substituent group (1). In general, the solubilities of CH_4, CO_2, and C_3H_8 in siloxane polymers decrease with the bulkiness of the functional side groups; however, values fall within a factor of only two (1,4). The decrease in gas solubility with increasing size of the substituent groups has been attributed to a decrease in free volume (4).

In this study, atomistic simulations are used to investigate gas diffusivity and solubility of four polysilanes. These are polyphenylmethylsilane (PPhMSi), polycyclohexylmethylsilane (PcHMSi), poly(n-propylmethylsilane) (PPrMSi), and polytrifluoropropylmethylsilane (PTFPrMSi). Structure and properties of these four polymers are given in Table I. Gases evaluated in this study are He, O_2, N_2, CO_2, and CH_4. For comparison, simulation results for PDMS and PDMSM are included and will be compared to experimental data and results from other simulations of small molecule diffusion in PDMS (13–19) and in PDMSM (15). In the most recent and extensive simulation study of gas permeability for PDMS, Charati and Stern (18) have reported values for free volume and gas diffusivity (He, O_2, N_2, CO_2, and CH_4) in three polysiloxanes that are structurally related to three of the four polysilanes selected for this study. The polysiloxanes included polypropylmethylsiloxane (PPrMS), polytrifluoropropylmethylsiloxane (PTFPrS), and polyphenylmethylsiloxane (PPMS). They found that diffusivity decreased with increasing T_g and decreasing fractional free volume (*f*) in the following order: PDMS≥PPrMS> PTFPrMS>PPhMS.

The force field selected for the simulation of polysilanes in this study is COMPASS,[*] an ab initio force field. COMPASS has been parameterized and validated for polysilanes *(20)*, alkanes and benzene compounds *(21)*, phosphazenes *(22)*, and fixed gases *(23)*. Very recently, Madkour et al. *(24)*, have used COMPASS to explore the relationship between persistence length and gas diffusivity for PDMS and one polysilane, poly(*n*-hexylmethylsilane). Over the past few years, our laboratory has reported the use of COMPASS in the atomistic simulation of the gas permeability of poly(organophosphazenes) *(25)* and poly(2,6-dimethyl-1,4-phenylene oxide) *(26)* and the glass transition (NPT dynamics) of poly(organophosphazenes) *(27)* and several asymmetrically substituted polysilanes *(28)*, including the four in the present study. Glass transition temperatures obtained by the previously reported simulation study compare favorably with experimental values (Table I). As the T_g values indicate, PPrMSi and PTFPrMSi are rubbery while PcHMSi and PPhMSi are glassy at room temperature.

In this study, we report and compare results of the atomistic simulation of gas diffusion (molecular dynamics) and sorption (Grand Canonical Monte Carlo) for representatives of the three major classes of silicon-based polymers — polysiloxanes (PDMS), polycarbosilanes (PDMSM), and polysilanes. The Transition-State Theory of Gusev and Suter *(29)* is used to determine the fractional free volume of these polymers. In addition, the effect of substituent groups on chain mobility is evaluated by use of a vectorial autocorrelation function (VACF) described in detail in an earlier communication *(28)*.

Computational Methods

COMPASS and the Insight II software package (version 4.0.0P, Molecular Simulation Inc.) were used for these simulations. Energy minimization was implemented using the default protocol provided by the Discover_3 program. For all MD simulations, the Andersen *(30)* method was used for temperature control and the Berendersen method *(31)* for pressure control. A cutoff distance of 10 Å and a spline width of 1.0 Å was used in all dynamics runs. As demonstrated in an earlier communication *(22)*, these constraints provide an adequate tradeoff between computational time and accuracy. Long-tail corrections were applied to the nonbonded dynamics calculations.

Amorphous Cell Construction and Characterization. A repeat unit of each polymer identified in Table I was built and divided into zero charge groups as identified in Table II. The repeat unit was then subject to 1000 steps of energy minimization. Using the Polymerizer module, the energetically minimized

---

[*] Condensed-phase Optimized Molecular Potentials for Atomistic Simulation Studies.

Table I. Structures and properties of silicon-based polymers included in this study

Polymer	Repeat Unit	T_g (K)	ρ g cm^{-3}	f
PDMS	CH₃ / ⁺Si—O⁺ₙ / CH₃	NA (150)[a]	1.037 (0.971)[a]	0.157 (0.219)[a] 0.33[c]
PDMSM	CH₃ / ⁺Si—CH₂⁺ₙ / CH₃	NA (181)[a]	0.983 (0.917)[a]	0.055 (0.134)[a]
PPrMSi	CH₃ / ⁺Si⁺ₙ / CH₂CH₂CH₃	242[b] (245)	0.9022	0.072
PTFPrMSi	CH₃ / ⁺Si⁺ₙ / CH₂CH₂CF₃	266[b] (270)[d]	1.2476	0.096
PcHMSi	CH₃ / ⁺Si⁺ₙ / (cyclohexyl)	338, 350[b] (366)	0.8885	0.121
PPhMSi	CH₃ / ⁺Si⁺ₙ / (phenyl)	400[b] (385–395)	1.0434	0.096

NOTE: Values are simulation results except otherwise indicated by parentheses. a. Group contribution calculations (4); b. Fried and Li (28); c. Charati and Stern (18); d. Fujino et al. (49).

Table II. Neutral charge groups selected for each silicon-based polymer included in this study

Polymer	Neutral Charge Group
PDMS	Si(CH₃)₂O
PDMSM	Si(CH₃)₂CH₂
PPrMSi	CH₃SiCH₂, CH₂, CH₃
PTFPrMSi	CH₃SiCH₂, CH₂, CF₃
PcHMSi	CH₃SiC, CH₂(cyclohexyl)
PPhMSi	CH₃SiC, CH(phenyl)

repeat units were assembled into a polymer chain of 300 repeat units for both PDMS and PDMSM and 200 repeat units for the four polysilanes. The choice of 200 repeat units for each of the four polysilanes resulted in a periodic cell, approximately 32 to 36 Å on a side. Based upon the studies of Wagner and coworkers (32,33), this box size is sufficiently large that simulation results for solubility, and especially diffusivity, should be insignificantly affected by size.

Insight II/Amorphous Cell was used to construct an amorphous cell for each polymer. The chain backbone was constructed in a bond-by-bond procedure. Due to the rigid backbone structure of the polysilanes, it was found necessary to construct the polymer chains with all backbone torsional angles fixed at the most probable conformation (i.e., all-*trans*) after the initial repeat unit was built and minimized. Details of specific steps used to insure an equilibrium conformational state are described elsewhere (34). An annealing process was used to refine the resulting cell. In this procedure, the cell was heated from 300 K to 600 K and cooled back to 300 K at increments of 50 K. At each temperature step, 50-ps NPT dynamics was run. This annealing cycle was followed by 60-ps NPT dynamics at 298 K. Next, 50 ps of NVT dynamics was run to further equilibrate the cell structure using the density averaged during the last 50 ps of the previous 60-ps NPT dynamics. In the case of PTFPrMSi, it was necessary to heat the cell to a higher temperature (up to 1000 K) during the annealing process and longer dynamics runs were required to remove unrealistic conformations.

To characterize the fully equilibrated cells following the procedures described above, an additional 60-ps NVT dynamics was run and the dynamics trajectories of the last 30 ps were recorded at intervals of 1 ps. Conformational and other parameters were determined from this dynamics run. Parameters included free volume and bond-length, bond-angle, and torsional-angle distribution.

Simulation of Gas Diffusion. From the NVT dynamics, a self-diffusion coefficient, D_α, was obtained from the mean-square displacement (MSD) of one gas molecule by means of the Einstein equation in the form (35)

$$D_\alpha = \frac{1}{6N_\alpha} \lim_{t \to \infty} \frac{d}{dt} \sum_{i=1}^{N} \left\langle \left| \mathbf{r}_i(t) - \mathbf{r}_i(0) \right|^2 \right\rangle \qquad (1)$$

In eq 1, N_α is the number of gas molecules of type α (i.e., He, O_2, N_2, CH_4, and CO_2), $\mathbf{r}_i(0)$ and $\mathbf{r}_i(t)$ are the initial and final positions of the center of mass of one gas molecule i over the time interval t, and $\left\langle \left| \mathbf{r}_i(t) - \mathbf{r}_i(0) \right|^2 \right\rangle$ is the MSD averaged over the ensemble. The Einstein relationship assumes a random walk for the diffusing species. For slow diffusing species, anomalous diffusion is sometimes observed and characterized by

$$\left\langle \left| \mathbf{r}_i(t) - \mathbf{r}_i(0) \right|^2 \right\rangle \propto t^n \qquad (2)$$

where $n < 1$ ($n = 1$ for Einstein diffusion regime). At sufficiently long time (i.e., the hydrodynamic limit), a transition from anomalous to Einstein diffusion ($n = 1$) may be observed.

To prepare for these MD simulations, six gas molecules were first randomly inserted into each cell. Only those gas molecules whose trajectories during the NVT dynamics run reached the Einstein regime (i.e., $n = 1$) were used to calculate D. Self-diffusion coefficients were obtained from 1-ns or longer time dynamics at 298 K for a maximum of 5 individual gas molecules (He, O_2, N_2, CO_2, and CH_4). In the case of the silicon-based polymers investigated in this study, Einstein diffusion regime was reached well before 1 ns. For example, a transition from anomalous to Einstein diffusion behavior for O_2 diffusion in PDMS and PDMSM was observed at about 100 ps.

Simulation of Gas Solubility. In this study, gas sorption isotherms were obtained by the Grand Canonical Monte Carlo (GCMC) method. This procedure employs a Metropolis *(36)* algorithm for sorbate insertion and deletion as well as for accepting or rejecting configurational moves (i.e., orientation and position changes of sorbate molecule). The solubility coefficient, S, is then calculated as

$$S = \lim_{p \to 0} \left(\frac{C}{p} \right) \qquad (3)$$

where C is the concentration of sorbate species in the unit of $cm^3(STP)/cm^3$ polymer and p is simulation pressure in atm.

For all the sorption simulations, the cutoff distance for nonbonded interactions was set to 10 Å and charge interactions were included. Calculations were made at 0.1 atm pressure intervals over the pressure range from 0 to 0.5 atm. At each pressure, 500,000 steps of GCMC calculations were performed with the first 200,000 steps used for equilibration and the last 300,000 steps for solubility computations.

Free Volume. The fractional free volume of each polymer was determined by the Transition State Theory (TST) method developed by Gusev and Suter *(29)* using He (Lennard-Jones parameters of $r = 2.9$ Å and $\varepsilon = 0.005$) as the probe molecule.

Chain Flexibility Analysis. A vectorial autocorrelation function (VACF) *(28)* was used to investigate main-chain flexibility. By defining a vector, $\mathbf{u}(t)$, that represents the orientation of a chain segment (backbone or side chain) at a given time t, the angle by which the orientation changes within a certain time period, t, is given as

$$m(t) = \left\langle \mathbf{u}(t_0) \bullet \mathbf{u}(t_0 + t) \right\rangle \qquad (4)$$

A value of one indicates a totally rigid structure over the time frame of the simulation while a value of zero indicates totally free bond rotation. For the polysilanes, the backbone vector selected was $Si(i)$–$Si(i+3)$ indicating a vector drawn from one Si-atom (i.e., the ith Si) to a Si-atom three atoms down the chain (i.e., the $i+3$ Si-atom). In case of PDMS, the backbone vector was $Si(i)$–$O(i)$–$Si(i+1)$–$O(i+1)$ indicating a vector connecting one Si-atom (i.e., the ith Si-atom) with the O-atom of the adjacent repeating unit. For PDMSM, it was defined as $Si(i)$–$C(i)$–$Si(i+1)$–$C(i+1)$, a vector connecting one Si-atom with the C-atom of the adjacent repeating unit. Vectors were also assigned to the five different substituent groups — methyl, n-propyl, trifluoropropyl, phenyl, and cyclohexyl. These sidechain vectors were $Si(i)$–$Si(i+1)$–C–H for methyl (i.e., a vector connecting one Si-atom with a H-atom of the methyl group of the adjacent Si-atom); $Si(i)$–$Si(i+1)$–$C(Ar)$–H for phenyl (connecting one Si-atom with the H-atom located at the p-position of the phenyl group attached to the adjacent Si-atom); $Si(i)$–$Si(i+1)$–$C(cH)$–H for cyclohexyl (connecting one Si-atom with the H-atom located at the p-position of the cyclohexyl group attached to the adjacent Si-atom); and $Si(i)$–$Si(i+1)$–$C(terminal)$–H for n-propyl (connecting one Si-atom with the H-atom located at the terminal C-atom of the propyl group attached to the adjacent Si-atom); and $Si(i)$–$Si(i+1)$–$C(terminal)$–F for trifluoropropyl (connecting one Si-atom with a F-atom located at the terminal position of the trifluoropropyl group attached to the adjacent Si-atom).

Results and Discussion

Diffusion and Solubility. Gas diffusion (D) and solubility (S) coefficients were obtained from molecular simulation of PDMS, PDMSM, and the four polysilanes. Gas diffusion coefficients were correlated with the square of the effective diameter of each gas molecule by the correlation *(37)*

$$\log D = K_1 - K_2 d_{eff}^2 \tag{5}$$

where the coefficients K_1 and K_2 are constants for a given polymer. The effective diameters were assigned by Teplyakov and Meares *(37)* from a comparison of diffusion coefficients for a number of gases with those of inert gases in different polymers *(38,39)*. Values are generally proportional but about 20% smaller than kinetic diameters. As shown in Figure 1, the correlation provides a good fit of the simulation data. PcHMSi and PDMSM have the lowest diffusion coefficients while PDMS has the highest value. Simulation values for PDMS agree generally well with reported experimental *(5,18)* and simulation *(13–19)* values. Diffusion coefficients for PPhMSi, PPrMSi, and PTFPrMSi fall between those of PcHMSi and PDMS. The order of decreasing diffusivity is as follows: PDMS>PPrMSi≈PTFPrMSi≈ PPhMSi>PDMSM>PcHMSi.

A correlation for gas solubility (S) uses the Lennard-Jones potential well depth, ε/k, a measure of gas condensability *(37,40)*

$$\log S = \log S^o + m\left(\frac{\varepsilon}{k}\right) \tag{6}$$

As an approximation, the slope m in eq 6 is independent of the nature of the sorbing medium and is in the order of $10^{-2}\,K^{-1}$. Simulation data for gas solubility in PDMS agrees reasonably well with reported experimental data *(1,5)*. Solubility coefficients obtained from GCMC simulation at 298 K for PDMS, PDMSM, and the four polysilanes are plotted in Figure 2. As shown, the correlation gives a good fit for the simulation data. Gas solubility is lowest for PDMS and PDMSM but highest for the two highest T_g (glassy) polymers — PPhMSi and PcHMSi. With respect to PDMS and PcHMSi, the order of solubility coefficients is opposite to those of the diffusion coefficients (i.e., PDMS with the higher diffusivity has the lower gas solubility).

Free Volume. Fractional free volumes, f, were calculated as

$$f = \frac{V_f}{V} \tag{7}$$

where V_f is the total free volume obtained by the TST method and V is the specific volume of the polymer. Fractional free volumes of PDMS, PDMSM, and the polysilanes are given in Table I. As the results show, PDMS has the largest free volume and diffusion coefficients. PDMSM, in contrast, has the smallest free volume and diffusion coefficients in agreement with the expected relationship between free volume and diffusivity *(41,42)*. For the polysilanes, f decreases in the following order: PcHMSi>PPhMSi≈PTFPrMSi>PPrMSi.

Chain Flexibility. The vectorial autocorrelation function (VACF) *(28)* has been used to investigate the relationship between chemical structure and main-chain flexibility for the silicon polymers included in this study. For purpose of discussion, PPhMSi was chosen as a representative polysilane; the other three polysilanes included in this study show similar results although the main-chain mobility of PPrMSi is the highest of the four. As shown by Figure 3, main-chain flexibility follows the following order: PDMS>PDMSM>PPhMSi. The main chain of PPhMSi exhibits very limited main-chain mobility over 1 ns of dynamics. The rigidity of the polysilane backbone as indicated by the VACF results is consistent with the conclusions of light scattering experiments *(43)* and other simulation studies *(20,24,44–46)*. The trend of the VACF results might suggest that PDMSM, with its more flexible chain, should have higher gas diffusivity than PPhMSi which is not the case (Figure 1); however, the higher diffusivity of PPhMSi is consistent with its higher free volume.

Further insight into chain dynamics is provided by comparing side-chain VACFs. As shown by Figure 4, methyl groups are highly mobile in agreement

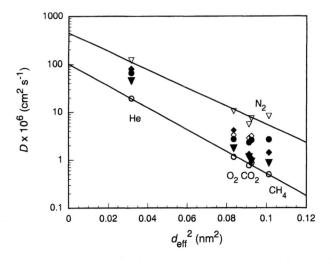

Figure 1. Semilog plot of self-diffusion coefficients obtained from molecular simulation (298 K) versus the square of the effective gas diameter (eq 5). (●) PPhMSi; (○) PcHMSi; (◆) PPrMSi;(◇) PTFPrMSi; (▽) PDMS; (▼) PDMSM. Lines represent best fit of simulation data for PcHMSi (lower boundary) and for PDMS (upper boundary).

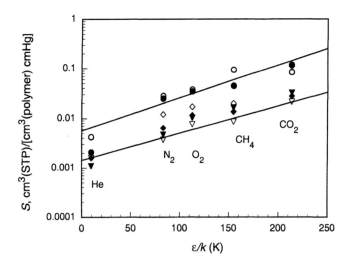

Figure 2. Semilog plot of gas solubility coefficients (GCMC, 298 K) against their Lennard-Jones potential well parameter, ε/k (eq 6). (●) PPhMSi; (○) PcHMSi; (◆) PPrMSi;(◇) PTFPrMSi; (▽) PDMS; (▼) PDMSM. Lines represent best fit of simulation data for PDMS (lower boundary) and for PcHMSi (upper boundary).

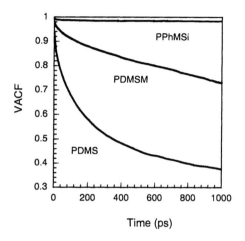

Figure 3. Plot of the vectorial autocorrelation function (VACF) of the main chain of PPhMSi, PDMSM, and PDMS.

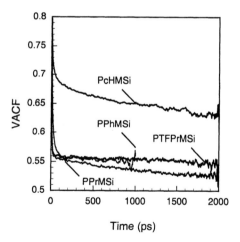

Figure 4. Plot of the methyl group VACFs.

with solid-state deuterium NMR results *(47)*. Of the four polysilanes, methyl group mobility is the most restricted for PcHMSi that also has the smallest diffusion coefficient. Comparison of the non-methyl groups VACFs in Figure 5 indicates that the phenyl group of PPhMSi and the cyclohexyl group of PcHMSi are the least mobile while the propyl group of PPrMSi is the most mobile. Simulation results show that the phenyl groups of PPhMSi are arranged in parallel, allowing maximum overlap *(46)*, while the cyclohexyl groups of PcHMSi reside in the chair conformation.

Permeability. The permeability of polysiloxanes decreases with the bulkiness of the side groups *(2)*. To compare gas permeability of the polysilanes, gas permeability coefficients (*P*) of the polysilanes were calculated as the product of the diffusion and solubility coefficients obtained from simulation. Oxygen and CO_2 permeabilities and O_2/N_2 and CO_2/CH_4 permselectivities

$$\alpha_{i,j} = \frac{P_i}{P_j} \tag{8}$$

of PDMS, PDMSM, and the four polysilanes are listed in Table III. For comparison, values for polytrifluoropropylmethylsiloxane (PTFPrMS) and two polyphosphazenes — poly[bis(trifluoroethoxy)phosphazene] (PTFEP) and poly[bis(*n*-butoxy)phosphazene] (PnBuP) — are included in Table III. Structures and properties of these three polymers are given in Table IV. In the case of PDMS and PDMSM, the current simulation values for permeability agree well with reported experimental data and other simulation results. Comparison of values show that substitution of the Si–O main-chain bond of PDMS with the stiffer Si–C bond of PDMSM results in a significant decrease in *P (7)*. The combination of moderate diffusivity and high solubility contribute to the overall high permeability of the polysilanes indicated by the simulation. It is noted that the simulation results suggest that the permeabilities of the polysilanes should be comparable to those of PDMS. Of the four polysilanes, PTFPrMSi may be the most interesting since it has particularly high CO_2 permeability and high CO_2/CH_4 permselectivity. High CO_2 permeability and permselectivity has been reported for other fluorinated polymers, including PTFPMS *(1,4)* and PTFEP *(48)*. In these cases, elevated CO_2 solubility has been attributed to an interaction between CO_2 and the electron-withdrawing fluorinated alkyl or alkoxy substituent groups. As shown by Table III, the CO_2 permeability and CO_2/CH_4 permselectivity of PTFPrMSi is higher than the corresponding values for either PTFPrMS and PTFEP.

Conclusions

Simulation studies have suggested that asymmetrically substituted polysilanes have gas permeabilities comparable to PDMS. These high

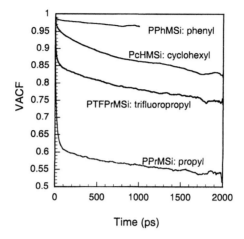

Figure 5. Plot of the non-methyl group VACFs.

Table III. Comparisons of simulated and experimental permeabilities

Polymer	T (K)	P(O₂) barrer	α_{O_2,N_2}	P(CO₂) barrer	α_{CO_2,CH_4}
PDMS	298	788	2.9	1180	1.7
	(308)[a]	(933)	(2.7)	(4550)	(3.4)
	(303)[b]	(781)	(2.1)		
PPhMSi	298	937	1.4	2788	2.3
PcHMSi	298	431	1.8	648	1.4
PPrMSi	298	483	7.0	371	1.9
PTFPrMSi	298	561	1.5	3181	11.1
PDMSM	298	179	4.0	372	2.6
	(308)[a]	(101)	2.3	(549)	(4.2)
	(303)[b]	(91)	(2.5)		
PTFPrMS	(308)[a]	(217)	(2.1)	(1388)	(4.2)
	(303)[b]	(183)	(2.2)		
PnBuP	(303)[b]	(128)	(2.3)	(647)	(3.3)
PTFEP[d]	(298)	(35)	2.4	(196)	(10.3)

NOTE: Experimental values are in parentheses.
1 barrer = 10^{-10} cm³(STP) cm/s-cm²-cmHg.; solubility units: cm³(STP)/(cm³ polymer cmHg). a. Stern et al. *(2)*; b. Lee et al. *(3)*; c. Hirose and Mizoguchi *(50)*; d. Hirose et al. *(48)*.

Table IV. Structures and properties of other inorganic polymers

Polymer	Repeat Unit	T_g (K)	ρ $g\ cm^{-3}$	f
PTFPrMS	$\left[Si-O\right]_n$ with CH_3 and $CH_2CH_2CF_3$	$(203)^a$	$(1.292)^a$	0.29^b
PTFEP[c]	$\left[P=N\right]_n$ with OCH_2CF_3 and OCH_2CF_3	(191)	(1.707)	—
PnBuP	$\left[P=N\right]_n$ with $OCH_2CH_2CH_2CH_3$ and $OCH_2CH_2CH_2CH_3$	$(165)^d$	$(1.047)^d$	0.097^e

NOTE: Experimental values are given within parentheses. a. Stern and Bhide (6); b. Charati and Stern (18); c. Sample with 60% crystallinity (48) d. Hirose and Mizoguchi (50); e. Fried and Ren (25).

permeabilities are due to a combination of very high gas solubility and high gas diffusivity. Although the main-chain structure of these polysilanes are highly rigid due to the high degree of substitution of the Si–Si backbone, side groups such as methyl and *n*-propyl have significant mobility and likely contribute to the high diffusivity of polysilanes. Of the four polysilanes investigated in this molecular simulation study, polytrifluoropropylmethylsilane is particularly attractive due to its potentially high CO_2 permeability and permselectivity.

References

1. Shah, V. M.; Hardy, B. J.; Stern, S. A. *J. Polym. Sci.: Part B: Polym. Phys.* **1986**, *24*, 2033–2047.
2. Stern, S. A.; Shah, V. M.; Hardy, B. J. *J. Polym. Sci.: Part B: Polym. Phys* **1987**, *25*, 1263–1298.
3. Lee, C.-L.; Chapman, H. L.; Cifiuentes, M. E.; Lee, K. M.; Merrill, L. D.; Ulman, K. L.; Venkataraman, K. *J. Membr. Sci.* **1988**, *38*, 55–70.
4. Shah, V. M.; Hardy, B. J.; Stern, S. A. *J. Polym. Sci.: Part B: Polym. Phys* **1993**, *31*, 313–317.
5. Merkel, T. C.; Bondar, V. I.; Nagai, K.; Freeman, B. D.; Pinnau, I. *J. Polym. Sci.: Part B: Polym. Phys* **2000**, *38*, 415-434.
6. Stern, S. A.; Bhide, B. D. *J. Appl. Polym. Sci* **1989**, *38*, 2131–2147.
7. Bhide, B. D.; Stern, S. A. *J. Appl. Polym. Sci.* **1991**, *42*, 2397–2403.
8. West, R. *J. Organomet. Chem.* **1986**, *300*, 327–346.
9. Miller, R. D.; Michl, J. *Chem. Rev.* **1989**, *89*, 1359–1410.
10. Mark, J. E.; Allcock, H. R.; West, R. *Inorganic Polymers*; Prentice Hall: Englewood Cliffs, NJ, 1992.
11. Sawan, S. P.; Ekhorutomwen, S. A. In *Polymeric Materials Encyclopedia*; Salamone, J. C., Ed.; CRC Press: Boca Raton, FL, 1996; pp 6722–6733.
12. Mark, J. E. In *Silicone-Based Polymer Science: A Comprehensive Resource*; Frearon, F. W. G., Ed.; ACS: Washington, DC, 1990; Vol. 224, pp 47–68.
13. Sok, R. M.; Berendsen, H. J. C.; Gunsteren, W. F. v. *J. Chem. Phys.* **1992**, *96*, 4699–4704.
14. Sok, R. M.; Berendsen, H. J. C. *Polym. Prepr.* **1992**, *33*, 641–642.
15. Taylor, G.; Galiatsatos, V. *Polym. Mater. Sci. Eng.* **1993**, *69*, 16–17.
16. Tamai, Y.; Tanaka, H.; Nakanishi, K. *Macromolecules* **1994**, *27*, 4498–4508.
17. Tamai, Y.; Tanaka, H.; Nakanishi, K. *Fluid Phase Equil.* **1995**, *104*, 363–374.
18. Charati, S. G.; Stern, S. A. *Macromolecules* **1998**, *31*, 5529–5535.
19. Hofmann, D.; Fritz, L.; Ulbrich, J.; Paul, D. *Comput. Theor. Polym. Sci.* **2000**, *10*, 419–436.
20. Sun, H. *Macromolecules* **1995**, *28*, 701–712.
21. Sun, H. *J. Phys. Chem. B* **1998**, *102*, 7338–7364.
22. Sun, H.; Ren, P.; Fried, J. R. *Comput. Theor. Polym. Sci.* **1998**, *8*, 229–246.

38

23. Yang, J.; Ren, Y.; Tian, A.; Sun, H. *J. Phys. Chem. B* **2000**, *104*, 4951–4957.
24. Madkour, T. M.; Mohamed, S. K.; Barakat, A. M. *Polymer* **2002**, *43*, 533–539.
25. Fried, J. R.; Ren, P. *Comput. Theor. Polym. Sci.* **2000**, *10*, 447–463.
26. Fried, J. R.; Sadat-Akhavi, M.; Mark, J. E. *J. Membr. Sci.* **1998**, *149*, 115–126.
27. Fried, J. R.; Ren, P. *Comput. Theor. Polym. Sci.* **1999**, *9*, 111–116.
28. Fried, J. R.; Li, B. *Comput. Theor. Polym. Sci.* **2001**, *11*, 273–281.
29. Gusev, A. A.; Suter, U. W. *Phys. Rev. A* **1991**, *43*, 6488–6494.
30. Andrea, T. A.; Swope, W. C., Anderson, H. C. *J. Chem. Phys.* **1983**, *79*, 4576–4584.
31. Berendsen, H. J. C.; Potsma, J. P. M.; von Gunsteren, W. F.; DiNola, A.; Haak. J. R. *J Chem. Phys.* **1984**, *81*, 3684–3690.
32. Cuthbert, T. R.; Wagner, N. J.; Paulaitis, M. E. *Macromolecules* **1997**, *30*, 3058–3065.
33. Cuthbert, T. R.; Wagner, N. J.; Paulaitis, M. E.; Murgin, H.; D'Aguanno, B. *Macromolecules* **1999**, *32*, 5017–5028.
34. Li, B. PhD Dissertation; University of Cincinnati, 2003.
35. Trohalaki, S.; Kloczkowski, A.; Mark, J. E., Rigby, D.; Roe, R. J. in *Computer Simulation of Polymers*; Roe, R. J. Ed.; Prentice Hall: Englewood Cliffs, NJ, 1991, pp. 220–232.
36. Metropolis, N.; Rosenbluth, A. W.; Rosenbluth, M. N.; Teller, A. H.; Teller, E. *J. Chem. Phys.* **1953**, *21*, 1087–1092.
37. Teplyakov, V.; Meares, P. *Gas Sep. Purif.* **1990**, *4*, 66–74.
38. Teplyakov, V. V.; Durgar'yan, S. G. *Polym. Sci. USSR* **1984**, *26*, 1678–1688.
39. Teplyakov, V. V.; Durgar'yan, S. G. *Polym. Sci. USSR* **1986**, *28*, 629–639.
40. Toi, K.; Morel, G.; Paul, D. R. *J. Appl. Polym. Sci.* **1982**, *27*, 1997–2005.
41. Fujita, H. *Adv. Polym. Sci.* **1961**, *3*, 1–47.
42. Frisch, H. L.; Klempner, D.; Kwei, T. K. *Macromolecules* **1971**, *4*, 237–238.
43. Kato, H.; Sassanuma, Y.; Kaito, A.; Tanigaki, N.; Tanabe, T.; Kinugasa, S. *Macromolecules* **2001**, *34*, 262–268.
44. Damewood, J. R.; West, R. *Macromolecules* **1985**, *18*, 159–164.
45. West, W. J.; DeBolt, L.; Mark, J. E. *Macromolecules* **1986**, *19*, 2978–2983.
46. Welsh, W. J.; Damewood, J. R.; West, R. C. *Macromolecules* **1989**, *22*, 2947–2951.
47. O'Connor, R. D.; Ginsburg, E. J.; Blum, F. D. *J. Chem. Phys.* **2000**, *112*, 7247–7259.
48. Hirose, T.; Kamiya, Y.; Mizoguchi, K. *J. Appl. Polym. Sci.* **1989**, *38*, 809–820.
49. Fujino, M.; Hisaki, T.; Fujiki, M.; Matsumoto, N. *Macromolecules* **1992**, *25*, 1079–1083.
50. Hirose, T.; Mizoguchi, K. *J. Appl. Polym. Sci* **1991**, *43*, 891–900.

Chapter 3

New Perspectives of Gas Sorption in Solution-Diffusion Membranes

N. F. A. van der Vegt[1,2]

[1]Membrane Technology Group, Department of Chemical Technology, University of Twente, P.O. Box 217, 7500 AE Enschede, The Netherlands
[2]Current address: Laboratory of Physical Chemistry, Swiss Federal Institute of Technology at Zürich, ETH-Hönggerberg, CH–8093 Zürich, Switzerland

This chapter discusses the thermodynamics of gas sorption in polymeric membranes from a molecular perspective. Its relation to molecular interactions is described theoretically and calculations based on Molecular Dynamics (MD) simulations of gas sorption in poly(ethylene) are reported to illustrate the feasibility of computing all thermodynamic quantities discussed. Solvation enthalpies and entropies are decomposed into contributions of *penetrant binding* and *polymer reorganization* to better understand the influence of the nature of the penetrant as well as the polymeric solvent on the molecular dissolution process. The contribution of penetrant binding and polymer reorganization energies to the gas solubility is discussed in detail.

© 2004 American Chemical Society

Studies of gas and vapor solubility in polymeric solvents (*1*) play an important role in rationalizing the separation performance of solution-diffusion membranes. Because gas solubility is easily accessible experimentally, it has been widely studied and correlated to gas phase properties and polymer characteristics. The role of polymer-penetrant interactions has frequently been invoked to explain trends observed with series of gases. A common qualitative approach is to identify the strength of these interactions with the partial heat of gas sorption which, in turn, is correlated to a gas phase property like the gas critical temperature. To account for polymer-penetrant interactions in a quantitative thermodynamic context, the designation "penetrant binding energy to the polymer" perhaps agrees best with our intuitive understanding of it. This binding energy contributes to the partial heat of penetrant sorption (solvation enthalpy), but, in equilibrium liquid solvents, never equates to it (*2*). If a penetrant dissolves in a liquid, several things happen. First, to accommodate the penetrant, solvent molecules need to reorganize. Solvent reorganization involves the formation of a cavity of sufficient size and shape and, once the penetrant takes advantage of it, rearrangement of solvent molecules in the solvation shell of the penetrant. These solvent rearrangements affect the average potential energy of the solvent and therefore contribute a so-called "solvent reorganization energy" to the solvation enthalpy. Dispersive (Van der Waals) and electrostatic interactions (dipolar interactions, hydrogen bonding, etc.) between penetrant and solvent molecules cause the penetrant to bind to the solvent and the average effect is referred to as the "penetrant binding energy", which contributes to the solvation enthalpy as well. A study of molecular dissolution processes in which these "solvent and solute contributions" to thermodynamic parameters can be distinguished clearly is advantageous in understanding the nature of penetrant sorption better. Experimentally, one cannot assess penetrant binding and solvent reorganization energies separately. Yet, theoretical methods can be invoked and molecular simulations (Molecular Dynamics or Monte Carlo simulations) are best equipped to suit this purpose as these describe molecular assemblies in full atomistic detail.

It is not the aim of this chapter to discuss simulation results on gas solubilities. Instead, I intend to discuss ways to obtain detailed information regarding solvent reorganization energies and polymer-penetrant interactions discussed above. This choice is motivated by the fact that membrane scientists commonly adopt MD simulations to *predict* diffusion and solubility coefficients because these two properties determine the gas permeability, which determines membrane performance. Such efforts are of great importance if they serve the development of improved, new membrane materials or if they are applied under "harsh process conditions" (high temperatures and/or pressures) in which experiments are no longer feasible. Despite their importance, simulations performed to predict such "macroscopic quantities" do not provide a priori information on the molecular scale dissolution process itself. To get this information, solvation energies must be studied instead and decomposed into the binding and solvent reorganization terms. MD studies of these quantities are rare in membrane science, yet they can be calculated by computational methods similar to the ones used to obtain the solubility (i.e., by performing test particle insertions in the system).

The present chapter is an extension of a detailed thermodynamic analysis of gas sorption in liquid dodecane, linear poly(ethylene), poly(vinyl chloride), and poly(vinyl alcohol) published recently (3). In this chapter, the thermodynamic theory described in reference (3) is extended further to illustrate that the enthalpy and entropy changes of an important molecular process in gas sorption, the creation of molecular sized cavity in the polymer fluid, exactly cancel out in the free enthalpy change of gas sorption. Because it is the free enthalpy change that determines the gas solubility, the energetics associated with creating molecular sized cavities in liquid (or rubbery) polymers do not *directly* affect gas solubility. The only energetics directly affecting gas solubility are those due to polymer-penetrant interactions. Yet, the cavity contribution does appear in derived quantities like the partial heat of gas sorption (which due to large cavity formation enthalpies may become endothermic (3)) and the sorption entropy. Although the above statements are exact results from a statistical mechanical treatment of solvation (see e.g. reference (10)), they remain difficult to understand, perhaps even conflicting with intuitive reasoning that tells us that formation of a cavity is the first thing must happen before a penetrant can dissolve. A hand-waving argument to explain this abstract thermodynamic fact is provided by realizing that the energy invested to create a cavity in the polymer fluid is distributed amongst a vast number of polymer degrees of freedom causing the fluid entropy to increase (the entropy of a molecular system increases with its energy content). Hence, the energetic cost of making a cavity is paid back for by an entropic gain such that the latter exactly compensates for the former when evaluating the free enthalpy change $\Delta G = \Delta H - T\Delta S$. Thus, those parts in ΔH and ΔS that account for cavity formation are compensating; what remains are the parts that account for polymer-penetrant interactions. In section 2.2, this thermodynamic description is presented in more detail. Perhaps, the most important consequence arising from this theory is that trends in solubilities of a series of penetrant gases can be understood solely on the basis of penetrant properties (like the penetrant Lennard-Jones force constant used in the early work of Michaels and Bixler (15, 16)) regardless of the fact that a more bulky penetrant takes advantage of the availability of larger cavities that require a larger energy penalty (a polymer property) for their formation. The relation between the penetrant solubility and its gas phase properties, of course, is well known from many more experimental studies. The theoretical justification above, however, has never been provided in the literature on this topic.

Above it was stated that the energetics of cavity formation does not *directly* affect gas solubilities because it does not explicitly appear in the free enthalpy change. Could there still be an indirect effect? The answer is yes. Clearly, a large penetrant with a Lennard-Jones force constant not significantly larger than that of a small penetrant will have a lower solubility because it will find fewer cavities big enough to occupy. This indirect effect occurs through the entropy. However, before we can venture any further on this topic, a detailed thermodynamic analysis is required and will be presented in the next section of this chapter.

In the "Theory" section of this article, I will discuss the statistical thermodynamics of gas sorption. This section is split into two parts. The first part briefly summarizes the test-particle insertion method, which is the method most commonly used to calculate gas solubility in molecular simulations. I will emphasize the solvation Gibbs energy, which directly relates to the solubility coefficient. In the second part, solvation enthalpies and entropies are discussed. In particular, I will pay attention to the physical interpretation of the polymer-penetrant interaction and cavity formation terms in the enthalpy and entropy expressions. In the section entitled "Thermodynamics of small permanent gases in liquid poly(ethylene)," some results from MD simulations of gas sorption in a poly(ethylene) melt will be discussed to illustrate the feasibility of calculating all thermodynamic quantities from molecular simulations. For a more detailed discussion of these simulations, the reader is referred to reference (3).

It is important to keep in mind that the discussion of the sorption thermodynamics in this chapter applies to systems in thermodynamic equilibrium. Nonetheless, the molecular interpretation of the thermodynamic quantities provides natural insights into what may happen when penetrants are dissolved in glassy polymer membranes. In the section entitled "Infinite dilution solubilities in glassy polymeric solvents," penetrant sorption in glassy polymeric solvents is briefly discussed. An elegant thermodynamic theory on gas sorption in non-equilibrium glassy systems is provided by F. Doghieri *et al.* and discussed elsewhere in this book.

Theory

Three thermodynamic functions are chosen to describe the gas sorption process: the solvation Gibbs energy and the corresponding solvation enthalpy and entropy. In the section on gas solubility in poly(ethylene), these quantities will be reported in the limit of infinite dilution, so penetrant-penetrant interactions will be ignored. The formalism below, however, is applicable at finite penetrant concentrations as well. Therefore, changes in these properties with penetrant loading can be studied using the same formalism.

The Solvation Gibbs Energy

The solvation Gibbs energy, ΔG_S^*, describes the change in Gibbs energy of the system associated with the process of transferring a penetrant molecule from a fixed point in the gas phase to a fixed point in the membrane at constant pressure P, temperature T, and composition (4, 5). This quantity is related to the Ostwald solubility L, which is defined as the ratio of the equilibrium (eq) penetrant number densities in the membrane phase and the ideal gas (ig) phase,

$$L \equiv \left(\rho_S / \rho_S^{ig} \right)_{eq} \tag{1}.$$

The solvation Gibbs energy reads (3),

$$\Delta G_S^* = -RT \ln L$$

$$= \mu_S^* - \mu_S^{*ig} \tag{2},$$

$$= -RT \ln \left[\left\langle V \exp(-B_S / RT) \right\rangle_{NPT} / \left\langle V \right\rangle_{NPT} \right]$$

where μ_S^* and μ_S^{*ig} are the pseudo chemical potentials (3,4, 5) of the penetrant at a fixed point in the membrane- and in the ideal gas phase, respectively. Hence, $\mu_S^* - \mu_S^{*ig}$ can be interpreted as the excess chemical potential of the penetrant, located at a fixed point, due to interactions with the membrane. This definition agrees with our notion of solvation and is therefore referred to as the solvation Gibbs energy. In eq 2, B_s is the potential energy difference between a system of N+1 particles and that containing only N particles (particle N+1 is the penetrant). Therefore, it will be referred to as the solute binding energy to the system. The averaging, denoted by the angle brackets $<\cdots>_{NPT}$, is over configurations of the N particle system (the neat solvent) at constant values of P and T, unperturbed by the presence of "test" particle N+1. In practice, the averaging is realized by randomly inserting the solute particle in the computer-generated solvent configurations. V is the solvent volume, which fluctuates around the average value $\langle V \rangle_{NPT}$ assumed by an ensemble of N solvent molecules at fixed values of P and T.

Solvation Enthalpies and Entropies

The most straightforward way to calculate the solvation enthalpy ΔH_S^* at infinite dilution is to compute the N+1 (the solution) and the N-body (the solvent) enthalpies directly by two independent simulations and to subtract the numerical values. This procedure, however, is very inaccurate and precludes any entropy determination, which can be obtained if the test-particle insertion method is used. To avoid this difficulty, ΔH_S^* is expressed in a form involving averaging over configurations of the N particle system (the neat solvent) only (3),

$$\Delta H_S^* = \frac{\left\langle B_S \, V \, e^{-B_S/RT} \right\rangle_{NPT}}{\left\langle V \, e^{-B_S/RT} \right\rangle_{NPT}} +$$

$$\frac{\left\langle (U_N + PV) \, V \, e^{-B_S/RT} \right\rangle_{NPT} - \left\langle (U_N + PV) \right\rangle_{NPT} \left\langle V \, e^{-B_S/RT} \right\rangle_{NPT}}{\left\langle V \, e^{-B_S/RT} \right\rangle_{NPT}} \qquad (3).$$

$$= \Delta U_{uv}^* + \Delta U_{vv}^*$$

The first term in eq 3 denotes the average solute binding energy to the solvent (the polymer-penetrant interaction, or, solute-solvent energy ΔU_{uv}^*). The second term denotes the solvent reorganization energy (the potential energy change of the solvent, or, solvent–solvent energy ΔU_{vv}^*) (6). Note that, despite being sometimes suggested in the literature, there is absolutely no a priori reason to assume this term to be negligible (7). Similar to the calculation of the solvation Gibbs energy in eq 2, eq 3 is evaluated by randomly inserting the solute particle into unperturbed solvent molecule configurations.

The solvation entropy,

$$T\Delta S_S^* = \Delta H_S^* - \Delta G_S^* \qquad (4)$$

is obtained by subtracting eq 3 from eq 2. Interestingly, the solvent reorganization energy, ΔU_{vv}^*, affects enthalpy and entropy values significantly, even changing their signs (3), but it does not affect the solubility in any way (10-13) because it does not appear explicitly in eq 2. Thus, as indicated in the introduction, the solvent reorganization energy is exactly counterbalanced by a term in the solvation entropy which causes it to cancel in the solvation Gibbs energy. To illustrate the cancellation of ΔU_{vv}^*, the solvation energy and entropy are written as (12,13)

$$\Delta U_S^* = \Delta U_{uv}^* + \Delta U_{vv}^*$$

$$\Delta S_S^* = \Delta S_{uv}^* + \left(\Delta U_{vv}^* / T\right) \tag{5},$$

to yield the solvation Gibbs energy

$$\Delta G_S^* = \Delta U_{uv}^* - T \Delta S_{uv}^* \tag{6},$$

which then contains an energetic and entropic term which both are functions of only the solute-solvent interaction, the solvent-solvent interactions occurring implicitly in the ensemble averaging (*10*). In eq 6, ΔU_{uv}^* is the solute binding energy introduced above, and ΔS_{uv}^* is a fluctuation term (*12*) defined by eq 5. This term expresses the entropic cost of inserting the solute (an excluded volume effect). The energetic term ΔU_{uv}^* is generally negative and tends to favor solubility whereas the entropic contribution ΔS_{uv}^* is always negative, which acts to inhibit solubility (*14*).

Note that there are two important features in determining the entropy change accompanying the dissolution process (see eq 5). First, the process of insertion requires the formation of a cavity of appropriate size and shape. At constant pressure, this process raises the solvent energy (solvent-solvent contacts must be broken and the system's volume increases) and causes a rise in the entropy, which is given by $\Delta U_{vv}^*/T$. Second, the process restricts the number of arrangements possible for the solvent molecules (the solvent molecules are excluded from the cavity). This causes a reduction in entropy, the magnitude of which is given by the ΔS_{uv}^* term.

In the introduction to this chapter, it was stated that the solvent energy change, ΔU_{vv}^*, of creating a cavity does not *directly* affect the gas solubility. Eq 6 is the basis of this statement. It was also stated that ΔU_{vv}^* may *indirectly* affect the solubility, because it is less probable to find larger cavities (which require a larger ΔU_{vv}^* for their formation) than smaller ones. The probability of finding a cavity large enough to accommodate a penetrant depends on the solvent density, which, in turn, is determined by the nature of the molecular interactions between the solvent molecules. In low-density solvents cavities are readily formed and ΔU_{vv}^* values are small. If the solvent density is larger, the opposite is true. The effect of ΔU_{vv}^* on the gas solubility thus appears implicitly by affecting the *probability* of finding a cavity that is big enough. The thermodynamic quantity expressing the magnitude of this probabilistic effect is ΔS_{uv}^*. The Ostwald solubility L (eq 2) can thus be written as a product of a probabilistic term $\exp(\Delta S_{uv}^*/R)$ and an energetic term $\exp(-\Delta U_{uv}^*/RT)$. The only energetics term directly influencing the solubility is the solute-solvent energy ΔU_{uv}^*. The energy change of the solvent affects the solubility only indirectly via the entropic term related to the probability of finding a cavity large enough to accommodate the solute molecule. This conclusion is important for developing microscopic theories on gas solubility.

Solubility, Enthalpy, and Entropy; Concluding Points

From the preceding discussion, a balance of enthalpic and entropic effects governs gas solubility. Test-particle insertions performed in polymer configurations generated by molecular simulation permit evaluation of all solvation quantities, and, in particular, polymer reorganization and penetrant binding terms, which cannot be obtained experimentally. Conclusions from this discussion are summarized below:

- Partial heats of gas sorption may become positive (endothermic) if the polymer-penetrant binding energy released upon penetrant insertion is smaller than the polymer reorganization energy invested to create a cavity.
- Solvation entropies contain a negative contribution resulting from a restriction in the number of arrangements possible for the polymer molecules due to the presence of the penetrant. Polymer reorganization provides a positive contribution to the solvation entropy because it reduces the cohesion amongst the repeat units, leading to an increase in their configurational freedom.
- The polymer reorganization energy invested does not affect penetrant solubility because it is exactly counterbalanced by the positive contribution to the solvation entropy mentioned in the previous point.
- Consequently, penetrant solubilities are fully determined by the penetrant binding energy (favoring solubility) and the entropy loss (inhibiting solubility) associated with keeping the polymer out of the region occupied by the penetrant. The latter is essentially an excluded volume effect.

Thermodynamics of Small Permanent Gases in Liquid Poly(ethylene)

To illustrate the formalism described above, I have calculated solvation properties of a series of gases in a poly(ethylene) melt at 300 and 400 K. In reality, high-density poly(ethylene) will crystallize below about 404 K. However, in these calculations, the poly(ethylene) melt is amorphous below the experimental melting temperature. This hypothetical amorphous poly(ethylene) should be a reasonable representation of nearly amorphous, branched poly(ethylenes) produced with metallocene catalysts, as well as ethylene propylene random copolymers at low propylene contents so that the propylene moieties distributed randomly along the chain backbone preclude crystallization.

The molecular models and simulation details have been described in reference (3). In that work the penetrant solvation behavior in poly(ethylene) has been compared to that in liquid dodecane to find out the differences between polymeric and low-molecular weight liquid solvents.

Tables I and II summarize solvation thermodynamic quantities in poly(ethylene) at 300 K and 400 K, respectively. Several interesting observations can be made. The solvation enthalpies (or partial heats of gas sorption) are endothermic for gases with a critical temperature, T_C, below approximately 150 K and exothermic for the higher T_C gases.

Table I. Solvation thermodynamic quantities (kJ/mol) for gases in poly(ethylene) at 300 K and 0.1 MPa

	T_C (K)	ΔG_S^*	ΔH_S^*	$T\Delta S_S^*$	ΔU_{vv}^*	ΔU_{uv}^*	$T\Delta S_{uv}^*$
He	5.1	7.7	4.9	−2.8	6.1 (0.2)	−1.2	−8.9
H_2	33.0	7.6	3.8	−3.8	8.7 (0.3)	−4.9	−12.5
Ne	44.3	6.8	3.1	−3.7	7.5 (0.3)	−4.4	−11.2
N_2	126.0	6.9	2.0	−4.9	14.6 (0.8)	−12.6	−19.5
Ar	150.8	4.4	−1.1	−5.5	12.1 (0.6)	−13.2	−17.6
O_2	154.8	5.3	0.0	−5.3	12.9 (0.7)	−12.9	−18.2
CH_4	191.0	3.7	−2.0	−5.7	15.0 (0.9)	−17.0	−20.7
Kr	209.4	2.2	−4.0	−6.2	13.7 (0.8)	−17.7	−19.9
Xe	289.4	0.3	−5.2	−5.5	19.4 (1.6)	−24.6	−24.9
CO_2	304.0	−0.8	−7.7	−6.9	14.7 (0.9)	−22.4	−21.6

NOTE: The gas critical temperatures are obtained from the gas Lennard-Jones model parameter ε_s/k used in this study (3). Averages were taken over 9.6 ns MD simulation during which 125000 test-particles were inserted every 0.2 ps. The numbers in parenthesis denote the statistical inaccuracy.

The solvation enthalpies are small (in absolute number) compared to the contributions of polymer reorganization and penetrant binding. The polymer reorganization energies are always positive, which is caused by the energetic penalty associated with creating a molecular sized cavity in the polymeric fluid. The penetrant binding energy to the polymeric fluid is always negative and, therefore, compensates for the former process to a large extent. Endothermic sorption heats are caused by a relatively large polymer reorganization energy involved in solvating penetrants that bind weakly. Since the partial heat of gas sorption is the temperature coefficient of the gas solubility, the helium, hydrogen, neon, and nitrogen solubilities in poly(ethylene) increase with increasing temperature, consistent with experimental observations (*15, 16*).

The solvation Gibbs energies are positive. Because the polymer reorganization energy (included in ΔH_S^* and $T\Delta S_S^*$) does not contribute to ΔG_S^*, its behavior can best be understood by inspection of the solute binding energy and the solute-solvent entropy, which are shown in the last two columns, respectively. Clearly, larger penetrants bind stronger (favoring solubility) and cause a stronger entropy reduction (inhibiting solubility). This behavior is due to the general tendency that larger penetrants possess stronger dispersive interactions with the polymeric fluid and cause polymer repeat units to be excluded from a larger region in the fluid. The unfavorable entropy leads to positive values of the solvation Gibbs energy.

Table II. Solvation thermodynamic quantities (kJ/mol) for gases in poly(ethylene) at 400 K and 0.1 MPa

	T_C (K)	ΔG_S^*	ΔH_S^*	$T\Delta S_S^*$	ΔU_{vv}^*	ΔU_{uv}^*	$T\Delta S_{uv}^*$
He	5.1	8.1	5.6	−2.5	6.5 (0.2)	−0.9	−9.0
H_2	33.0	8.2	4.5	−3.7	8.7 (0.3)	−4.2	−12.4
Ne	44.3	7.4	3.9	−3.5	7.7 (0.3)	−3.8	−11.2
N_2	126.0	7.7	2.1	−5.6	13.3 (0.6)	−11.2	−18.9
Ar	150.8	5.4	−0.6	−6.0	11.1 (0.5)	−11.7	−17.1
O_2	154.8	6.2	0.3	−5.9	11.8 (0.5)	−11.5	−17.7
CH_4	191.0	4.8	−2.0	−6.8	13.2 (0.6)	−15.2	−20.0
Kr	209.4	3.5	−3.6	−7.1	12.1 (0.6)	−15.7	−19.2
Xe	289.4	1.7	−6.8	−8.5	15.3 (0.9)	−22.1	−23.8
CO_2	304.0	0.8	−7.5	−8.3	12.6 (0.6)	−20.1	−20.9

NOTE: Averages were taken over 6.2 ns MD simulation during which 125000 test-particles were inserted every 0.2 ps. The numbers in parenthesis denote the statistical inaccuracy.

If we do not consider He, H_2 and Ne, increasing temperature causes more negative solvation entropies, smaller polymer reorganization energies, and weaker (less negative) penetrant binding to the polymeric solvent. The first effect signifies a stronger polymer ordering around the solute at elevated temperature. The second effect expresses the tendency that creating a molecular sized gap in hot poly(ethylene) is easier than in cold poly(ethylene). The third effect expresses the tendency that at higher temperature, the penetrant more frequently visits positions that are energetically less favorable.

Figure 1 displays the solvation Gibbs energy, solvation enthalpy, and solvation entropy of a series of gases as a function of their critical temperatures. The correlation of ΔG_s^* with T_C is similar to presenting gas solubility versus T_C on a log-scale, which has been done in some experimental studies (see e.g. (15, 16)). Note again that ΔG_s^* is determined by the sum of the penetrant binding energy ΔU_{uv}^* and the solute-solvent entropy $-T\Delta S_{uv}^*$ (eq 6), which both, as indicated in Table I and II, correlate well with solute size. The fact that larger penetrants, in general, have higher T_C's justifies the correlation in this figure. The reduction of ΔG_s^* with increasing T_C (i.e., larger penetrants generally have higher solubility) means that the concomitant reduction in entropy (inhibiting solubility) is balanced by increasingly favorable polymer-penetrant binding energetics.

Some more interesting features appear in this representation. First, solvation entropies tend to become more negative as T_C increases. As indicated above, this trend expresses the tendency that polymer-ordering effects around larger penetrants are more significant. The partial heat of solution becomes more exothermic as penetrant T_C increases. This correlation expresses the general feature that the heat of sorption becomes more exothermic as the condensability of the gas (as measured by T_C) increases (15, 16). The solvation entropies and enthalpies partially compensate yielding smaller absolute values of the solvation Gibbs energy.

Figure 2 displays the penetrant binding energy to poly(ethylene) versus the solute surface area weighted energy $\sigma_S^2\sqrt{\varepsilon_S}$ (3), where σ_s is the penetrant Lennard-Jones size and ε_s the penetrant Lennard-Jones force constant. This choice is based on the fact that the penetrant binding energy is determined by the molecular interactions between the penetrant and the methylene units in its solvation shell. The penetrant-methylene pair interaction energy is proportional to $\sqrt{\varepsilon_S \cdot \varepsilon_{CH_2}}$, while the number of penetrant-methylene contacts is assumed to be proportional to the penetrant surface area ($= \pi\sigma_S^2$).

This correlation describes the penetrant binding process surprisingly well. In terms of the parameter $\sigma_S^2\sqrt{\varepsilon_S}$, Ar, O_2, and N_2 lump together, the same applies to Kr and CH_4. If the penetrant binding energy is plotted versus T_C or ε_s/k, the penetrants in these groups scatter along the correlation line strongly (3). It is worthwhile to note that the partial heat of penetrant sorption in glassy polymeric solvents is expected to correlate just as well with this quantity (see next section).

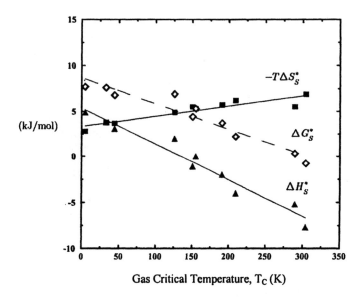

Figure 1. Solvation Gibbs energy ΔG_S^* (open diamonds), enthalpy ΔH_S^* (solid triangles), and entropy $-T\Delta S_S^*$ (solid squares) in poly(ethylene) (kJ/mol) at 300 K and 0.1 MPa presented versus the gas critical temperature T_C.

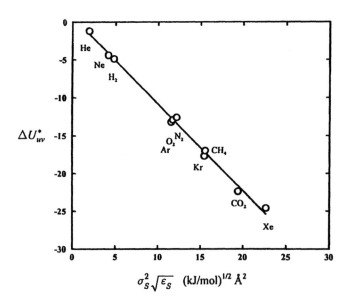

Figure 2. Solute binding energy, ΔU_{uv}^* (kJ/mol), in poly(ethylene) at 300 K and 0.1 MPa as a function of solute energy parameter $\sigma_S^2 \sqrt{\varepsilon_S}$.

Infinite Dilution Solubilities in Glassy Polymeric Solvents

An important distinction between glassy and equilibrium rubbery polymeric solvents is that, in the latter, molecular-sized cavities appear and disappear due to the irregular thermal motions of the polymer. In the equilibrium rubbery state one therefore cannot count them and the question "how many cavities are there?" stated with the intention to relate that information to the solubility is meaningless. Therefore, the concept of "free volume" applied to rationalize solubility in rubbery polymers is of no use; it says no more than that solubility depends on the polymer density. As irregular random motions of the polymer repeat units are responsible for the formation of transient cavities one may surmise that the statistical fluctuations of the polymer fluid's potential energy can be related to the polymer reorganization energy associated with the cavity formation process. This is exactly what eq 3 tells us: polymer reorganization only contributes to the partial heat of gas sorption if the statistical fluctuations of the polymer's potential energy U_N are correlated with the solute insertion probability $\exp(-B_S / RT)$ (*17*). This correlation exists if a fluctuation in the polymeric solvent causing the formation of a cavity at the same time increases the potential energy U_N of the solvent. That is, to create a cavity, polymer contacts must be broken or polymer repeat units must be pushed aside. If this cavity is of sufficient size and shape, it can accommodate the solute, hence the insertion probability in this polymer configuration assumes an increased value as well.

In glassy polymers, the microscopic picture is different. The polymer matrix, including cavities in it, is frozen in a metastable configuration. If the cavity is of sufficient size and shape to accommodate a penetrant its number decreases by one once the penetrant takes advantage of it. Therefore, the number of free volume elements, and, in particular, the distribution of their sizes and shapes strongly impact penetrant solubility. In view of the discussion above, the fact that cavities are static entities (at least on the time scale relevant to a sorption experiment) which may be occupied by penetrants that fit into them, suggests that at infinite dilution the glassy solvent neither reorganizes nor shows a correlation between its potential energy and the penetrant insertion probability. Consequently, the partial heat of gas sorption is fully determined by the local binding interactions between the penetrant and the polymer atoms surrounding the cavity. Because this binding interaction is always exothermic (see Tables I and II), gas sorption in glassy polymeric solvents is always exothermic. In reference (*3*) it was shown that this picture is indeed valid and may cause minima in temperature dependent (helium) sorption data at the glass transition temperature of several polymers. Below the T_g, the penetrant binding process causes exothermic sorption behavior (solubility decreases with increasing temperature). Above T_g solvent reorganization prevails, causing endothermic sorption behavior (solubility increases with increasing temperature).

A consequence of the assertions above is that (at infinite dilution) the partial heat of gas sorption in glassy polymers expresses the process of penetrant binding and thus the nature of polymer-penetrant interactions. Hence, it may make physical sense to express ΔH_S^* of gases in glassy polymers versus $\sigma_S^2 \sqrt{\varepsilon_S}$. Of course, this will be a physically reasonable exercise only if the penetrant molecule in the glassy state is surrounded by a number of polymeric units proportional to the penetrant's surface area. In ultra-high free volume polymers, such as PTMSP, for which experiments indicate the existence of continuous interconnected free volume elements, this assumption will obviously not hold.

Summary and Conclusions

In this chapter the thermodynamics of penetrant sorption are discussed from a molecular perspective. The discussion is based on the theory of statistical mechanics because only this theory provides a direct relation between macroscopic thermodynamic properties and detailed interactions between the molecules comprising the system.

Only polymer-penetrant interactions *directly* affect solubility through terms that occur explicitly in the sorption enthalpy and entropy. The energetic penalty associated with creating a molecular sized cavity in the polymer fluid has no direct influence on the solubility (it cancels with a term in the solvation entropy) but, indirectly, affects the probability of finding a cavity of suitable size to accommodate the penetrant. This *probability*, which determines the value of the sorption *entropy*, decreases as the penetrant gets larger (larger penetrants require bigger cavities), thereby inhibiting sorption. A very important consequence of this conclusion is that trends in the sorption of a series of gases can be fully rationalized on the basis of the penetrant's gas phase properties only; i.e. the penetrant's Lennard-Jones force constant and the penetrant's size determine the values of the polymer-penetrant interaction energy and entropy, respectively. No polymer-specific properties (like the energy required to create a cavity) are needed to explain such trends as is well known from experimental studies.

To illustrate the feasibility of calculating the relevant thermodynamic quantities from molecular simulations, I have summarized some calculations performed for gas sorption in poly(ethylene). For a more detailed description of these simulations the reader is referred to reference (*3*).

Because the above description applies to thermodynamic equilibrium systems only (polymer systems at a temperature $T>T_g$), I have briefly discussed what happens in glassy polymers. The energy required to create a cavity in a glassy polymer is negligible as long as the frozen-in cavities in these systems are big enough to accommodate the penetrant. This condition is met for penetrants like helium and hydrogen in any case. The cavity formation enthalpy causes an endothermic partial heat of sorption for small, weakly interacting gases (He, H_2) in the rubbery state. In the glassy state, the absence of the endothermic cavity formation enthalpy causes exothermic heats of sorption. As a consequence, temperature dependent sorption data of small, weakly interacting gases may pass through a minimum at the T_g of the polymer.

References and Notes

1. The word "solvent" used throughout this chapter refers to a penetrant-free polymer membrane, which acts as a solvation medium for the penetrant (the "solute").
2. In equilibrium (rubbery) polymeric solvents, spatial rearrangement of polymer segments induced by penetrant sorption results in an endothermic contribution to the partial heat of sorption in addition to the (exothermic) penetrant binding energy. In glassy polymeric solvents, at infinite dilution, this "solvent contribution" vanishes as will be shown below.
3. van der Vegt, N.F.A. *J. Membrane Sci.* **2002**, 205, 125-139.
4. Ben-Naim, A. *Statistical Thermodynamics for Chemists and Biochemists*; Plenum: New York, 1992.
5. Ben-Naim, A. *Solvation Thermodynamics*, Plenum: New York, 1987.
6. The second term equates to the sum of the solvent reorganization energy and a contribution, $P\Delta V^*_S$, which is negligible ($\sim 10^{-3}$ kJ/mol) at ambient pressures. Hence, I will refer to this term as the solvent reorganization energy instead of enthalpy.
7. The solvent reorganization term contributes only if the fluctuations of the solvent potential energy are correlated with the penetrant insertion probability $exp(-B_S/RT)$. This correlation is significant, especially in dense liquids. In several studies reported in the literature (*8, 9*), including work performed by this author, the solvent reorganization term has unjustly been neglected. Interestingly, in glassy polymeric solvents at infinite dilution, solvent reorganization plays a negligible role (*3*); therefore partial heats of gas sorption in glassy polymers are fully determined by the penetrant binding energy, which is exothermic.
8. Xiang, T.-X.; Anderson, B.D. *J. Chem. Phys.* **1999**, 110, 1807-1818.
9. van der Vegt, N.F.A. *Macromolecules* **2000**, 33, 3153-3160.
10. Yu, H.-A.; Karplus, M. *J. Chem. Phys.* **1988**, 89, 2366-2379.
11. Guillot, B.; Guissani, Y. *J. Chem. Phys.* **1993**, 99, 8075-8094.
12. Gallicchio, E.; Kubo, M.M.; Levy, R.M. *J. Phys. Chem. B* **2000**, 104, 6271-6285.

13. Lee, B. *Biopolymers* **1985**, 24, 813-823.
14. Note that ΔS_{uv}^{*} (as well as ΔS_{S}^{*}, ΔH_{S}^{*}, and ΔG_{S}^{*}) is an excess contribution measuring the effects of molecular interactions. Hence it excludes all ideal gas contributions. If one views sorption as a form of mixing, one may expect a positive mixing entropy to enter the equations in addition to the negative ΔS_{uv}^{*}. The positive entropy of mixing appears only implicitly in eqs 1 and 2: The penetrant has a chemical affinity, μ_{S}^{*}, for each phase, gas and polymer. The penetrant also has a translational entropy in each phase, represented by a term $RT\ln C$, where C is some measure of the solute concentration in that phase. The solute partitioning ("mixing process") between the two phases reaches equilibrium when the difference in chemical affinities, $\Delta G_{S}^{*} = \mu_{S}^{*} - \mu_{S}^{*ig}$, balances the entropy gain $-RT\ln(\rho_{S}/\rho_{S}^{ig})$ upon mixing, $\Delta G_{S}^{*} = -RT\ln L$ (eq 2).
15. Michaels, A.S.; Bixler, H.J. *J. Polym. Sci. L* **1961**, 393-412.
16. Merkel, T.C.; Bondar, V.; Nagai, K.; Freeman, B.D.; Yampolskii, Y.P. *Macromolecules* **1999**, 32, 8427-8440.
17. The numerator in the last term in eq 3 reads $\langle XY \rangle - \langle X \rangle \langle Y \rangle$ with $X = U_N + PV$ and $Y = V\exp(-B_S/RT)$. If X and Y are uncorrelated, $\langle XY \rangle = \langle X \rangle \langle Y \rangle$ and the entire term cancels.

Chapter 4

Solubility of Gases and Vapors in Glassy Polymer Blends

F. Grassia, M. Giacinti Baschetti, F. Doghieri, and G. C. Sarti

**Dipartimento di Ingegneria Chimica, Mineraria e delle Tecnologie
Ambientali, Alma Mater Studiorum, Università di Bologna, Viale
Risorgimento 2, 40136 Bologna, Italy**

The solubility of gases and vapors in miscible polymer blends
in the glassy state has been considered. The sorption isotherms
are calculated using an extension of the NELF model to
polymer blends for both nonswelling and swelling penetrants.
The swelling coefficient for the blend is given in terms of the
swelling coefficients for the pure polymers. In all cases, the
isotherms for the polymer blends can be completely predicted
based on the parameters needed for the solubility in the pure
polymer. The comparison between the model predictions and
experimental data for PS-TMPC blends are shown in detail
and indicate the reliability of the model.

Introduction

Several technical applications require knowledge of the thermodynamic and
transport properties of low molecular weight solutes in solid polymers, either in
the rubbery or the glassy state. Special attention is devoted to this field for the
development of membrane separation processes, packaging materials, and
barrier protections from different environmental contamination. For such
applications, it is important to describe the solubility of gases and vapors in solid
polymers, both in the rubbery and in the glassy state.

© 2004 American Chemical Society

Rubbery phases are typically in thermodynamic equilibrium so that their description simply requires the use of classical equilibrium thermodynamic arguments and the application of an appropriate model for the Gibbs free energy of polymer-penetrant mixtures. The same procedure cannot be applied to glassy phases due to their departure from thermodynamic equilibrium; specific approaches have therefore been developed to describe phase equilibria involving a glassy phase.

For glassy polymers, the representation of sorption isotherms is generally offered by a well-established empirical correlation, the dual-mode (DMS) model (1,2), which invokes a combination of dissolution and adsorption and results in a correlation typically rather accurate and simple to use. According to its formulation, the model contains three parameters which are endowed with a physical meaning, within the rather simplified framework of the model. These parameters are ultimately treated only as adjustable constants for curve-fitting the model to experimental data, and their values cannot be predicted based on the knowledge of pure component properties; the parameter values are rather used to essentially test internal consistency and to obtain qualitative correlations with polymer structure.

This model has been applied successfully and extensively over the years to represent the typical convex shape of solubility isotherms of glassy polymers; nonetheless it is characterized by important limitations briefly recalled hereafter. Its physical assumptions present an oversimplified picture of the physico-chemical factors affecting the sorption process. The three model parameters can only be used in the range of temperature and pressure over which the regression is performed and for the specific penetrant-polymer pair for which the experimental sorption data were obtained. In this regard, an extensive study of the dependence of the dual-mode parameters on the pressure range used for the fitting procedure was presented by Bondar et al. (3) These observations imply that the DMS model cannot be used for predictive purposes. Moreover, there are cases in which the basic physical interpretation of the dual mode equation cannot be used even qualitatively, to describe the behavior of polymer-penetrant pairs. For instance, the solubility of alcohols in glassy PTMSP exhibits an S-shaped sorption isotherm, which intersects the "*Henry line*" drawn through the origin (4). Consequently, in such cases the contribution to the penetrant concentration due to the adsorption into the microvoids of the glassy matrix would be negative in the low pressure range, which is clearly physically unacceptable.

The physical model, albeit oversimplified, offers a guideline for a useful extension of DMS to penetrant mixtures (5) when the model parameters for single penetrants are known. Some extensions to polymer blends have also been proposed for the DMS, based either on specific volumetric data and T_g of the glassy blend (6) or simply on additivity of the contributions of the single polymers (7).

In the last decade several alternative models have been developed to describe gas solubility in glassy polymers. These models are generally constructed on a more fundamental basis than the DMS model (8-10). In this work, we consider the NELF (Non Equilibrium Lattice Fluid) model proposed by Doghieri and Sarti (10-12) to describe sorption in glassy polymers. It is based on a non-equilibrium thermodynamic analysis and on the resulting relationship between equilibrium and non-equilibrium free energy. The model suitably applies the lattice fluid equation of state to represent the thermodynamic properties of glassy phases and is quite appropriate to calculate, and even to predict, solubility isotherms in such systems. Input information for the NELF model includes equilibrium P-V-T properties of both polymeric and penetrant species and, more significantly, the polymer density at each sorption pressure. When this information is available, sorption isotherms can be reliably predicted for different temperatures over a rather broad pressure range (11-13). In the more frequent case in which the polymer density is experimentally known only for the penetrant-free polymer, the NELF model can reliably calculate the solubility isotherms for swelling penetrants in the range of low penetrant concentrations or pressures (12). The sorption isotherm for swelling gases and vapors can be calculated over a broad pressure range only by accounting for the volume dilation induced on the polymeric matrix. In the absence of direct experimental data on the volume isotherm, the effects of dilation on the sorption isotherm can be estimated through the correlation procedure recently proposed (14), which requires the additional knowledge of only a single solubility data point.

The model has been tested thus far for the solubility of pure penetrants and mixed gases in pure polymers, while numerous technical applications concern the use of polymer blend and copolymers. The theoretical basis of the model appears rather solid: all the coefficients have a precise physical meaning and are subject to independent measurements. Therefore, the model extension to polymer blends and copolymers seems promising and straightforward. Indeed, the thermodynamic approach indicates that the solubility isotherms of nonswelling gases in miscible polymer blends would simply require the application of the mixing rules of the model, while for swelling penetrants the only additional information concerns the relationship between the matrix dilation and the swelling of the pure polymers comprising the blend.

Our present aim is to test the reliability of the NELF model to describe solubility in glassy blends of miscible polymers of nonswelling as well for swelling gases and vapors. For the sake of simplicity, we will consider only binary polymer blends, although the method is easily generalized to multicomponent blends.

After the proper formulation of the model, we will consider the set of parameters required as input to the NELF model for the cases of both non-

58

swelling and swelling penetrants. Notably, we will see that for the latter case, the NELF model offers a complete prediction of the solubility in polymer blends even when the volume dilation data of the blend are not experimentally available.

Overview of the NELF Model

In the glassy state a polymer is in non-equilibrium conditions, and thus the Gibbs free energy does not attain the minimum possible value at the temperature and pressure of the system. Therefore, the usual equilibrium equations of state cannot be applied for the description of the material properties. The NELF model relies on the hypothesis that the Gibbs free energy of a glassy polymer can be described by considering the density of the polymer as an internal state variable whose value, given by separate experimental information, is a measure of the departure from equilibrium. It is thus possible to extend to the nonequilibrium domain the use of equilibrium expressions of the system free energy to adequately describe the properties of the glassy mixture. The model properly uses the Gibbs free energy expression for polymer mixtures proposed by the lattice fluid theory by Sanchez and Lacombe (SL) (15-18), and reduces to the SL model at equilibrium. The gas solubility is calculated from the phase equilibrium condition, which requires equality between the chemical potential of the penetrant species in the glassy polymer and in the external gaseous phase.

In the following sections, we briefly recall the SL expression for the Gibbs free energy of a multicomponent solution, the related mixing rules and the extension to non-equilibrium glassy phases. Based on the system's bulk rheology, the density of the polymer matrix will be recognized as the appropriate internal state variable for the glassy mixture. The corresponding thermodynamic analysis will lead to the expression for the penetrant chemical potential in the glassy phase. From the phase equilibrium condition, a procedure will be obtained to calculate the solubility of low molecular weight penetrants in glassy polymers, provided that the value of the polymer density during sorption is available. Finally, we will examine the use of the model for the estimation of the solubility isotherms of swelling penetrants in glassy polymer blend over a wide range of pressures, even when no dilation data are experimentally available.

Sanchez-Lacombe Gibbs Free Energy and Mixing Rules

Sanchez and Lacombe (15-18) developed a lattice fluid (LF) model which is suitable for describing the thermodynamic properties of polymers and polymeric mixtures. For a mixture of N components, the theory offers the following expression for the total Gibbs free energy (16):

$$G = RT^* \left(\sum_{i=1}^{N} r_i \, n_i \right) \cdot \left\{ -\tilde{\rho} + \frac{\tilde{p}}{\tilde{\rho}} + \frac{\tilde{T}}{\tilde{\rho}} \left[(1-\tilde{\rho}) \ln(1-\tilde{\rho}) + \tilde{\rho} \sum_{i=1}^{N} \frac{\phi_i}{r_i} \ln(\phi_i \, \tilde{\rho}) \right] \right\}$$

(1)

In eq 1, R is the ideal gas constant, r_i is the number of lattice sites occupied by each of the n_i moles of species i; ϕ_i is the volume fraction of species i, defined in terms of the mass fraction ω_i and the characteristic densities of the pure components ρ_i^* as: (14)

$$\phi_i = \frac{r_i n_i}{\sum_{j=1}^{N} r_j n_j} = \frac{\omega_i / \rho_i^*}{\sum_{j=1}^{N} \omega_j / \rho_j^*} = \frac{\omega_i \rho^*}{\rho_i^*}$$

(2)

The quantities $\tilde{\rho}, \tilde{p}, \tilde{T}$ represent the reduced density, pressure, and temperature of the mixture, defined as follows:

$$\tilde{\rho} = \rho / \rho^*, \quad \tilde{p} = p / p^*, \quad \tilde{T} = T / T^*.$$

(3)

The characteristic lattice parameters ρ^*, p^* and T^* of the mixture are calculated from the corresponding pure component values using the following mixing rules:

i) *density* ρ^*:

$$\frac{1}{\rho^*} = \sum_{j=1}^{N} \frac{\omega_j}{\rho_j^*}$$

(4)

ii) pressure p*:

$$p^* = \sum_{j=1}^{N} \phi_j p_j^* - \frac{1}{2} \sum_{j,k=1}^{N} \phi_j \phi_k \Delta p_{j,k}^*$$

(5)

The binary parameters $\Delta p_{j,k}^*$ are constant with respect to composition and temperature and are determined from experimental data of different binary mixtures containing species j and k, and representing either volumetric behavior or mutual solubility.

iii) *temperature* T^*:

$$T^* = \frac{p^* v^*}{R} \tag{6}$$

where the site molar volume of the mixture lattice v^* is

$$\frac{1}{v^*} = \sum_{i=1}^{N} \frac{\phi_i}{v_i^*} \tag{7}$$

The characteristic volumes v_i^* of each pure species are related to the characteristic densities as follows:

$$r_i^0 v_i^* = \frac{M_i}{\rho_i^*} \tag{8}$$

The symbol r_i^0 indicates the number of lattice sites occupied by a mole of species i in the pure component lattice. The mixing rules are completed by introducing the following relationship between r_i^0 and the number of lattice sites r_i occupied by a mole of species i in the mixture:

$$v^* = \frac{r_i^0 v_i^*}{r_i} = \frac{M_i}{r_i \rho_i^*} \tag{9}$$

The only binary parameters entering the mixing rules are $\Delta p_{j,k}^*$, which can equivalently be expressed in the following alternative way, through the dimensionless binary parameters $\psi_{j,k}$:

$$\Delta p_{j,k}^* = p_j^* + p_k^* - 2 \cdot \psi_{j,k} \sqrt{p_j^* p_k^*} \tag{10}$$

so that the characteristic pressure becomes:

$$p^* = \left(\sum_{j=1}^{N} \phi_j \sqrt{p_j^*} \right)^2 + \sum_{j,k=1}^{N} \left[\phi_j \phi_k \left(\psi_{j,k} - 1 \right) \sqrt{p_j^* p_k^*} \right] \tag{11}$$

As a reasonable first approximation, one can set the binary parameter $\psi_{j,k}$ equal to unity or, equivalently (15-18):

$$\Delta p_{j,k}^* = \left(\sqrt{p_j^*} - \sqrt{p_k^*} \right)^2 \quad \Leftrightarrow \quad \psi_{jk} = 1 \tag{12}$$

This first order approximation is entirely equivalent to considering the pair energy potential between unlike molecules as the geometric mean of the energy potential between like species.

From the above expression of the Gibbs free energy, one can obtain the density of the mixture under equilibrium conditions by minimizing the Gibbs free energy at T, p, and ω_i, with respect to the density:

$$\left(\frac{\partial G}{\partial \rho}\right)_{T,p,n_j} = 0 \quad \text{at equilibrium} \tag{13}$$

Equation 13 yields the well-known Sanchez-Lacombe equation of state, which has been used successfully to describe the thermodynamic properties of many equilibrium polymeric solutions.

The density of a glassy mixture, however, cannot be calculated from eq 13, since the system is not at equilibrium. In this case, temperature, pressure and density are treated as independent variables, and it can be shown that eq 1 provides the proper and unique nonequilibrium Gibbs free energy for a lattice fluid in which the bulk rheology is governed by a Voigt model (*11*). In the following discussion, we use eq 1 and the mixing rules presented above as the model for the Gibbs free energy of a glassy mixture.

Polymer Density as An Internal Variable of State

The density of a glassy mixture is typically smaller than the equilibrium density at the temperature, pressure and composition considered and thus offers a quantitative measure of the departure from equilibrium. In general terms, equivalent information can also be obtained by considering other physical quantities, such as the fractional free volume or the fraction of empty lattice sites; indeed, any quantity characterizing the departure of the glassy state from equilibrium can be used as an alternative order parameter. The polymer density is one of the possible order parameters for a glassy phase; however, among all the possible choices, the polymer partial density is the only order parameter which is an internal state variable for the glassy phase (*11*), i.e. it is the only quantity whose time rate of change is itself a function of the state of the system. Such a result is obtained based on the bulk rheology of the glassy mixture during sorption (*10, 11*). When the Voigt model describes the bulk rheology of a polymeric system, the time rate of change of the volume is a unique function of the state of the system, and this leads to (*11*):

$$\frac{d\rho_p}{dt} = f\left(T, p, \omega_i, \rho_p\right) \quad i = 1, 2, ..., N-1 \tag{14}$$

Equation 14 is the local expression for the time rate of change of the polymer density $\rho_p = m_p/V$ during the approach to the final pseudo-equilibrium condition.

For the same mixture, the nonequilibrium Gibbs free energy in the glassy phase is given by:

$$\frac{G}{\sum_{j=1}^{N} n_j} = g(T, p, \omega_i, \rho) = a(T, p, \omega_i, \rho) + \frac{p}{\rho} = a\left(T, p, \omega_i, \frac{\omega_p}{\rho_p}\right) + \frac{p\omega_p}{\rho_p} \tag{15}$$

with $\quad i = 1, 2, ..., N-1$

where a is the molar Helmholtz free energy of the mixture. The last equality in eq 15 is obtained in view of the properties of the polymer densities ρ_p, and comes directly from its definition :

$$\rho_p \equiv \frac{m_p}{V} = \omega_p \rho$$

Based on Eqs 14 and 15, the time rate of change of both polymer density ρ_p and the molar Gibbs free energy are function of the same variables indicated above and thus, by definition, ρ_p is an internal state variable for the non-equilibrium lattice fluid. One can thus apply the thermodynamics of systems endowed with internal state variables to describe the non-equilibrium behavior of a glassy mixture and, by using well-established theoretical results for such systems (*11, 12, 19*), the chemical potential of the low molecular weight penetrant, species 1, in the glassy phase is obtained as follows:

$$\mu_1 = \left(\frac{\partial G}{\partial n_1}\right)_{T, p, \rho_p, n_{j \neq 1}} \tag{16}$$

In the pseudo-equilibrium state of a glassy polymer, the relaxation of polymeric density with time is ultimately kinetically hindered; its value becomes asymptotically constant in time, even though it is not equal to the true equilibrium value. Under these conditions, the time rate of change of ρ_p becomes negligibly small, though not exactly equal to zero, so that for the polymer density, eq 14 asymptotically becomes:

$$\frac{d\rho_p}{dt} = f\left(T, p, \omega_1, \rho_p\right) \approx 0 \qquad (17)$$

Even when the time rate of change of polymer density become imperceptibly small, eq 13, representing true thermodynamic equilibrium conditions, is not obeyed. Equation 17 implies that the viscoelastic contribution to the mixture volume reaches a pseudo-equilibrium value that can be considered asymptotically steady over the time frame of the experimental solubility determinations. Equation 17 is satisfied for a finite range of polymer density and does not constrain ρ_p to a unique value. This feature is consistent with the experimental observation that, at fixed T, p, and ω_1, different values of pseudo-equilibrium polymer density are observed for samples of different mechanical, thermal or sorption history. For example, lower polymer density values are often observed when higher cooling rates are applied to obtain the glass from above T_g; another example is the well-known sorption/desorption hysteresis of polymer density. However, by considering only isothermal sorption histories with pressure values changed after suitably long time intervals, we can say that the asymptotic polymer density reduces to a function of pressure, which varies with the penetrant used:

$$\rho_p^\infty = \rho_p^\infty(p) \qquad (18)$$

Chemical Potential Under Pseudo-Equilibrium Conditions

By considering the case of a polymer mixture formed by N species the chemical potential of penetrant i in the glassy polymeric phase can now be obtained from eq 1, through eq 16:

$$\frac{\mu_i^{(S)}}{RT} = \ln\left(\tilde{\rho}\phi_i\right) - \left[r_i^0 + \frac{\left(r_i - r_i^0\right)}{\tilde{\rho}}\right]\ln\left(1 - \tilde{\rho}\right) - r_i - \tilde{\rho}\frac{2r_i^0 v_i^*}{RT}\left[p_i^* + \sum_{k=1}^{N}\phi_k\left(p_k^* - \Delta p_{i,k}^*\right)\right] =$$

$$= \ln\left(\tilde{\rho}\phi_i\right) - \left[r_i^0 + \frac{\left(r_i - r_i^0\right)}{\tilde{\rho}}\right]\ln\left(1 - \tilde{\rho}\right) - r_i - \tilde{\rho}\frac{2r_i^0 v_i^*}{RT}\left[p_i^* + \sum_{k=1}^{N}\phi_k\left(-p_i^* + 2\psi_{i,k}\sqrt{p_i^* p_k^*}\right)\right] \qquad (19)$$

Equation 19 for the pseudoequilibrium chemical potential of a penetrant in a glassy mixture is the main result of the NELF theory (10, 11).

In eq 19, the last term represents the energetic contribution to the chemical potential, while the first three terms are due to entropic effects. The value of the energetic contribution to the chemical potential is affected by the binary parameters, which characterize energetic interactions between unlike species in the mixture.

Pseudoequilibrium Solubility

The phase equilibrium requirement for a glassy mixture in contact with a fluid phase containing a single penetrant j leads to the expected condition that the penetrant chemical potential both in the solid and in the external fluid phase must be equal (10-12)

$$\mu_j^S \left(T, p, \omega_i, \rho_p^\infty (p) \right) = \mu_j^E \left(T, p, y_i \right) \tag{20}$$

where y_i is the molar fraction of component i in the external phase contiguous to the polymeric glass. Equation 20 represents the phase equilibrium condition when, due to hindered relaxation, the asymptotic polymer density is ρ_p^∞, rather than its equilibrium value. In eq 20 one must use the explicit expression offered by eq 19 for the penetrant chemical potential in the solid glassy phase.

In principle, to evaluate the pseudoequilibrium solubility in a polymer in contact with a external gas phase containing penetrant j at a given T and p, we solve simultaneously the equilibrium condition (eq 20), and the evolution equation for the order parameter, which provides for the function $\rho_p(t)$:

$$\begin{cases} \dfrac{d\rho_p}{dt} = f(T, p, \omega_1, \rho_p) \\ \rho_p = \rho_o \qquad at \ \ t = 0 \end{cases} \tag{21}$$

NELF Correlations for Swelling Penetrants: Polymer Dilation

The analysis of several experimental measurements on polymer dilation (21-23) indicates that the pseudo equilibrium density of the single polymer, ρ_p^∞, follows a linearly decreasing function of penetrant pressure p, in the isothermal sorption experiments in pure polymer matrices. It is thus convenient to introduce the swelling coefficient k defined as:

$$k = \left[\frac{\rho_p^0 - \rho_p^\infty(p)}{p} \right] \frac{1}{\rho_p^0} \tag{22}$$

where ρ_p^∞ represents the polymer density asymptotically reached after exposure to the penetrant at pressure p. According to the available evidence, for the case of glassy polymers, the coefficient k can be considered constant in a relatively wide pressure range, and the linear dependence on penetrant pressure p of the pseudo equilibrium density ρ_p^∞ of the single polymer can be expressed equivalently as follows:

$$\rho_p^\infty(p) = \rho_p^0 \left(1 - kp\right) \tag{23}$$

This observation is the basis for the NELF model correlation (14) for the solubility of swelling penetrants. In the absence of direct experimental evidence of the function $\rho_p^\infty(p)$, or of the swelling coefficient k, one can use an experimental solubility data point, at a sufficiently high pressure where dilation is appreciable. Based on that single point, use of the phase equilibrium condition, eq 20, with the NELF model chemical potential, eq 19, allows a calculation of the polymer density ρ_p^∞ at the pressure p used for the solubility datum; then eq 22 offers an estimate of the swelling coefficient and eq 23 provides the polymer density at all pressures of interest. Based on these values, one can calculate the solubility isotherm at all pressures. As already discussed previously (14), the procedure is rather stable and reliable.

The extension of the above correlation procedure for swelling penetrants to the case of miscible polymer blends requires a proper mixing rule for k, i.e. a relationship between the swelling coefficients of the pure polymers and the swelling coefficient of the resulting blend.

NELF Model for Binary Polymer Blends

Let us consider for simplicity sake the case of a binary blend of miscible polymers, species 2 and 3, in the glassy state, in contact with an external phase containing the penetrant 1. Before sorption, the blend composition is characterized by the mass ratio $\Omega_{2,3} = m_2/m_3$, which remains unchanged after sorption, since the masses m_2 and m_3 of the two polymers remain constant. During sorption, therefore, the following relationship holds true for the mass fractions of the polymeric species:

$$\frac{\omega_2}{\omega_3} \equiv \frac{m_2}{m_3} = \Omega_{2,3} \tag{24}$$

As a consequence, for any polymer blend, the composition of the ternary polymeric phase is identified simply by the mass fraction ω_1 of the penetrant species and by the polymer mass fraction $\Omega_{2,3}$, which remains constant during sorption/desorption.

In view of the properties of the partial polymer densities ρ_2 and ρ_3 deriving from their definitions we also have:

$$\frac{\rho_2}{\rho_3} \equiv \frac{m_2}{m_3} = \Omega_{2,3} \tag{25}$$

so that the two polymer densities are not independent of one another for any fixed blend; the knowledge of each polymer density ρ_2 or ρ_3 is entirely equivalent to the knowledge of the overall polymer density ρ_p, in view of eq 25 and of the following:

$$\rho_p \equiv \frac{m_2 + m_3}{V} = \rho_2 + \rho_3 = (\omega_2 + \omega_3)\rho = (1 - \omega_1)\rho \tag{26}$$

Therefore, even though ρ_2, ρ_3 and ρ_p are all internal state variables, they are uniquely related to one another and the polymer blend is characterized by one single independent internal state variable. In the following discussion, we will choose ρ_p as the internal state variable. Equations 14 and 15 become:

$$\frac{d\rho_p}{dt} = f\left(T, p, \omega_1, \rho_p\right) \tag{27}$$

$$\frac{G}{n_1 + n_2 + n_3} = g(T, p, \omega_1, \rho_p) = a(T, \omega_1, \rho_p) + \frac{p(1 - \omega_1)}{\rho_p} \tag{28}$$

For the ternary mixture under consideration eq 1 can be simplified as follows, since r_2 and r_3 are much larger than r_1:

$$G = R T^* \left(r_1 n_1 + r_2 n_2 + r_3 n_3\right) \left\{ -\tilde{\rho} + \frac{\tilde{p}}{\tilde{\rho}} + \frac{\tilde{T}}{\tilde{\rho}}\left[(1 - \tilde{\rho})\ln(1 - \tilde{\rho}) + \frac{\phi_1}{r_1}\tilde{\rho}\ln(\phi_1\tilde{\rho})\right]\right\} \tag{29}$$

The corresponding mixing rules are immediately obtained from the general expressions reported above. In particular for the characteristic pressure we have:

$$p^* = \phi_1 p_1^* + \phi_2 p_2^* + \phi_3 p_3^* - \phi_1\phi_2 \Delta p_{12}^* - \phi_1\phi_3 \Delta p_{13}^* - \phi_2\phi_3 \Delta p_{23}^* \qquad (30)$$

so that the binary parameters contained in the characteristic pressure are Δp^*_{12}, and Δp^*_{13} (or equivalently ψ_{12} and ψ_{13}) relative to the binary mixtures formed by penetrant 1 and either pure polymer 2 or pure polymer 3, and Δp^*_{23} (or ψ_{23}) relative to the binary polymeric blend of the two polymers 2 and 3. No additional parameters are needed for the thermodynamic properties of the ternary mixture.

The chemical potential of penetrant 1, based on eq 19, becomes:

$$\frac{\mu_1^{(S)}}{RT} = \ln(\tilde{\rho}\phi_1) - \left[r_1^0 + \frac{(r_1 - r_1^0)}{\tilde{\rho}} \right] \ln(1-\tilde{\rho}) - r_1 - \tilde{\rho}\frac{2r_1^0 v_1^*}{RT} \left[\phi_1 p_1^* + \phi_2 \psi_{1,2}\sqrt{p_1^* p_2^*} + \phi_3 \psi_{1,3}\sqrt{p_1^* p_3^*} \right]$$

$$(31)$$

The solubility isotherm can thus be calculated considering the phase equilibrium equation, eq 20. For a non swelling penetrant no further information is needed but the density of the unpenetrated glassy blend. For a swelling penetrant the unpenetrated polymer density allows the solubility calculation in the low pressure range only, while polymer dilation, or the swelling coefficient, is required to construct the entire isotherm. In the absence of direct experimental information about the volume dilation of the blend, the swelling coefficient of the blend itself can be obtained from the swelling coefficients of the single polymers 2 and 3, through a proper mixing rule. Based on a volume additivity guideline, the mixing rule proposed for the dilation coefficient is based on the volume average between the dilation coefficients of the pure polymers, that is:

$$k = \varphi_2 k_2 + (1 - \varphi_2)k_3 \qquad (32)$$

where ϕ_2 and $(1 - \phi_2)$ are the volume fractions of polymer 2 and 3 in the pure unpenetrated blend, respectively.

It is worth noting that the swelling coefficient is indeed a fitting parameter for the solubility isotherm in a pure polymer matrix, as already discussed. On the contrary, based on the mixing rule embodied by eq 32, the swelling coefficient for a polymer blend is a known value once the pure component's terms have been determined; as a consequence, the solubility isotherms do not require any further adjustable parameters and the model can be used in a predictive mode both for non-swelling as well as for swelling penetrants.

Comparison with Experimental Data for Polymer Blends.

A comparison of the model with experimental data will be performed using the solubility isotherms in polymer blends formed by polystyrene (PS) and tetra-methylpolycarbonate (TMPC). Methane is selected as a nonswelling gas, and carbon dioxide is selected as a swelling penetrant. The experimental data available, obtained by Muruganandam and Paul (24), report the behavior observed for the pure polymers as well as for their blends.

The solubility isotherms at 35 °C for methane in pure PS and pure TMPC are shown in Figure 1. The dotted lines indicate the NELF model prediction based on the first order approximation for the binary parameter, while the

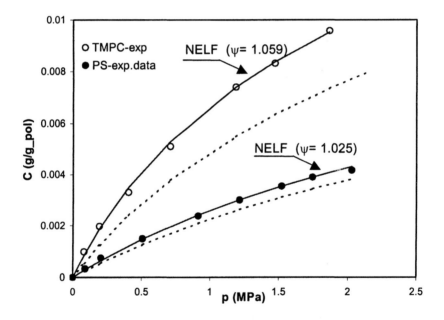

Figure 1 Solubility isotherms of methane in pure PS and in pure TMPC. Data from ref. (24); dotted lines are NELF model predictions based on $\psi_{ij}= 1$, continuous lines are NELF model calculations based on fitted binary parameters.

continuous lines indicate the best fit of the NELF model obtained with the binary parameter $\psi_{12}= 1.059$ for the pair $CH_4 - TMPC$ and with $\psi_{13}= 1.025$ for CH_4 in PS. One can appreciate that for the case of PS, the default binary parameter is rather appropriate; in any case we will use the adjusted value of 1.025 for a better representation.

Once the two binary parameters have been obtained, the solubility isotherm of CH_4 in all the PS-TMPC blends are calculated by the model in an entirely predictive way. The corresponding results are reported in Figure 2, which clearly indicates the ability of the model to calculate the solubility isotherms for all the blends inspected; the maximum deviation is 15%.

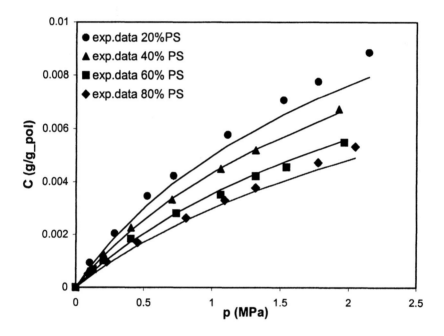

Figure 2 Solubility isotherms of methane in PS – TMPC blends. Data from ref. (24); continuous lines are NELF model predictions based on the binary parameters obtained for the pure polymers.

For the swelling penetrant, CO_2, the solubility isotherms at 35 °C in pure PS and pure TMPC are shown in Figure 3. The dotted lines indicate the NELF model calculations based on the first order approximation for the binary parameter, and on a zero value for the swelling coefficient. The continuous lines indicate the best fit of the NELF model obtained with the binary parameter ψ_{12}= 1.060 and the swelling coefficient of k_{12}=0.0155 MPa^{-1} for the pair CO_2 – TMPC and with ψ_{13}= 1.054 and the swelling coefficient of k_{13}=0.0057 MPa^{-1} for CO_2 in PS. For the case of PS, the first order approximation for both binary parameter and swelling coefficient is rather appropriate in the pressure range of interest. In any case, we will use the adjusted values for a better representation. For the determination of the binary parameters the fitting procedure considered only the low pressure portion of the solubility isotherm, while the higher pressure range was used to determine the swelling coefficients.

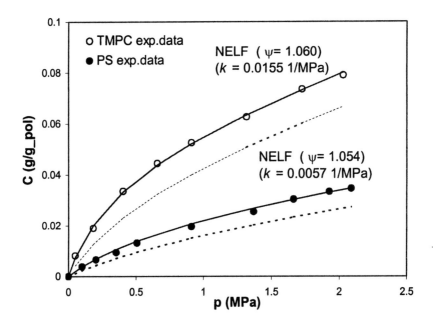

Figure 3 Solubility isotherms of CO_2 in pure PS and pure TMPC. Data from ref. (24); dotted lines are NELF model predictions based on ψ_{ij}= 1 and k_{ij} =0, continuous lines are NELF model calculations based on fitted binary parameters and swelling coefficients.

The parameter values determined based on the solubility of CO_2 in pure PS and in pure TMPC are sufficient to determine in a predictive way the solubility isotherms in all of the PS-TMPC blends, in view of Eqs. 30 and 32. In Figure 4 the results obtained are compared with the experimental data reported by Muruganandam and Paul (24). It is remarkable to notice that also for the case of swelling penetrants the model can represent the solubility isotherms in an entirely predictive way for all the blends considered, based on the parameters needed to describe the solubility isotherms in the pure polymers. The maximum deviation between the experimental data and the model is 13%.

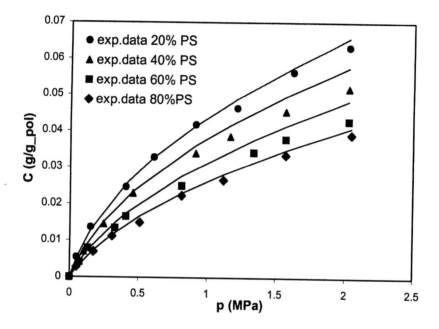

Figure 4 Solubility isotherms of CO$_2$ in PS – TMPC blends. Data from ref. (24); continuous lines are NELF model predictions based on the binary parameters and swelling coefficients obtained for the pure polymers.

Conclusions

The extension of the NELF model to calculate the solubility isotherm in glassy polymer blends is rather direct, both for the case of nonswelling gases as well as for swelling penetrants. In the former case, it is sufficient to obtain the binary parameters for the penetrant and each of the pure polymers, then the model becomes predictive for all the polymer blends considered. In many cases

72

the first order approximation of ψ_{jk}=1 was satisfactory. For swelling penetrants, the swelling coefficient in each pure polymer contained in the blend is required in addition to the binary mixing parameters. The mixing rule proposed for the swelling coefficient of the blend, based on the average swelling coefficient weighted by the volume fraction of each polymer in the blend, was found appropriate and effective for all blend compositions. The method has been applied to several blends with rather satisfactory predictions. Detailed examples have been explicitly reported for the case of methane and carbon dioxide solubility in PS-TMPC blends.

Acknowledgements

This work has been supported by the Italian Ministry of Education University and Research, *40% funds*, 2001.

References

1. Barrer, R.M.; Barrie, J.A.; Slater, J. *J. Polym Sci*, **1958**, *27*, 177-197.
2. Michaels, A.S.; Vieth, W.R.; Barrie, *J.A. Appl. Phys.*, **1963**, *34*, 1-12.
3. Bondar, V. I.; Kamiya, Y.; Yampol'skii, Y. P. *J. Polym. Sci.: Part B : Polym. Phys.*, **1996**, *34*, 369-378.
4. Doghieri, F.; Biavati, D.; Sarti, G. C. *Ind. Eng. Chem. Res.*, **1996**, *35*, 2420-2430.
5. Koros, W.J.; *J. Polym. Sci., Polym. Phys. Ed.*, **1980**, *18*, 981-992.
6. Masi, P.; Paul, D.R.; Barlow, J.W. *J. Polym. Sci.: Polym. Phys. Ed.*, **1982**, *20*, 15-26.
7. Kato, S.; Tsujita, Y.; Yoshimizu, H.; Kinoshita, T. *Polymer*, **1997**, *38*, 2807-2811.
8. Conforti, R. M.; Barbari, T. A.; Vimalchand, P.; Donohue, M. D. *Macromolecules*, **1991**, *24*, 3388-3394.
9. Wissinger, R.G.; Paulaitis, M.E. *Ind. Eng. Chem. Res.*, **1991**, *30*, 842-851.
10. Doghieri, F.; Sarti, G. C. *Macromolecules* **1996**, *29*, 7885-7896.
11. Sarti, G. C.; Doghieri, F. *Chem. Eng. Sci.* **1998**, *53*, 3435-3447.
12. Doghieri, F.; Sarti, G. C. *J. Membr. Sci.* **1998**, *147*, 73-86.
13. De Angelis, M. G.; Merkel, T. C.; Bondar, V. I.; Freeman, B. D.; Doghieri, F.; Sarti, G. C., *J. Polym. Sci.: Part B : Polym. Phys.*, **1999**, *37*, 3011-3026.

14. Giacinti, M.; Doghieri, F.; Sarti, G.C. *Ind. Eng. Chem. Res.* **2001**, *40*, 3027-3037.
15. Sanchez, I. C.; Lacombe, R. H. *J. Phys. Chem.*, **1976**, *80*, 2352-2362.
16. Lacombe, R. H.; Sanchez, I. C. *J. Phys. Chem.*, **1976**, *80*, 2568-2580.
17. Sanchez, I. C. ; Lacombe, R. H. *J. Polym. Sci.: Polym. Letts. Ed.*, **1977**, *15*, 71-75.
18. Sanchez, I. C.; Lacombe, R. H. *Macromolecules*, **1978**, *11*, 1145-1156.
19. Astarita, G. *Thermodynamics*. Plenum Press, New York, USA, **1989**.
20. Doghieri, F.; Canova, M.; Sarti, G.C. in *Polymer Membranes for Gas and Vapor Separation;* B.D.Freeman-I.Pinnau Eds., Am. Chem. Soc. Symp. Series 733, ACS, Whashington D.C., 1999, pp. 179-193.
21. Fleming, G. K.; Koros, W. J. *Macromolecules,* **1990**, *23*, 1353-1360.
22. Koros, W. J., Paul, D. R.; Rocha, A. A. *J. Polym. Sci. Polym. Phys. Ed.* , **1976**, *14*, 687-702.
23. Koros, W. J.; Paul, D. R.; *J. Polym. Sci. Polym. Phys. Ed.*, **1978**, *16*, 1947-1963.
24. Muruganandam, N.; Paul, D.R. *J. Polym. Sci.: Part B : Polym. Phys.*, **1987**, *25*, 2315-2329.

Chapter 5

Predicting Gas Solubility in Glassy Polymers through Nonequilibrium EOS

F. Doghieri[1], M. Quinzi[1], D. G. Rethwisch[2], and G. C. Sarti[1]

[1]Dipartimento di Ingegneria Chimica, Mineraria e delle Tecnologie
Ambientali, Università di Bologna, Viale Risorgimento 2,
40136 Bologna, Italy
[2]Department of Chemical and Biochemical Engineering, University of Iowa,
4139 Seamans Center, Iowa City, IA 52242–1527

The non-equilibrium thermodynamics approach to describing
properties of glassy polymers, earlier applied in the so-called
NELF model using Lattice Fluid Theory, is generalized to the
use of arbitrary equations of state. Application to Statistical
Associating Fluid Theory (SAFT) is discussed in detail and
use of this approach to predict or correlate gas solubility in
glassy polymers is presented for the case of low-pressure
sorption. Prediction procedures rely on volumetric data for the
pure polymer and gaseous species, while for data correlation a
temperature independent binary interaction coefficient is used
as an adjustable parameter. Results from model calculations
are compared with literature data for gas solubility in
polycarbonate and polysulfone. Analysis of solubility
coefficient variations above and below the glass transition
temperature indicates that the non-equilibrium approach is the
proper procedure to extend an EoS description of
thermodynamic properties to the glassy state.

© 2004 American Chemical Society

Satisfactory descriptions of the properties of amorphous polymer-solvent mixtures above the glass transition temperature can be obtained through several well-established, thermodynamic models based on activity coefficients (*1, 2*) or equation of state approaches (*3-7*). These models provide reliable predictions, or at least meaningful correlations, for gas solubility in rubbery polymers. This is valuable for understanding the performance of pervaporation or gas separation membranes prepared from such materials.

On the contrary, empirical or semi-empirical tools are typically used to discuss and correlate gas or vapor solubility data in glassy polymers. For example, in analyzing experimental solubility data by the Dual Mode model (*8*), which is by far the most popular of these tools, good correlations and useful insights into the polymer structure can be obtained, but no predictive use can be made of these models. The lack of models able to estimate the solubility coefficient in glassy polymers is a serious limitation to the development of effective tools for the design of materials for gas separation membranes (*9*).

Development of tools for the description of thermodynamic properties in polymeric glasses is complicated by the need to address the non-equilibrium nature of the glassy state. Below the glass transition temperature, samples of the same material prepared according to different protocols can display quite different properties though identical external conditions prevail. In the simplest view, the hindered mobility of polymer chains in the glassy state does not allow the system to access all possible microstates and ultimately prevents the structure from reaching the volume that would minimize its Gibbs free energy.

From the modeling point of view, departure from equilibrium conditions does not allow direct use of relations from either classical or statistical thermodynamics of mixtures. In fact, the physical picture described above for departure from equilibrium inspired the development of several approaches (*10-13*) to calculate thermodynamic properties of glassy mixtures, all aimed at overcoming the limitations of the semi-empirical tools.

Among the most effective methods in predicting gas solubility in polymers or polymer blends is the Non-Equilibrium Lattice Fluid (NELF) model (*12, 14, 15*). Initially applied within the framework of Lattice Fluid Theory, it represents a general procedure for the extension of relations for thermodynamic properties established for equilibrium conditions to non-equilibrium states attainable by polymeric phases in glassy state.

In this work a generalization of the procedure is proposed so that it can be applied to thermodynamic relations for equilibrium conditions resulting from an arbitrary equations of state. In consideration of the remarkable results obtained in describing phase equilibria above the glass transition temperature, the statistical-associating-fluid theory (*4*) was selected for a detailed discussion.

Background

The basic assumptions and results of the NELF approach to the description of thermodynamic properties of glassy polymeric phases, which has been reported in detail in previous publications (12, 16), are summarized below. This model considers the glassy phase to be a homogeneous, isotropic and amorphous system whose properties not only depend on composition and externally imposed conditions such as temperature and pressure, but also on convenient order parameters which describe the out-of-equilibrium degree of the system.

In general, several order parameters could be used to properly describe the non-equilibrium states Σ of the system. However, in the present study, the specific volume of the polymer network is used to describe the out of equilibrium condition of the system. According to this view, which is a simplification of a more general picture, the state of the system Σ for a polymer-solute binary mixture can be represented by the following set of variables:

$$\Sigma = \left\{ T, p, \Omega_{sol}, \rho_{pol} \right\} \tag{1}$$

where T, p and Ω_{sol} are the temperature, pressure and solute-to-polymer mass ratio, while ρ_{pol} is the polymer mass per unit volume. In eq 1 ρ_{pol} is an order parameter that characterizes the degree of departure from equilibrium conditions by comparison to ρ_{pol}^{EQ}, its value at equilibrium. The latter depends on system composition, temperature and pressure and it can be identified through the condition of minimum Gibbs free energy for the system:

$$\rho_{pol} = \rho_{pol}^{EQ}\left(T, p, \Omega_{sol} \right) \quad \Leftrightarrow \quad \left(\frac{\partial G}{\partial \rho_{pol}} \right)_{T, p, \Omega_{sol}} = 0 \tag{2}$$

According to this simplified picture, the density of the polymer network ρ_{pol} is the only information necessary to represent the out-of equilibrium degree of the system. Thus, in this model two distinct glassy polymer samples with different thermal, mechanical or sorption histories, are predicted to have the same thermodynamic properties at a given temperature, pressure and composition, provided that they are characterized by the same mass density.

The second key assumption of the model is that the order parameter ρ_{pol} evolves in time according to a rate that only depends upon the state of the system, i.e. the following equation holds:

$$\frac{d\,\rho_{pol}}{d\,t} = f\left(T, p, \Omega_{sol}, \rho_{pol}\right) \qquad (3)$$

Indeed, in this view the pseudo-equilibrium states for the glassy phase are recognized as those states in which the function f on the right hand side of eq 3 becomes negligible, though the resulting ρ_{pol} values are appreciably different from ρ_{pol}^{EQ}.

Eq 3 states that ρ_{pol} is assumed to be an internal state variable for the system (17), and, thus, the rest of the analysis can be performed using the tools of thermodynamics for systems endowed with internal state variables. Indeed, after the assumptions stated in Eqs 1 and 3, simple and meaningful results for the dependence of the Helmholtz free energy density a^{NE} and solute chemical potential μ_{sol}^{NE} upon the state of the system can be obtained after application of the second law of thermodynamics. Only the results are indicated here, while the reader is referred to a previous publication (16) for details of the analysis:

$$\left(\frac{\partial\,a^{NE}}{\partial\,p}\right)_{T,\Omega_{sol},\rho_{pol}} = 0 \qquad (4)$$

$$\mu_{sol}^{NE} = \left(\frac{\partial\,a^{NE}}{\partial\,\rho_{sol}}\right)_{T,p,\rho_{pol}} \qquad (5)$$

In eqs 4 and 5 a^{NE} and μ_{sol}^{NE} represent the non-equilibrium Helmholtz free energy density and solute chemical potential for the system, respectively, and should not be mistaken for the corresponding, more familiar, equilibrium functions.

Eqs 4 and 5 are key results which show that, under the assumptions stated in eqs 1 and 3, both non-equilibrium free energy and solute chemical potential depend only upon temperature and component densities, and they are unaffected by the imposed pressure. Thus, the following relations hold:

$$a^{NE}\left(T, p, \Omega_{sol}, \rho_{pol}\right) = a^{EQ}\left(T, \Omega_{sol}, \rho_{pol}\right) \qquad (6)$$

$$\mu_{sol}^{NE}\left(T, p, \Omega_{sol}, \rho_{pol}\right) = \mu_{sol}^{EQ}\left(T, \Omega_{sol}, \rho_{pol}\right) \qquad (7)$$

In the above equations, a^{EQ} and μ_{sol}^{EQ} are the equilibrium values of the Helmholtz free energy density and solute chemical potential in the polymer-solute mixture at a given temperature and species density. Eqs 6 and 7 express a simple but non-trivial result indicating that the free energy and solute chemical potential can be evaluated in all accessible non-equilibrium states of the system starting from knowledge of their value on the restricted equilibrium domain.

Solubility Calculation

Results of the above analysis are important for the calculation of the pseudo-equilibrium solute content in a glassy polymer sample in contact with a pure gaseous phase at temperature T and pressure p. To make a clear distinction between equilibrium and pseudo-equilibrium calculations for phase equilibria in polymeric solution, first recall the procedure for calculating the solute content in a polymeric mixture under conditions of true thermodynamic equilibrium. In this case, the phase equilibrium problem is described by the following set of equations, where G indicates the Gibbs free energy of the mixture and superscripts (s) and (g) indicate properties for the condensed and gaseous phase, respectively:

$$\begin{cases} \mu_{sol}^{EQ\,(s)}\left(T, \Omega_{sol}^{EQ}, \rho_{pol}^{EQ}\right) = \mu_{sol}^{EQ\,(g)}\left(T, p\right) \\ \left(\dfrac{\partial G}{\partial \rho_{pol}}\right)_{T, p, \Omega_{pol}} = 0 \end{cases} \qquad (8)$$

Once input values for the temperature T and pressure p are assigned, the equilibrium solute mass per polymer mass Ω_{sol}^{EQ} and polymer density ρ_{pol}^{EQ} are determined by solving eq 8.

The pseudo-equilibrium solute mass per polymer mass Ω_{sol}^{PE} results from the solution of a phase equilibrium problem which only requires a common value of the solute chemical potential in the gaseous and condensed phases:

$$\mu_{sol}^{NE\,(s)}\left(T, p, \Omega_{sol}^{PE}, \rho_{pol}^{PE}\right) = \mu_{sol}^{EQ\,(g)}\left(T, p\right) \qquad (9)$$

In the pseudo-equilibrium formulation stated above, the thermodynamic equilibrium constraint of minimum Gibbs free energy has been relaxed and the resulting problem can be solved to obtain Ω_{sol}^{PE} only after a proper value for the pseudo-equilibrium polymer density ρ_{pol}^{PE} has been input. Indeed, independent measurement of the pseudo-equilibrium polymer density for the sorption condition is needed to provide a complete formulation of the pseudo-equilibrium problem in this picture. While this necessity can be seen as a serious drawback for the application of the model in a purely predictive way, ρ_{pol} is the yardstick for the out-of-equilibrium degree of the system, and it allows the possibility of different solubility coefficient values for samples of the same material treated according to different protocols (12). The remaining question, however, is: what is the appropriate method to evaluate the pseudo-equilibrium polymer density ρ_{pol}^{PE} in eq 9 to obtain the pseudo-equilibrium solute content Ω_{sol}^{PE} for the case of interest?

When low-pressure sorption is considered, the pure glassy polymer density ρ_{pol}^{0} provides a good estimate of ρ_{pol}^{PE}. Indeed, at low gas pressure, polymer swelling due to gas sorption can be neglected, and the system volume does not change significantly with respect to its initial value. The same is also true at moderate pressures for non-swelling agents, such as many permanent gases. However, in the general case of higher pressures and condensable gases, the non-equilibrium value of the density characterizing the sample of interest should be experimentally evaluated or estimated from the value measured for a sample prepared according to the same protocols.

As only low-pressure solubility calculations are discussed in this work, we will hereafter refer to the pseudo-equilibrium problem as that described by the following condition:

$$\mu_{sol}^{NE(s)} \left(T, p, \Omega_{sol}^{PE}, \rho_{pol}^{0} \right) = \mu_{sol}^{EQ(g)} \left(T, p \right) \qquad (10)$$

A discussion of convenient procedures to address the high-pressure pseudo-equilibrium solubility of swelling agents is given in a previous publication (14).

Eqs 6 and 7 provide a general result that was derived under the simplifying assumptions used for the order parameters and its evolution equation (see Eqs 1 and 3), but they are not limited to any specific thermodynamic model for the representation of equilibrium properties for the polymer-solute mixture. This means that, given any reliable representation of the Helmholtz free energy or the solute chemical potential of the mixture in terms of temperature and species densities, expressions for the same quantities in non-equilibrium states are obtained from results of the above analysis. Low-pressure phase equilibrium calculations can then be performed by solving eq 10 once the dry polymer density ρ_{pol}^{0} for the samples of interest is specified.

In this work the SAFT EoS, which has already proven to be valuable in representing phase equilibria for simple fluids and polymeric systems in the melt phase, was considered. In the next section, a brief description of the characteristics of this EoS will be given before presenting results for gas solubility in glassy polymers obtained from its use in this framework.

SAFT EoS

SAFT belongs to the class of Tangent Hard Sphere Chain models in which molecules are represented as chains of spherical segments with an assigned mass and a temperature dependent volume. Segments are connected to adjacent segments in the chain, and they interact with other segments according to a proper pair interaction potential of spherical symmetry. In the SAFT model, unlike the case of other EoS in the same class, the free energy contribution due to specific hydrogen bonding interactions between segments is treated separately, requiring a preliminary identification of associating sites in each molecule. In the calculations that follow, however, only non-associating species have been considered and representation of specific hydrogen bonding interactions in SAFT has no role in the present discussion. Similar to other models in this class, the SAFT EoS relies on statistical thermodynamics arguments and take advantage of proper simplifications. It is not the aim of this section to offer an exhaustive presentation of its characteristic features but rather to introduce the model parameters and to indicate the pure component and mixture properties that are needed for their retrieval. The reader is referred to the original papers for details.

Several versions of the SAFT equation of state have been proposed in the literature after the initial works by Chapman and coworkers (*18, 19*). In the present work we use one of the earliest versions, which is described in detail by Huang and Radosz (*4*). In the SAFT model, the residual Helmholtz free energy of the system a^{res}, defined as the free energy difference with respect to the corresponding ideal gas mixture at the same temperature and volume, results from the sum of the *hard sphere, dispersion, chain* and *association* contributions:

$$a^{res} = a^{hs} + a^{disp} + a^{chain} + a^{assoc} \qquad (11)$$

The first two terms in eq 11 refer to the segment-segment hard sphere and mean-field interactions, respectively. The chain term accounts for the free energy increment due to permanent bonding between segments in the chain, while the last contribution refers to specific hydrogen bonding between associating sites. The dispersion term uses the power series expression developed by Chen and

Kreglewski (*20*), after fitting accurate PVT data for argon. For non-associating species only three component parameters, besides the molar mass (*MM*), appear in the pure component free energy expression of SAFT: the sphere volume (v^{00}), the sphere mass (*MM/m*) and the characteristic energy of the interactions represented in the dispersion contribution (u_{ii}^{0}).

The equilibrium free energy expression is extended to solutions by using mixing rules where additional binary parameters appear. We will discuss here only the role of the binary interaction parameter k_{ij}, which characterizes the relation between the characteristic interaction energy for pairs of unlike and like segments:

$$u_{ij}^{0} = \left(1 - k_{ij} \right) \sqrt{u_{ii}^{0} \, u_{jj}^{0}} \qquad (12)$$

The default value $k_{ij} = 0$ can be used to recover the usual geometric mean rule for the characteristic interaction energy.

Pure component parameters for SAFT are typically evaluated by comparing EoS predictions with equilibrium PVT data for the given chemical species. For low molecular weight components, phase equilibrium data such as vapor pressure and saturated liquid density over a relatively large temperature interval are used in most cases. Specific volume (above the glass transition temperature) as a function of temperature and pressure is typically used to determine pure component parameters for polymeric species.

As all pure component parameters for SAFT have precise interpretations in terms of a microscopic picture of the species molecule, their evaluation would also be possible, in principle, using group contribution methods based on simple information about the chemical structure. A reliable procedure of this kind, however, should be based on an extensive analysis of group contributions that has not yet been completed for any EoS in this class. Thus, at present, proper experimental PVT data or, as an alternative, computer simulation results are still needed to safely estimate pure component SAFT parameters.

Evaluation of the binary interaction parameter requires binary phase equilibrium data. Typically, to obtain a good representation of binary VLE or LLE data, the binary interaction parameters must vary with temperature. In this work, however, for the sake of simplicity, k_{ij} will be regarded as temperature independent. Reasonable values for k_{ij} in Eqs 12 or 13 can sometimes be estimated even in the absence of specific binary data, as it is expected to be close to zero for a pair of chemically similar species and only small variations should result when changing one species in a pair with another one of the same class.

NE-SAFT Model for Pseudo-Equilibrium Gas Solubility in Glassy Polymers

Once the necessary pure component and binary interaction parameters have been determined or estimated, a map of the solute chemical potential over the entire domain of non-equilibrium states is obtained from the corresponding equilibrium map using eq 7. For SAFT, a pseudo-equilibrium solubility model results, hereafter called the non-equilibrium SAFT [NE-SAFT] model, which for the case of low pressure has the following form:

$$\mu_{sol}^{NE(s)} \begin{pmatrix} T,\, p,\, \Omega_{sol}^{PE},\, \rho_{pol}^{0}; \\ MM_{sol},\, V_{sol}^{0},\, \left(MM/m\right)_{sol},\, u_{sol}^{0}, \\ V_{pol}^{0},\, \left(MM/m\right)_{pol},\, u_{pol}^{0}, \\ k_{ij} \end{pmatrix} = \mu_{sol}^{EQ(g)} \left(T,p\right) (13)$$

where the left-hand-side represents the chemical potential of the gas dissolved in the polymer phase and the right-hand-side is the chemical potential of the gas in the vapor phase. In eq 13, the role of the molar mass of solute component MM_{sol} on its non-equilibrium chemical potential is explicitly indicated. No indication is reported, however, for the molar mass of polymeric species MM_{pol}, as its effect on solute chemical potential becomes negligible for sufficiently high values and all calculations done for working this study refer to the limit of infinite MM_{pol}.

Solubility Coefficient Calculation and Comparison with Experimental Data

In this section, the reliability of calculations from the pseudo-equilibrium solubility model, NE-SAFT, described in the previous section is evaluated for low-pressure gas solubility coefficients in glassy polycarbonate (PC) and polysulfone (PSF). To illustrate the relation with a description of thermodynamic properties of polymeric mixtures above the glass transition temperature, corresponding equilibrium solubility calculations from the SAFT EoS will be considered. Results for low pressure equilibrium and pseudo-equilibrium solubility are reported in terms of the infinite dilution solubility coefficient S_0, which is defined as

$$S_0 = \lim_{p \to 0} \frac{C(p)}{p} \qquad (14)$$

where $C(p)$ indicates molar gas concentration in the polymeric mixture at pressure p.

A comparison of results from the two approaches with experimental data in a sufficiently large temperature range, embracing both rubbery and glassy regions, will be particularly meaningful. In this regard, experimental values for infinite dilution solubility coefficients of CO_2 or Ar in PC and CO_2 in PSF, as measured by Wang and Kamiya (22, 23), were considered.

For either equilibrium or pseudo-equilibrium solubility calculations, pure component SAFT parameters for PC and PSF are first needed. They were evaluated using volumetric data for PC and PSF reported by Zoller (21). SAFT parameters for both polymeric species were determined from the best fit of the experimental volumetric data in a temperature range about 150 K wide above the glass transition and for pressures up to 150 MPa. A more than satisfactory representation of the specific volume of rubbery PC or PSF was obtained through the SAFT EoS by tuning just two pure component parameters, namely sphere mass and interaction energy, while sphere volume was set to the value characteristic of high molecular weight hydrocarbons (4). Results of these fitting procedures are shown in Figure 1 and parameters values retrieved this way are reported in Table I. In the same table, pure component SAFT parameters used in this work for gaseous species are also indicated, which were obtained from the technical literature (4).

In Figure 2, the infinite dilution equilibrium solubility coefficients of CO_2 in PSF calculated using the SAFT EoS are compared with data by Wang and Kamiya (23). The data span a relatively large temperature range. The solubility coefficients calculated using a purely predictive procedure, (i.e. by setting the binary parameter $k_{ij} = 0$) are indicated by the dashed line. The solubility S_0 for temperatures above the glass transition, where the equilibrium SAFT model can be consistently applied, was clearly over predicted.

A remarkably good correlation for the temperature dependent gas solubility coefficient above the glass transition temperature can be obtained by adjusting the binary interaction parameter k_{ij} of the SAFT EoS for CO_2-PSF pair to 0.045 (values shown in the dashed-dotted line). The assumption of a temperature independent binary interaction parameter in the SAFT EoS allows, in this case, for the correct representation of the variation of S_0 with temperature above the glass transition. On the other hand, not surprisingly, the equilibrium SAFT model fails to represent the gas solubility coefficient measured below the glass transition temperature, as well as its temperature variation. A comparison of the latter experimental data with the pseudo-equilibrium solubility which can be calculated using the NE-SAFT model is in order.

In fact, the phase equilibrium problem in eq 13 can be solved for the pseudo-equilibrium solubility coefficient once the dry polymer density ρ_{pol}^0 is evaluated at the proper temperature. As already mentioned, this parameter

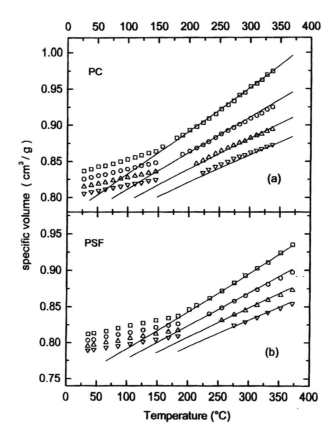

Figure 1. Results of the fitting procedure for PVT polymer data with SAFT EoS; experimental data are symbols, calculated data are lines. (a) Data for PC from (21), pressures: 0.1 (□), 59 (○), 118(△), 177 MPa(▽);(b) Data for PSF from (21), pressures: 0.1 (□), 49 (○), 98(△), 149 MPa(▽).

Table I. Pure Component Parameter for SAFT EoS

Species	MM [g/mol]	v^{00} [L/kmol]	(MM/m) [g/mol]	u^{0}/k [K]	reference
PC		12.00	25.00	371.0	this work
PSF		12.00	25.67	410.0	this work
CO_2	44.01	13.58	31.06	216.1	(4)
Ar	39.95	16.29	39.95	150.9	(4)

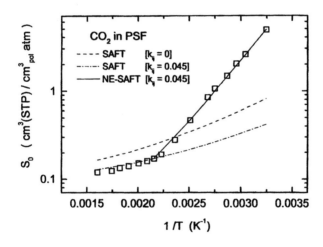

Figure 2. Comparison of experimental infinite dilution solubility coefficient for CO_2 in polysulfone from (22) [symbols] with predictions and correlations from SAFT and NE-SAFT models

should be measured for the sample of interest, as it can be different for materials treated according to different preparation procedures, ultimately resulting in different solubility coefficients. In this work, however, typical values were estimated for the specific volume of glassy polysulfone samples as a function of temperature, assuming that the PSF mass density at T_g (= 180 °C) is equal to 1.195 kg/L and that the cubic dilation coefficient in the glassy state is 2.0×10^{-4} K^{-1} (21). The binary interaction parameter k_{ij} for the penetrant-polymer pair was set to the value determined from the best fit of equilibrium solubility coefficient from the EoS above T_g. Results for pseudo-equilibrium solubility coefficients calculated this way using NE-SAFT are also reported in Figure 2 (solid lines). Calculated values for S_0 below T_g agree remarkably well with experimental data, and the change in the temperature dependence of the solubility coefficient across the glass transition temperature is correctly predicted.

Similar calculations have been repeated for the infinite dilution solubility coefficients of CO_2 and Ar in polycarbonate, and the results have been compared with the experimental data by Wang and Kamiya (22) in Figure 3. Also in these cases, the purely predictive estimation by the SAFT EoS (dashed lines in the figure) leads to an over prediction of the equilibrium solubility coefficient above the glass transition temperature. Quite satisfactory correlations for S_0 above T_g are obtained using the SAFT EoS for both CO_2 and Ar in PC by adjusting the binary interaction parameter k_{ij} to 0.06 and 0.05, respectively, (dashed-dotted lines in the figure). It must be stressed that the correct representation of the temperature variation of the solubility is obtained using a temperature independent value of k_{ij} both for the case of an exothermic mixing process, as in that of the CO_2-PC pair, and for an endothermic case (i.e., Ar-PC).

As is evident from the data in Figure 3, gas solubility coefficients measured below the glass transition temperature in PC are definitely higher than those predicted by the SAFT equation of state for equilibrium conditions. However, when pseudo-equilibrium solubility is calculated from the corresponding non-equilibrium version of the same EoS, namely NE-SAFT, remarkably good predictions are obtained for S_0 below T_g (solid lines in the figure) using the same binary interaction parameter k_{ij} determined from the optimal representation of the solubility coefficient above T_g. The polymer mass density of PC in the glassy state was evaluated as a function of temperature based on a density of 1.155 kg/L at T_g (= 150°C) and a cubic dilation coefficient in the glassy state of 2.8×10^{-4} K^{-1} (23). These results are obtained for CO_2 in PC, which exhibits a decrease in solubility with increasing temperature in both the rubbery and glassy states, and for Ar solubility in PC, which shows the opposite effect of temperature on S_0 above T_g.

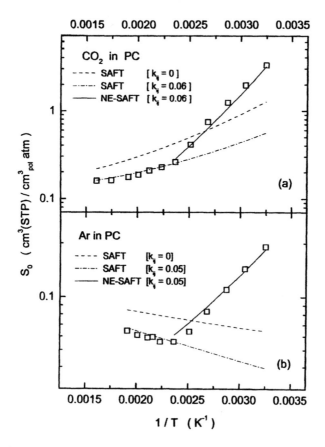

Figure3 Comparison of experimental infinite dilution gas solubility coefficient in polycarbonate from (23) [symbols] with predictions and correlations from SAFT and NE-SAFT models: CO_2 (a) and Ar (b).

Summary and Conclusions

A general procedure has been described for the derivation of thermodynamic models for pseudo-equilibrium gas solubility in glassy polymers from an arbitrary "equilibrium" equation of state. This procedure was originally applied to the Lattice Fluid Theory, resulting in what is known as the Non-Equilibrium Lattice Fluid (NELF) model and it is here extended to the Statistical Associating Fluid Theory (SAFT) to derive its "non-equilibrium" version (NE-SAFT). Using the assumption that the mass density of the polymer network is the only order parameter needed to represent the out-of-equilibrium degree of the system and that it evolves in time as described by a Voigt viscoelastic model for bulk rheology, simple results are obtained which provide an unambiguous framework to extend free energy and solute chemical potential in a polymeric mixture from equilibrium conditions to the entire domain of non-equilibrium states. Derivation of a pseudo-equilibrium solubility model is then straightforward, at least for those conditions in which the polymer density at pseudo-equilibrium can be evaluated from independent information. That is the case, for example, in low-pressure solubility problems, in which sorption induced swelling of the glassy matrix can be neglected and the pure polymer glassy density closely approximates pseudo-equilibrium values in sorption conditions.

Calculation procedures for estimating equilibrium gas solubility coefficients in polymers from equations of state have been discussed and compared with those for pseudo-equilibrium solubility values from their non-equilibrium versions. Examples of such calculations have been offered for the case of infinite dilution gas solubility in polycarbonate and polysulfone. Comparison of equilibrium solubility coefficient for CO_2 or Ar in PC and for CO_2 in PSF calculated using the SAFT EoS with experimental data measured above the glass transition temperature have been performed to determine an optimal value for the penetrant-polymer interaction parameter. NE-SAFT was then used for the calculation of pseudo-equilibrium low pressure gas solubility curves for the same penetrant/polymer pairs below T_g, after accounting for the variations of pure polymer glassy density with temperature. Quite accurate predictions have been obtained this way for CO_2 and Ar solubility in glassy PC, as well as for CO_2 in PSF, despite the different effect of temperature on solubility observed for the two gases in the rubbery phase.

Within the proposed scheme, the higher gas sorption capacity of glassy polymeric materials, with respect to that of the corresponding rubbery phases, is ultimately addressed by accounting for the actual value of polymer density in glassy state relative to that predicted for equilibrium conditions. The latter is, indeed, the most relevant difference between the pseudo-equilibrium and equilibrium solubility problems described in this work, although it is not the only one. Excess specific volume of the polymer below the glass transition temperature can be easily appreciated for PC and PSF based on the data shown in Figure 1. This excess volume is ultimately responsible for the difference between the pseudo-equilibrium gas solubility coefficient calculated from NE-SAFT and the equilibrium solubility estimated from the SAFT EoS using the same pure component and binary parameters (solid versus dashed-dotted lines in Figures 2 and 3).

Indeed, excellent representations of gas solubility curves in rubbery and glassy regions were obtained by using the same description for the basic thermodynamic properties of the penetrant/polymer pair. This is a clear demonstration that the approach proposed in this work represents the proper way to extend a given equilibrium thermodynamic model to the non-equilibrium domain of glassy states. Success in predicting infinite dilution gas solubility coefficients in polymers and their change over a wide temperature interval also relies, of course, on the use of a good description of the system thermodynamic properties in the original equation of state. For the cases examined in this work, pure component volumetric properties of linear chain polymeric species (i.e., PC and PSF) were satisfactorily represented by the SAFT EoS, essentially adjusting two pure component model parameters. Tuning of just one temperature independent binary interaction parameter for each penetrant/polymer pair yielded a correct representation of the infinite dilution gas solubility curve in the rubbery region for both exothermic or endothermic mixing processes. Accounting for the proper extension of thermodynamic properties to glassy state, an accurate representation of the pseudo-equilibrium gas solubility below the glass transition temperature was finally obtained for the same systems by using a straightforward procedure based on a consistent and solid thermodynamic foundation.

Acknowledgments

This research was partially supported by the Italian Ministry for Education, University and Research.

References

1. Oishi, T.; Prausnitz, J. M. *Ind. Eng. Chem. Res.* **1978**, *17(3)*, 333-339.

2. Elbro, H. S.; Fredenslund, A.; Rasmussen, P. *Macromolecules* **1990**, *23*, 4707-4714.
3. Sanchez, I. C.; Lacombe, R. H. *Macromolecules* **1978**, *11*, 1145-1156.
4. Huang, S. H.; Radosz, M. *Ind. Eng. Chem. Res.* **1990**, *29*, 2284-2294.
5. Song, Y.; Hino, T.; Lambert, S. M.; Prausnitz, J. M. *Fluid Phase Equilib.* **1996**, *117*, 69-76.
6. Hino, T.; Prausnitz J. M. *Fluid Phase Equilib.* **1997**, *138*, 105-130.
7. Kang, J. W.; Lee, J. H.; Yoo, K. P.; Lee, C. S. *Fluid Phase Equilib.* **2002**, *194*, 77-86.
8. Michaels, A. S.; Vieth, W. R.; Barrie, J. A. *J. Appl. Phys.* **1963**, *34*,1-12.
9. Freeman B. D.; Pinnau I. In *Polymer Membranes for Gas and Vapor Separation: Chemistry and Materials Science*; Freeman B. D.; Pinnau I. Eds.; ACS Symposium Series 733; American Chemical Society: Washington, D.C., 1999, 1-27.
10. Wissinger, R. G; Paulaitis, M. E. *J. Polym. Sci. Polym. Phys. Ed.* **1987**, *25*, 2497-2510.
11. Barbari, T. A.; Conforti, R. M. *J. Polym. Sci. Polym. Phys. Ed.* **1992**, *30*, 1261-1271.
12. Doghieri, F.; Sarti G. C. *Macromolecules* **1996**, *29*, 7885-7896.
13. Boudouris D.; Panayiotou C. *Macromolecules* **1996**, *31*, 7915-7920.
14. Giacinti Baschetti, M.; Doghieri, F.; Sarti G. C. *Ind. Eng. Chem. Res.* **2001**, *40*, 3027-3037.
15. Doghieri, F.; Canova, M.; Sarti G. C. In *Polymer Membranes for Gas and Vapor Separation: Chemistry and Materials Science*; Freeman B. D.; Pinnau I. Eds.; ACS Symposium Series 733; American Chemical Society: Washington, D.C., 1999, 179-193.
16. Sarti, GC.; Doghieri F. Chem. Eng. Sci. 1998, 53, 3435-3447.
17. Coleman, B.D.; Gurtin M.E. J.Chem.Phys. 1967, 47, 597-613.
18. Chapman, W. G.; Gubbins, K. E.; Jackson, G.; Radosz, M. *Fluid Phase Equilib.* **1989**, *52*, 31-38.
19. Chapman, W. G.; Gubbins, K. E.; Jackson, G.; Radosz, M. *Ind. Eng. Chem. Res.* **1990**, *29*, 1709-1721.
20. Chen, S. S.; Kreglewski, A. Ber. Bunsen-Ges. *Phys. Chem.* **1977**, *81*, 1048-1052.
21. Zoller, P. *J. Polym. Sci. Polym. Phys. Ed.* **1978**, *16*, 1261-1275.
22. Wang, J.-S.; Kamiya, Y. *J. Polym. Sci. Polym. Phys. Ed.* **2000**, *38*, 883-888.
23. Wang, J.-S.; Kamiya, Y. *J. Membrane Sci.* **1995**, *98*, 69-76.

Chapter 6

Nanostructure of Free Volume in Glassy Polymers as Studied by Probe Methods and Computer Simulation

Yu. Yampolskii[1], V. Shantarovich[2], D. Hofmann[3], and M. Heuchel[3]

[1]A. V. Topchiev Institute of Petrochemical Synthesis, 29 Leninsky Drive, 119991 Moscow, Russia
[2]N. N. Semenov Institute of Chemical Physics, 4, Kosygine ul 117977, Moscow, Russia
[3]GKSS Research Center, Kantstrasse 55, D–14513, Teltow, Germany

Several experimental methods are currently available for investigation of nanostructure of free volume in polymers: positron annihilation lifetime spectroscopy (PALS), inverse gas chromatography (IGC), [129]Xe-NMR, and some others. These methods differ by physical principles, size of the probe, and methodical problems that can complicate interpretation of the results. A systematic study of those methods resulted recently in two important observations: (i) a reasonable agreement is reached between the data of different methods for a given polymer and (ii) the results of probe methods can be verified by computer simulation. In this work, the results of two probe methods (PALS, IGC) are compared with each other and with the predictions based on molecular simulation of the nanostructure of two groups of polymers: (i) high free volume materials, i.e. poly(trimethylsilyl-1-propyne) and amorphous Teflons AF) and (ii) conventional glassy polymers (Si- and F-containing polystyrene derivatives).

© 2004 American Chemical Society

Introduction

The free volume model is the most efficient concept to describe transport phenomena in condensed matter and, particularly, in glassy polymers. However, the classical work by Frenkel (1) and Cohen-Turnbull (2) has often been interpreted only as an abstract notion. On the other hand, new experimental probe methods made it possible to assess free volume directly in various polymers. The generic features of different probe methods are: species of different size and nature are introduced into a polymer under investigation. It is assumed that the probe molecules are accommodated in sufficiently large free volume elements (FVE). By monitoring their behavior in the FVE one can obtain information about average size and, sometimes, size distribution of free volume. A list of several probe methods is given in Table I. The most important parameters used in monitoring free volume are: (i) positron annihilation lifetimes, (ii) thermodynamic potentials of gas sorption, and (iii) NMR chemical shifts. The literature on probing free volume in polymers is abundant, and the references given in Table I serve only as examples. The results of the probe methods have been compared with the data of scattering methods (e.g. X-ray or neutron scattering) and, in general, a good agreement between the various techniques is obtained.

However, free volume is also a complex physical object involving topology, connectivity, shape, and size distribution. Every experimental method probes, strictly speaking, only a part of the size distribution. Specifically, only those free volume elements that are larger than the intrinsic probe size can accommodate the probe.

Table I. Probe methods for investigation of free volume

Method	Probe	Parameter monitored	Reference
Positron annihilation lifetime spectroscopy	o-Positronium (e^{+}-e^{-} pair)	Lifetimes	(3)
Inverse gas chromatography	n-Alkanes C_3-C_{12}	Partial molar potential of mixing	(4)
Spin probe method (ESR)	TEMPO and other nitroxile radicals	Correlation times	(5)
Electrochromism	Aromatic azo-compounds	Rotation diffusion coefficient	(6)
Photochromism	Aromatic compounds	Photoisomerization rate	(7)
^{129}Xe-NMR	Xe atom	Chemical shift	(8)

Thus, positron annihilation lifetime spectroscopy provides information on FVE larger than an o-positronium (intrinsic size ~ 1Å), whereas very large photochromic probes sense only "the tail" of the size distribution of free volume in the range of 5-10 Å. In spite of the fact that for highly permeable polymers, such as poly(trimethylsilyl-1-propyne) (PTMSP), high free volume was found using different methods *(6)*, a quantitative comparison of different methods is complicated and has never been made in a systematic manner.

Molecular modeling investigations can provide fundamental insights to the problem of free volume characterization. These techniques have been widely used during the past decade to obtain a better understanding of the structure and the transport behavior of amorphous glassy materials for nonporous membranes. General results of these studies can be found in a number of reference and feature articles *(9,10)*. A great advantage of this approach is that it provides size distribution of FVE and also gives some insight in connectivity of free volume elements, both critical parameters for diffusion of small molecules in polymers. However, very few attempts have been made to compare the results of computer simulations with the data of the same polymer using experimental probe methods. Therefore, it is relevant to make a quantitative comparison of free volume sizes found using different probe methods.

In this work we focus on recent findings of the aforementioned approaches and compare the results of experimental methods and computer simulations. Two groups of amorphous glassy polymers were investigated (Table II). PTMSP as well as the copolymers of 2,2-bis(trifluoromethyl)-4,5-difluoro-1,3-dioxole and tetrafluoroethylene (amorphous Teflons AF, Du Pont Co.) were studied as typical high free volume, high permeability polymers. On the other hand, two polystyrene derivatives of the general formula $-CH_2-CH(p-C_6H_4R)-$, where R = $Si(CH_3)_3$ (PTMSS) or $Si(CH_3)_2(CH_2CH_2CF_3)$ (PFPDMSS), were taken as representatives of "conventional", i.e. low free volume glassy polymers. Gas permeability of all the polymers considered varied over more than 2 orders of magnitude.

Positron Annihilation Lifetime Spectroscopy

It has been shown that the o-positronium lifetime spectrum consists of two components, that is, the size distribution of FVE is bimodal for glassy polymers *(12)*. The radii of FVE in the polymers studied (Table III) are in the range 6.8 –2.6 Å (the finite term program PATFIT *(13)* was used in treatment of the primary lifetime spectrum). Assuming spherical symmetry of FVE one has to conclude that the volumes of larger FVE are in the range 74 –1,320 Å^3. The volume of smaller FVE of high permeability polymers falls in a much narrower range of 80-170 Å^3, typical values for conventional glassy polymers. Similar results were obtained when the continuous lifetime spectrum was obtained using the CONTIN program *(14)*. These results will be discussed later.

Table II. Structure and properties of various glassy polymers

Polymer	Structure	Density (g/cm^3)	FFV (%)	P(O2) (Barrer)	Tg (°C)
PTMSP	(structure: —[C=C]ₙ— backbone with CH₃ and H₃C–Si–CH₃ / CH₃)	0.75	33	7,700	>200
AF2400 (n=0.87)	(structure: —[CF–CF]ₙ[CF₂–CF₂]₁₋ₙ— with O O / CF₃ CF₃)	1.75	32	1,140	240
AF1600 (n=0.65)		1.82	28	170	160
PTMSS	(structure: —[CH–CH₂]ₙ— with phenyl ring and H₃C–Si–CH₃ / CH₃)	0.965	20	56	135
PFPDMSS	(structure: —[CH–CH₂]ₙ— with phenyl ring and H₃C–Si–CH₂–CH₂ / CH₃ CF₃)	1.127	18	38	62

The lifetime spectra enable an estimation of the concentrations N of FVE in the polymers studied (15). It was shown that, in spite of large differences in the sizes of FVE and transport properties of polymers, the concentrations N do not differ significantly and are $0.55\text{-}1.6 \times 10^{20}$ cm^{-3}. The concentration of smaller FVE (e.g. in PTMSS) is somewhat higher than that of high free volume, highly permeable polymers such as PTMSP. If we suppose that spherical FVE are distributed uniformly in the polymer matrix, the N values allow an estimation of the mean distance l between neighboring FVE (16). The l values are also presented in Table III. It is obvious that the l values are comparable or in some cases even smaller than the diameter of larger FVE ($2R_4$). These values are also of the same order as the diffusion jump lengths estimated in computer simulation studies.(17)

The fractional free volume found via annihilation parameters $FFV=N(4\pi/3)R_4^3$ is in the range of 2-8% and correlates reasonably well with the gas permeability of the polymers. It is several times smaller than the fractional free volume determined via polymer density according to Bondi's method, as shown in Table II. There are several reasons that can cause these discrepancies. On one hand, the Bondi method is based on several very rough approximations: constant ratio of van der Waals and occupied volume and the universal increments in the van der Waals volume. It is likely that the additive scheme does not account for variations of bond lengths and dihedral angles between the groups that are incorporated in repeat units. Other shortcomings of Bondi's additive scheme have been addressed earlier by Park and Paul *(18)*. On the other hand, as the Bondi free volume is evaluated via specific and occupied volumes, it accounts for the whole size distribution of free volume, whereas the PALS method "feels" only those FVE that can accommodate o-positronium. Indeed, molecular simulation showed *(19)* that, as the size of a probe molecule decreases, the fractional free volume sensed by these probes increases rapidly.

Table III. Parameters of free volume size distribution in glassy polymers

Parameter	PTMSP	AF2400	AF1600	PTMSS	PFPDMSS
R_3, Å	3.4	2.7	-	2.7	2.6
R_4, Å	6.8	6.0	4.9	3.7	3.7
$N\ 10^{20}$ cm^{-3}	0.75	0.55	0.55	1.6	1.4
V_{4f}, Å^3	1320	880	490	220	210
l, Å	10	14	16	11	12
FFV,%	8.3	3.4	1.9	2.9	2.1

Inverse Gas Chromatography

For a series of solutes with increasing molecular dimensions (*n*-alkane homologs), it has recently been shown that the partial molar enthalpies of mixing found using the IGC method pass through minima for all glassy polymers studied by this technique *(20)*. The coordinates of these minima (e.g. critical or van der Waals volumes of the corresponding solutes) were assumed to be equal to the mean size of FVE. The same approach was employed in the IGC study of the two AF Teflons having different content of 2,2-bis-trifluoro-4,5-difluoro-1,3-dioxole co-monomer: 87 % (AF2400) and 65% (AF1600). The partial molar enthalpies of mixing with these copolymers also pass through a minimum when the solute size (critical volume V_c) increases for the series of C_4-C_{13} alkanes (Figure 1a). The coordinates of these minima ($V_{c,min}$) correlate with the parameters of size distribution as found by the PALS method (Figure 1b). Thus, in AF1600 the limiting radius of FVE that can accommodate a solute

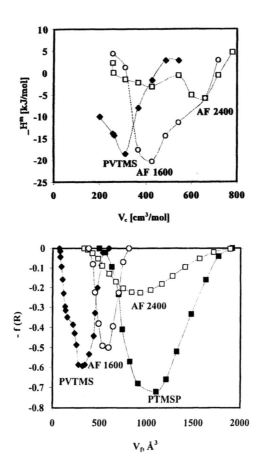

Figure 1. Estimation of the size of free volume elements in glassy polymers: (a) partial molar enthalpy of mixing (ΔH_m) as a function of critical volume of solute (V_c) according to IGC; (b) probability density function f(R) in polymers with different sizes of free volume elements according to PALS.

molecule (C_7) is about 5 Å, whereas FVE in AF2400 can accommodate a much larger molecule (C_{11}) with a radius of 6.0 - 6.4 Å. Both values are in reasonable agreement with the R_4 values presented in Table IV, where the radii of FVE were obtained using CONTIN positron lifetime data treatment. This result shows that independent probe methods give very similar evaluation of the FVE size in polymers.

Table IV. Radii of spherical free volume elements (Å)

Polymer	Method	
	IGC	PALS
Poly(vinyltrimethylsilane) (4)	5.3	4.4
Teflon AF1600	5.5	5.2
Teflon AF2400	6.4	6.0
PTMSP	-	6.4

Molecular Modeling

The main focus of the molecular simulation was scaling of the size of amorphous packing models to about 10^4 atoms and utilization of the newly developed COMPAS force field (21, 22). For three polymers - PTMSP, PTMSS, and PFPDMSS several independent atomistic bulk models were realized utilizing the Amorphous Cell module of the InsightII/Discover Software of Molecular Simulations Inc.(23). The basic techniques used were published in a previous paper (10). For PTMSP the initial packing procedure was performed with a 50:50 probability for the occurrence of monomers with cis- and trans-configuration at a density of 0.75 g/cm³. Cubes with an edge of 45-50 Å were constructed, which contained polymer chains equilibrated to attain the experimental densities.

Figures 2 and 3 illustrate free volume distribution for packing models of PTMSS and PTMSP, respectively, as series of slices cut perpendicular to the respective z-axis and separated by about 3 Å. In total, three models were constructed for both polymers. Qualitatively, they were very similar to those shown in Figures 2 and 3. The former polymer shows a widely homogeneous distribution of free volume which is quite typical for conventional amorphous polymers with small and medium amounts of free volume (10,24). The same tendency was observed for PFPDMSS (19). Besides much smaller total free volume in the case of PTMSS (if compared to PTMSP), it can be noted that microcavities have shapes much closer to spherical symmetry. In most cases, a microcavity shown in one slice disappears on the next one, that is, the average size of a free volume element is often comparable to the distance between slices, i.e. about 3 Å.

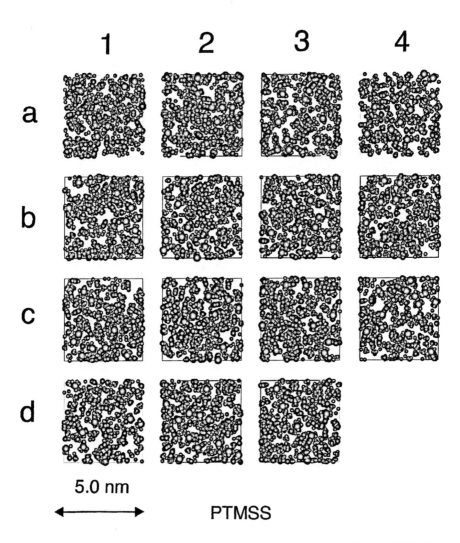

Figure 2. Representation of the free volume of completely equilibrated PTMSP packing models as series of 3.1 Å slices cut perpendicular to the z-axis. 1a is the first and d4 is the last slice in each case.

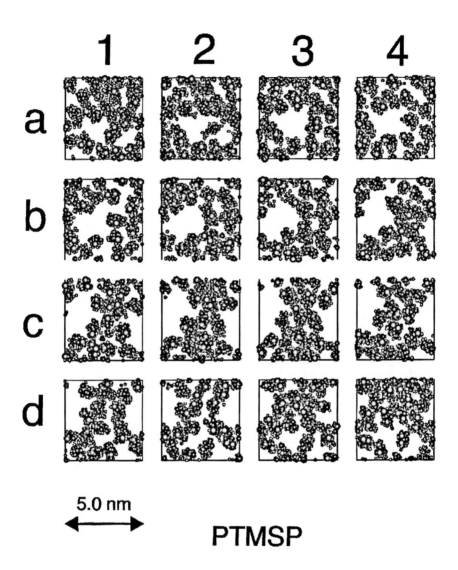

Figure 3. Representation of the free volume of completely equilibrated PTMSS packing model as a series of 3.0 Å thick slices cut perpendicular to the z-axis. 1a is the first and d4 is the last slice in each case.

An analysis of the packing model for PTMSP indicates that this material contains regions of high segmental packing density where the free volume distribution resembles the one typically observed in low and medium free volume polymers. On the other hand, rather large voids are present with a tendency to a partly continuous hole phase with lateral void-widths of about 5-20 Å. A more attentive examination of the slices of the PTMSP cube shown in Figure 3 reveals that some larger holes have shapes that differ dramatically from spherical symmetry. For example, a hole shown in the slice a1 (PTMSP-1) with a cross-section about 18 Å extends along the z-axis of the cube to a depth of at least 50 Å (down to the slice d4), indicating that larger microcavities have distinctly elongated shape. One can observe branching of these "pores" in several cases. The local density can be quite large in certain areas shown in some slices.

These features of the nanostructure of PTMSP are consistent with numerous observations made in studies of gas transport in this polymer. Recently, it has been shown that mixed-gas selectivity in separation of vapors/permanent-gas mixtures is significantly higher than the ideal, pure-gas selectivities for the same gases (25,26). The result was explained by condensation of vapors in the pores and blocking of transport of light gases. The studies with PTMSP revealed unusually low activation energies of diffusion and negative activation energies of permeation for light gases (27), which can be explained by extremely low energy barriers for mass transfer through the pores in this material. We believe that visualization of free volume in PTMSP can be considered as the first validation of these assumptions based on kinetic arguments.

The size distribution of free volume was obtained by treating the models shown in Figures 2 and 3. It is clear that the resulting distribution depends on the size of the test particle which is used for scanning the free volume regions. In order to compare the simulation results with those of the PALS method, a cutoff radius of about 1.1 Å, which is the size of an o-positronium, was used in these calculations. The free volume of the simulated packing models was determined by superimposing a grid of a grid-step-width $\delta = 0.7$ Å with a probe molecule radius of 1.1 Å. While considering the connectivity of the "free" grid points and connecting "free" grid points into groups, an important decision regarding the nano-structure of free volume elements had to be made. In the first approach (named V_connect) several neighboring subgroups formed and extended free volume elements, as shown in Figure 4. Alternatively, in the second approach (named R_max, see Figure 4) larger FVE of elongated or highly complex shape were split into smaller ones. These two approaches gave dramatically different results in the case of PTMSP.

The resulting distribution functions for PTMSS and PTMSP calculated with the R_max approach and statistically weighted with the volume of the respective hole are presented in Figures 5 and 6. For comparison, Figure 7 shows the respective FVE size distribution based on the V_connect approach for PTMSP. The distribution for PTMSS shows one major, nearly symmetrical peak in the

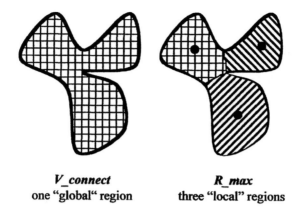

V_connect
one "global" region

R_max
three "local" regions

Figure 4. Principal view of the two approaches to connect free grid points in a specific free volume region. In V_connect (left) all connected "free" grid points belong to one region. In the R_max approach (right) first points of local maxima of "shortest distances" to polymer atoms are determined (dots), and, then the grid points are assigned. The result is a decomposition into three more local regions.

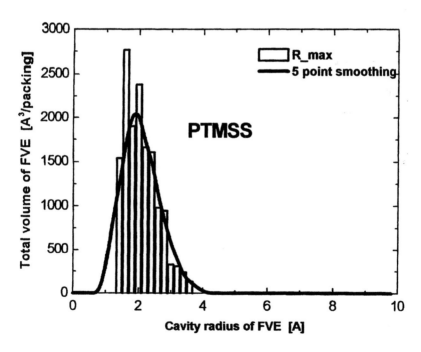

Figure 5. Size distribution of free volume elements for PTMSS as a function of cavity radius of spheres of equivalent volume averaged over three packing models. R_max was applied in this case.

range 1.5-3.5 Å with a maximum at about 2 Å. Similar results were obtained for PFPDMSS.

PTMSP has a much wider range of cavity sizes (radius: 1.5 Å to about 9 Å). Furthermore, the computer simulation predicts for PTMSP (cf. Figure 6) an asymmetric distribution with at least two peaks, one at about 3-4 Å and one (appearing as a shoulder) between about 6 Å and 8 Å. This result reflects again the intuitive image from observing slice pictures of PTMSP packing models (cf. Fig. 1), which showed a large number of small voids in this polymer, whereas other regions were composed of very large free volume elements.

Figure 7, which presents the distribution of complete holes (approach V_connect), underlines the fact that in PTMSP most of the free volume is organized in a continuous hole phase. This is indicated by the dominating, very large peak at 19 Å. The position of this peak increases with increasing model size. This clearly emphasizes the highly non-spherical shape of this free volume phase. It also provides an additional argument for using a description of the free volume distribution which splits larger lengthy holes of complex geometry into smaller more compact ones.

Thus, computer modeling confirmed qualitatively, and in some instances quantitatively, many results obtained by other experimental methods, namely: (i) existence of much larger free volume elements in PTMSP, (ii) narrow size distribution of free volume in conventional glassy polymers and a much wider distribution in high permeability polymers, like PTMSP, (iii) a tendency to bimodal size distribution of free volume in PTMSP (Figures 6 and 7), (iv) reasonable agreement between the radii of FVE as obtained from computer modeling (CM) and PALS: 1.5-3.5 Å (CM) and 2.6-3.7 Å (PALS) in conventional glassy polymers and 1.5-9 Å (CM) and 3.4-6.8 Å (PALS), and (v) elongated, seemingly open pores found by CM in PTMSP are in amazing agreement with the results obtained by gas permeation studies of this polymer.

It would be desirable if molecular modeling could be applied to design improved membrane materials as compared to state-of-the-art polymers. However, molecular modeling does not advance such a goal: this is an objective of computer-aided molecular design, a field just emerging. It can be speculated though that advanced membrane materials should combine the following features: large free volume like that in PTMSP or amorphous Teflons and thin, size-selective "walls" of free volume elements. The results of several recent papers (16,28) indicated that permeation properties are controlled not only by free volume size and size distribution but also by energy barriers that can be well characterized by cohesion energy density.

Another generic feature of highly permeable glassy membrane materials is their strong non-equilibrium state. All glassy polymers are non-equilibrium materials. However, the deviations from equilibrium in highly permeable polymers are much more pronounced than in conventional glassy polymers. PTMSP is an interesting example to elucidate this point. It is known that this polymer is prone to very fast physical aging: intrinsic aging times are on the order of 500 h (29). The ultra-high free volume state of PTMSP is far from any

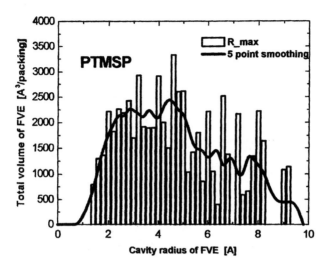

Figure 6. Size distribution of free volume elements for PTMSP as a function of cavity radius of spheres of equivalent volume averages over three packing models. R_max was applied in this case.

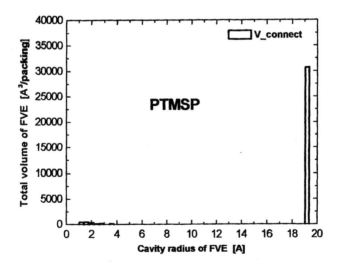

Figure. 7 Size distribution of free volume elements for PTMSP as a function of cavity radius of spheres of equivalent volume averages over three packing models. V_connect was applied in this case.

equilibrium as can be seen from the observed fast aging. Molecular modeling, on the other hand, attempts to find "equilibrium" structures (to the extent, where this term can be applied to glassy polymers.). That is, if a model for PTMSP with no density as an input parameter is used under standard procedures, the predicted density would be that of aged PTMSP (ρ=0.94 g/cm^3) *(30)*. This problem has been addressed earlier *(10)*. The aged PTMSP is in fact just a conventional glassy polymer and, therefore, not of particular interest for commercial membranes. In other words, the extreme properties of as-cast PTMSP films and their nanostructure of free volume result from a complex temporary influence of the processing conditions, and only via this connection are they related with specific structural properties of PTMSP. It may even possible that the processing conditions are in some cases more important than the structural effects. Indeed, variations in film pretreatment conditions for some polyimides *(31)* resulted in changes of O_2/N_2 selectivity in the range of 4-12 with virtually no variation of oxygen permeability.

In summary, it is clear that molecular modeling alone cannot be used for design of better membrane materials. Only a combination of modeling work and a thorough thermodynamic analysis of the applied macroscopic processing conditions, i.e. an interdisciplinary approach, can lead to success.

References

1. Frenkel, Ya. I. *Kinetic Theory of Liquids*; Nauka: Moscow-Leningrad, 1945.
2. Cohen, M.H.; Turnbul, D. *J. Chem. Phys.,* **1959**, *31*, 1162.
3. *Positron and Positronium Chemistry;* Scrader, D.M.; Jean, Y.C. (Eds.), Elsevier: Amstredam, 1988.
4. Yampolskii, Yu. P.; Kaliuzhnyi, N. E.; Durgaryan, S. G. *Macromolecules,* **1986**, *19*, 846-851.
5. Wasserman, A. M.; Kovarskii, A. L. *Spin Probes and Kabels;* In *Physical Chemistry of Polymers;* Nauka: Moscow, 1986.
6. Yampolskii, Yu. P.; Shantarovich V. P.; Chernyakovskii, F. P.; Kornilov A. I.; Plate N. A. *J. Appl. Polym. Sci.,* **1993**, *47*, 85-92.
7. Victor J. G.; Torkelson, J. M. *Macromolecules,* **1987**, *20*, 2241-2250.
8. Nagasaka, B.; Eguchi, T.; Nakayama, H.; Nakamura, N.; Ito, Y. *Radiation Phys. Chem.* **2000**, *58*, 581-585.
9. Müller-Plathe, F. *Acta Polym.* **1994**, *45*, 259-293.
10. Hofmann, D.; Fritz, L.; Ulbrich, J.; Schepers, C.; Böhnung, M. *Macromol. Theory Simul.* **2000**, *9*, 293-327.
11. Bondi, A. *Physical Properties of Molecular Crystals, Liquids, and Glasses;* Wiley: New York, 1968.
12. Shantarovich, V. P.; Kevdina I. B.; Yampolskii Yu.P; Alentiev A. Yu. *Macromolecules,* **2000**, *33*, 7453-7466.

105

13. Kirkegaard, P.; Pedersen, N.J.; Eldrup, M. PATFIT-88: A Data Processing System for Positron Annihilation Spectra on Mainframe and Personal Computers, Risoe-M-2740.
14. Provensher, S. W. *Comput. Phys. Commun.* **1982**, *27*, 229.
15. Shantarovich, V. P.; Goldanskii, V. I. *Hyperfine Interact.* **1998**, *116*, 67.
16. Alentiev, A. Yu.; Yampolskii, Yu. P. *J. Membr. Sci.* **2002**, *206*, 291-306.
17. Gusev, A. A.; Arizzi, S.; Suter, U. W. *J. Chem. Phys.,* **1993**, *99*, 2221.
18. Park, J. Y.; Paul, D. R. *J. Membr. Sci.*, **1997**, *125*, 23.
19. Hofmann, D.; Heuchel, M.; Yampolskii, Yu.; Khotimskii, V.; Shantarovich V. *Macromolecules,* **2002**, *35*, 2129-2140.
20. Plate, N. A.; Yampolskii, Yu. P. In: *Polymeric Gas Separation Membranes,* Paul, D. R.; Yampolskii, Yu. P., Eds.; CRC Press: Boca Raton, 1993, p.155.
21. Sun, H.; Rigby, D. *Spectochimica Acta,* **1997**, *53A*, 1301.
22. Rigby, D.; Sun, H.; Eichinger, B.E. *Polymer Intern.*, **1997**, *44*, 311.
23. *Polymer User Guide, Amorphous Cell Section, Version 4.0.0*: Molecular Simulations: San Diego (1996).
24. Nagel, C.; Schmidtke, E.; Günther-Schade, K.; Hofmann, D.; Fritsch, D.; Strunskus, T.; Faupel, F. *Macromolecules* **2000**, *33*, 2242-2248.
25. Srinivasan, R.; Auvil, S. R.; Burban, P. M. *J. Membr. Sci.,* **1994**, *86*, 67-86.
26. Pinnau, I.; Toy, L. G. *J. Membr. Sci.,* **1996**, *116*, 199-209.
27. Masuda, T.; Iguchi, Y.; Tang, B.-Z.; Higashimura, T. *Polymer,* **1988**, *29*, 2041-2049.
28. Nagel, C.; Guenther-Schade, D.; Fritsch, D.; Strunkus, T.; Faupel, F. *Macromolecules*, **2002**, *35*, 2071-2077.
29. Robeson, L. M.; Burgoyne, W. F.; Langsam, M.; Savoca, A.C.; Tien, C.F. *Polymer*, **1994**, *35*, 4970-4978.
30. Ichiraku, Y.; Stern, S. A.; Nakagawa, T. *J. Membr. Sci.,* **1987**, *34*, 5-18.
31. Alentiev, A. Yu.; Yampolskii, Yu. P., unpublished results.

Chapter 7

Fluoropolymer–Hydrocarbon Polymer Composite Membranes for Natural Gas Separation

Rajeev S. Prabhakar and Benny D. Freeman

Chemical Engineering Department, University of Texas at Austin, Austin, TX 78758

Hydrocarbon-based polymer membranes used to remove CO_2 from natural gas streams can lose selectivity due to plasticization caused by sorption of higher hydrocarbon compounds present in natural gas. Fluoropolymers have a weak tendency to sorb higher hydrocarbons, but commercially available fluoropolymers do not possess CO_2/hydrocarbon selectivities as high as those exhibited by hydrocarbon polymers. In this paper, a simple model is presented to evaluate the conditions under which coating a hydrocarbon-based polymer membrane with a fluoropolymer could reduce the sorption of higher hydrocarbons into the hydrocarbon polymer, thereby protecting the hydrocarbon polymer from plasticization by these compounds. Based on this analysis, an effective plasticization-resistant coating should have a lower ratio of higher hydrocarbon to CO_2 solubility than that of the hydrocarbon polymer and be as strongly size-sieving as possible. Model cases are presented to illustrate the possibilities and limitations of this approach.

© 2004 American Chemical Society

The world market for natural gas is approximately US$ 22 billion annually *(1)*. Removal of CO_2 from natural gas is an important processing operation in converting raw natural gas to pipeline quality gas. While this separation has traditionally been accomplished by amine absorption, polymeric membrane systems are now proving to be an attractive alternative, especially for treating small to moderate size gas streams *(2)*. Table I presents the composition of a natural gas stream *(3)*. While the stream is composed primarily of CH_4, it also contains a wide variety of higher hydrocarbons. CO_2 is smaller than all the hydrocarbons in this stream, so it has a higher diffusion coefficient in the membrane than the hydrocarbons. CO_2 is also more condensable than CH_4, and therefore it is more soluble than CH_4. Larger hydrocarbons, however, are more condensable than CO_2 and therefore, significantly more soluble than CO_2 in polymers. Therefore, to remove CO_2 from natural gas using polymer membranes, the strategy has been to design materials with strong size sieving abilities *i.e.*, with diffusivity selectivity values strongly favoring CO_2 diffusion relative to that of the hydrocarbon components. Hydrocarbon-based polymers with high CO_2 permeabilities and high CO_2/CH_4 selectivities have been developed for this separation (*cf.* Table II). The results in Table II show the importance of high diffusivity selectivity in obtaining overall selectivity.

Unfortunately, the high solubility coefficients of higher hydrocarbons in hydrocarbon-based polymer membranes introduce complications in this separation process. Upon sorbing into a polymer, these higher hydrocarbons can act as plasticizers. They increase polymer chain mobility and decrease the size-sieving ability (or diffusivity selectivity) of the polymer. Figure 1 presents results from an experimental study of the effect of plasticization on diffusion coefficients in poly(vinyl chloride), which is a glassy, rigid, strongly size-sieving polymer in the unplasticized state *(4)*. As indicated in Figure 1, in the unplasticized polymer, diffusion coefficients decrease dramatically with increasing penetrant size. For example, as penetrant critical volume increases from 65 cm^3/mol to 370 cm^3/mol (from hydrogen to *n*-hexane), diffusivity decreases by more than 10 orders of magnitude. However, in plasticized poly(vinyl chloride), diffusivity decreases by only two orders of magnitude over the same penetrant size range, thus losing its size-sieving ability to a very large extent.

An example of the negative effect of higher hydrocarbon sorption on CO_2/CH_4 separation is presented in Figure 2 *(5)*. In the presence of toluene or hexane, the polyimide membrane exhibits a significant reduction in CO_2/CH_4 mixed-gas selectivity. For a CO_2 removal membrane unit, as CO_2/CH_4 selectivity decreases, more of the desired methane product appears in the low pressure permeate stream, which either forces the use of a second membrane stage to recover the permeated methane and repressurize it to pipeline conditions or results in larger losses of methane from the separation system. Both of these options increase the cost of purifying the gas.

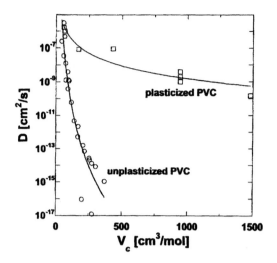

Figure 1. Diffusion coefficients of gases in poly(vinyl chloride) (4).

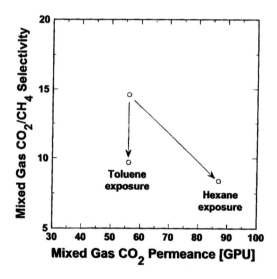

Figure 2. Mixed gas CO_2 permeance and CO_2/CH_4 selectivity of a polyimide membrane. The feed gas was composed of 10 mol% CO_2 and 90 mol% CH_4, and the experiments were performed at 48°C using a feed pressure of 1000 psi. To obtain the results for membranes exposed to hydrocarbons, the CO_2/CH_4 feed stream was saturated with 0.055 vol% toluene or 0.23 vol% hexane (5). 1 GPU = 1×10^{-6} $cm^3 (STP)/(cm^2 \cdot s \cdot cmHg)$.

Table I. Composition of a natural gas stream processed for CO_2 removal. The gas stream is a blend from 15 wells in the Pailin field in the Gulf of Thailand *(3)*.

Compound	Composition (mole%)
CO_2	32.79
N_2	2.89
C1	48.46
C2	8.22
C3	4.45
iC4	1.22
nC4	1.04
iC5	0.40
nC5	0.23
C6+benzene	0.18
C7+toluene	0.095
C8+xylenes	0.012
C9	0.002
C10	0.001
C11	0.0009
C12	0.0011
C13	0.0001
C14	0.0001
C15+	0.0002

NOTE: The numeral after "C" indicates the number of carbon atoms per molecule in the hydrocarbon compound. The letters 'i' or 'n' preceding "C" refer to 'iso' and 'normal', respectively. Benzene, toluene and xylenes are grouped with other compounds having the same number of carbon atoms.

Table II. Pure gas CO_2 and CH_4 permeation data at 35°C in cellulose acetate (degree of substitution = 2.45) *(6)* and an aromatic polyimide *(7)*.

Polymer	Pressure (atm)	CO_2 Permeability (Barrer)	CO_2/CH_4 Selectivity	CO_2/CH_4 Solubility selectivity	CO_2/CH_4 Diffusivity selectivity
Aromatic polyimide	6.8	7.50	93	2.9	32
Cellulose acetate	10	4.75	36.5	8	4.6

Reports of experimentally determined solubilities of large hydrocarbons in polymers, especially the strongly size-sieving ones considered for natural gas separations, are extremely rare due to the long times needed to measure solubility or diffusivity of large penetrants in such polymers (8, 9). However, penetrant solubility in polymers often scales with simple measures of penetrant condensability, such as critical temperature; so values of higher hydrocarbon solubility in polymers may be estimated (10). For example, Figure 3 presents permanent gas and C1-C5 hydrocarbon solubility in low-density polyethylene (LDPE). If the trendline through the experimental data is extrapolated to, for example, n-decane (T_c=617.7K), its solubility would be about 2 orders of magnitude larger than that of CO_2. The linear relation between lnS and T_c in Figure 3 has a slope of 0.019 K^{-1}, which is similar to values for a wide range of hydrocarbon polymers and organic liquids (cf. Table III). This linear relation for hydrocarbon sorption in hydrocarbon polymers and liquids can be predicted by straightforward thermodynamic considerations (11). Since the slope values are similar in a variety of polymers, the extent of higher hydrocarbon sorption relative to, for example, CO_2 solubility is expected to be similar in these polymers. Even though higher hydrocarbons are present in minute amounts in natural gas streams (cf. Table I), their sorbed concentration in the polymer can be appreciable because they have very high solubility coefficients and low vapor pressures.

Fluorinated liquids (12, 13) and fluoropolymers (14) display a much weaker dependence of solubility on penetrant condensability than hydrocarbon polymers and organic liquids. An example of this behavior is presented in Figure 4, which shows gas and vapor solubility in n-heptane and its fluorinated analog (12, 15). While fluoroheptane has higher solubility than n-heptane for helium, increases in solubility coefficient with increasing critical temperature are markedly less in fluoroheptane than in n-heptane. The slope of the line through the fluoroheptane data in Figure 4 is approximately half that of the line through the n-heptane data. Extrapolating to critical temperatures representative of higher hydrocarbons such as n-decane suggests a one and a half orders of magnitude difference in n-decane solubility coefficients, with the fluorinated liquid having much lower solubility coefficients than its hydrocarbon analog.

One potential approach to address the issue of membrane plasticization due to sorption of higher hydrocarbon contaminants would be to use polymers, such as fluoropolymers, with inherently low solubility coefficients for these compounds. However, currently available commercial fluoropolymers have much lower CO_2/CH_4 selectivities than hydrocarbon polymers currently used in these separations (cf. Table IV). Systematic structure-property studies, which might reveal materials with high CO_2 permeabiltiy and high CO_2/CH_4 selectivity, are not widely available for fluoropolymers. Such studies might represent a good long-term approach to develop optimized fluoropolymers for gas separations (16, 17).

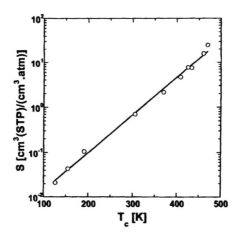

Figure 3. Infinite dilution solubility coefficients for permanent gases and hydrocarbons in low density polyethylene (19). The best fit line through the data is: lnS (cm³(STP)/(cm³ atm)) = -6.17 + 0.019T_c (K).

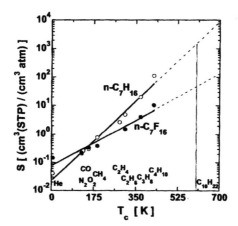

Figure 4. Solubility coefficients of various gases in liquid n-heptane (o) and perfluorinated n-heptane (•) at 25°C and 1 atm partial pressure (12, 15).

Table III. Slope values for the correlation of gas solubility with critical temperature (cf. Eq. 3) in liquids, rubbery polymers and glassy polymers.

Category	Solvent	b (K^{-1})	Source
Liquids	Benzene[a]	0.017	(15)
	n-heptane[a]	0.018	
Rubbers	Natural rubber[a]	0.018	(18)
	Amorphous Polyethylene[a]	0.016	
	Hydropol[a]	0.017	
	Poly(dimethylsiloxane)[b]	0.017	(19)
Glasses	Polysulfone[c]	0.017	(20)
	Poly(phenylene oxide)[d]	0.016	(21)
	Poly(ethylene terephthalate)[e]	0.019	(22)

[a] 25°C and 1 atm

[b] 35°C

[c] 35°C and 10 atm for all gases except $n-C_4H_{10}$ which is at infinite dilution

[d] 35°C and infinite dilution

[e] 24-45°C and infinite dilution

Table IV. Pure gas CO_2/CH_4 selectivity of fluoropolymers.

Polymer	CO_2/CH_4 selectivity	Reference
Polytetrafluoroethylene	13	(23)
Poly(fluoroethylene-co-propylene)	17	(24)
AF2400[a]	5.6	(25)
AF1600[a]	5.9	(26)
Hyflon AD 80	13	This work

[a] AF2400 and AF1600 are random copolymers of tetrafluoroethylene (TFE) and 2,2-bistrifluoromethyl-4,5-dilfluoro-1,3-dioxole containing 13 mol% and 35 mol% of TFE, respectively.

A shorter term strategy to consider is coating the selective hydrocarbon polymer with a thin fluoropolymer layer that has low higher-hydrocarbon solubility. This approach might lower the effective higher-hydrocarbon partial pressure to which the underlying hydrocarbon polymer layer is exposed, thereby reducing plasticization. This approach has obvious tradeoffs. The fluoropolymer coating would reduce gas flux due to the extra mass transfer resistance that it imposes on all penetrants. In addition, the composite membrane selectivity could be adversely affected if the selectivity of the coating layer was less than that of the hydrocarbon layer. In this article, a theoretical analysis is used to assess the ability of a fluoropolymer coating to reduce the exposure of an underlying hydrocarbon membrane to higher hydrocarbons and the penalty associated with having an extra resistance to mass transfer. Some results from this model have been previously reported *(27)*; the current article provides a complete derivation of the model and illustrates the possibilities and limitations of this approach with the aid of model cases.

Theory

The flux of a penetrant A through a polymer membrane, N_A, is *(10, 28)*:

$$N_A = \frac{P_A(p_{up,A} - p_{down,A})}{l} \tag{1}$$

where $p_{up,A}$ and $p_{down,A}$ are the penetrant partial pressures in the gas phase contiguous to the upstream and downstream faces of the membrane, respectively, l is the membrane thickness and P_A is the permeability of the penetrant through the polymer. Permeability is often expressed in units of Barrer where 1 Barrer = 1×10^{-10} $(cm^3(STP) \cdot cm)/(cm^2 \cdot s \cdot cmHg)$.

The solution-diffusion model describes gas transport in polymers *(29)*. Using this model, when the downstream pressure is much less than the upstream pressure, penetrant permeability can be expressed in terms of penetrant solubility and diffusivity in the polymer:

$$P = S \times D \tag{2}$$

As indicated previously, gas solubility is often correlated with gas critical temperature, T_c, as follows *(10)*:

$$S = a \exp(b \times T_c) \tag{3}$$

Additionally, gas diffusion coefficients are commonly understood to scale with some measure of penetrant size, such as critical volume, V_c *(30)*:

$$D = \frac{\tau}{V_c^{\eta}} \qquad (4)$$

In Equations 3 and 4, a, b, τ and η are adjustable constants. b characterizes the increase in penetrant solubility in the polymer with increasing penetrant critical temperature. η is a measure of the polymer's size selectivity for gaseous penetrants. Polymers with larger values of η have higher diffusivity selectivity than polymers with lower η values. Liquids do not pose a significant diffusion resistance to dissolved gas molecules and hence the size selectivity of liquids is very low. For example, low molar mass organic liquids like hexane, heptane and benzene have η values of only about 0.45 *(30)*. Rubbery polymers like polydimethylsiloxane have low size-sieving capabilities due to the high mobility of polymer chains above the polymer's glass transition temperature. PDMS has an η value of 2.2 *(30)*. In comparison, glassy polymers like PSF and poly(vinyl chloride) are strongly size-sieving due to low chain mobility below the glass transition temperature and have η values of 8.4 and 10.5 respectively *(30)*.

Problem Definition

Figure 5a presents the cross section of a hydrocarbon polymer membrane of thickness l^{HP} used for removing CO_2 from natural gas. The membrane is exposed, on its upstream side, to higher hydrocarbons having a partial pressure of $p_{up,HC}$ [†]. Figure 5b shows a cartoon of the proposed approach of applying a fluoropolymer layer to the hydrocarbon polymer. In this scenario, the overcoated hydrocarbon polymer membrane is now exposed to a hydrocarbon partial pressure p^*_{HC} which is lower than the upstream partial pressure of the hydrocarbon, $p_{up,HC}$ due in part to low solubility and diffusivity of the higher hydrocarbon in the fluoropolymer coating. The objective is to use the fluoropolymer coating to achieve a large reduction in p^*_{HC} relative to $p_{up,HC}$ with a minimal loss in CO_2 flux and CO_2/CH_4 selectivity provided by the original hydrocarbon membrane. Mathematically, these criteria can be expressed as:

$$\left(\frac{p^*}{p_{up}} \right)_{HC} \to 0, \quad while \quad \left(\frac{N^c}{N} \right)_{CO_2} \to 1 \quad and \quad \left(\frac{\alpha^c}{\alpha} \right)_{CO_2/CH_4} \to 1 \qquad (5)$$

[†] The subscript 'HC' stands for hydrocarbon and will be used later to indicate the name of the specific higher hydrocarbon under consideration.

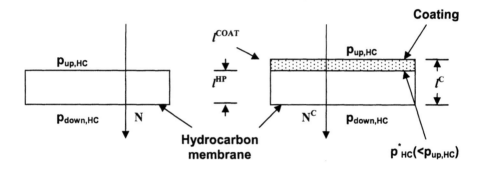

(a) Hydrocarbon polymer membrane

(b) Composite membrane

Figure 5. Schematic diagram of (a) a hydrocarbon polymer membrane and (b) a composite membrane. The subscript 'HC' denotes hydrocarbon gas.

where N and α are the membrane gas flux and selectivity, respectively, and the subscript HC refers to higher hydrocarbons (*e.g.*, hexane, octane, decane, aromatic compounds, *etc.*). The composite membrane properties are denoted by a superscript 'c'.

Analysis

Flux condition

Using the definition of permeability given in Eq. 1 and modeling the composite membrane as two polymer layers in series, the steady state flux of penetrant A through the composite membrane is:

$$N_A^c = \frac{P_A^{COAT}(p_{up,A} - p_A^*)}{l^{COAT}} = \frac{P_A^{HP}(p_A^* - p_{down,A})}{l^{HP}} = \frac{P_A^c(p_{up,A} - p_{down,A})}{l^c} \quad (6)$$

where the superscript 'COAT' refers to the fluoropolymer coating layer. This equation sets the hypothetical interfacial partial pressure of the penetrant, p_A^*, equal in the two polymers at the polymer-polymer interface, which is equivalent to equating chemical potential of the penetrant in the two polymers at the interface. Equation 6 can be recast as follows:

$$N_A^c = \frac{(p_{up,A} - p_{down,A})}{\dfrac{l^{COAT}}{P_A^{COAT}} + \dfrac{l^{HP}}{P_A^{HP}}} \quad (7)$$

The ratio of the membrane thickness to its gas permeability coefficient represents the mass transfer resistance of the membrane layer to permeation of gas A. From Eqs. 1 and 7, the flux condition in Eq. 5 becomes:

$$\left(\frac{N^c}{N}\right)_{CO_2} = \frac{\dfrac{l^{HP}}{P_{CO_2}^{HP}}}{\dfrac{l^{HP}}{P_{CO_2}^{HP}} + \dfrac{l^{COAT}}{P_{CO_2}^{COAT}}} \to 1 \quad (8)$$

which implies that the fluoropolymer coating layer resistance to CO_2 transport should be as low as possible to maintain CO_2 flux in the composite membrane as close as possible to that in the uncoated polymer:

$$\frac{\left(\dfrac{l^{COAT}}{P_{CO_2}^{COAT}}\right)}{\left(\dfrac{l^{HP}}{P_{CO_2}^{HP}}\right)} \to 0 \qquad (9)$$

Partial Pressure Condition

From Eq. 6,

$$p_A^* = p_{up,A} - N_A^c \frac{l^{COAT}}{P_A^{COAT}} \qquad (10)$$

Substituting the expression for N_A^c from Eq. 7 and assuming that the downstream penetrant partial pressure is negligible relative to the upstream penetrant partial pressure, the partial pressure condition of Eq. 5 can be rewritten as

$$\left(\frac{p^*}{p_{up}}\right)_{HC} = \frac{\dfrac{l^{HP}}{P_{HC}^{HP}}}{\dfrac{l^{COAT}}{P_{HC}^{COAT}} + \dfrac{l^{HP}}{P_{HC}^{HP}}} \to 0 \qquad (11)$$

which implies that the resistance of the coating layer to higher hydrocarbon transport should be as large as possible:

$$\frac{\left(\dfrac{l^{COAT}}{P_{HC}^{COAT}}\right)}{\left(\dfrac{l^{HP}}{P_{HC}^{HP}}\right)} \to \infty \qquad (12)$$

Eqs. 9 and 12 may be combined to yield the following expression:

$$\frac{\left(\dfrac{P_{CO_2}}{P_{HC}}\right)^{COAT}}{\left(\dfrac{P_{CO_2}}{P_{HC}}\right)^{HP}} \to \infty \qquad (13)$$

Equation 13 depends only on the permeation properties of the materials used in the coating and hydrocarbon polymer separating layer, so it can be used to provide materials selection guidelines. Using the solution diffusion model (Eq. 2) and the solubility and diffusivity correlations in Eqs. 3 and 4, the above condition can be expressed as:

$$\exp[(b^{HP} - b^{COAT})(T_{c_{HC}} - T_{c_{CO_2}})]\left(\frac{V_{c_{HC}}}{V_{c_{CO_2}}}\right)^{\eta^{COAT} - \eta^{HP}} \to \infty \; (>>1) \quad (14)$$

From a practical viewpoint, this condition is modified to the inequality shown in parenthesis in Eq. 14 with the understanding that the higher the value of the left hand side of the inequality, the better will be the performance of the composite membrane. As the higher hydrocarbon critical temperature and critical volume are greater than those of CO_2, this inequality is satisfied when:

$$b^{HP} > b^{COAT} \text{ and } \eta^{COAT} > \eta^{HP} \qquad (15)$$

Based on these conditions, for optimal performance, the fluoropolymer coating should have a lower ratio of higher hydrocarbon to CO_2 solubility and a higher size-selectivity than the hydrocarbon polymer. In other words, ideally, the coating material should pose a large resistance to higher hydrocarbon permeation. The conditions in Eq. 15 provide guidelines for appropriate materials selection of the coating material to achieve a large reduction in the higher hydrocarbon partial pressure to which the hydrocarbon membrane is exposed without a large sacrifice in membrane flux.

Analysis of the condition on CO_2/CH_4 selectivity is presented in Appendix A. The analysis highlights the tradeoff between maintaining high CO_2/CH_4 selectivity while minimizing the transport of higher hydrocarbons to the hydrocarbon membrane. Based on these results, with existing fluoropolymer membranes, which do not have exceptionally high CO_2/CH_4 selectivity, CO_2/CH_4 selectivity will be reduced by overcoating a hydrocarbon polymer to protect it from higher hydrocarbons. However, as will be seen from the model cases, as long as the conditions of Eq. 15 are satisfied, the selectivity loss can be quite small.

Model Cases

The validity of the materials selection guidelines in Eq. 15 was tested by contrasting the performance of two fluoropolymer-coated hydrocarbon membranes, one that satisfies the conditions in Eq. 15 and one that does not. For both of these hypothetical composite membranes, a commercial fluoropolymer, Hyflon AD 80, served as the coating layer. The solubility and permeability of N_2, CO_2, CH_4, C_2H_6 and C_3H_8 in this polymer were determined experimentally over a range of pressures. For the calculations in this study, the solubility and permeability coefficients were determined at infinite dilution by extrapolation from the experimentally determined values at higher pressures. In the case of permeability coefficients, the infinite dilution values are close to the permeability values in the low-pressure range of 1-2 atm. The experimental conditions of these data are not representative of those that might be experienced by a membrane being used to treat natural gas. However, we did not have sufficient mixed gas solubility and permeability data at high pressure to enable a more realistic study. The results presented here using low pressure pure gas experimental data are therefore only qualitatively indicative of the benefits and tradeoffs of the proposed approach.

Solubility coefficients of Hyflon AD 80 are presented in Figure 6. Diffusion coefficients shown in Figure 7 were calculated from the ratio of infinite dilution permeability to solubility coefficients. The two hydrocarbon polymers are ethyl cellulose and polysulfone. The transport properties of these polymers were obtained from the literature (23, 31-33) and are displayed in Figures 6 and 7. The data in the two figures are for He, N_2, O_2, CO_2 and hydrocarbon penetrants up to C_3H_8 or C_4H_{10}, depending on the polymer. These data were used to find least-square best fit values of the coefficients a, b, τ and

120

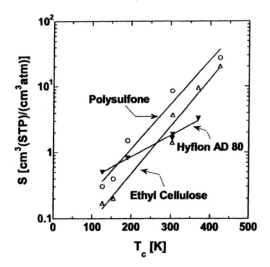

Figure 6. Infinite dilution solubility coefficients in polysulfone (o) (30), ethyl cellulose (Δ) (23) and Hyflon AD 80 (▼) at 35°C as a function of penetrant critical temperature. (Reproduced with permission from Elsevier Science, B.V.).

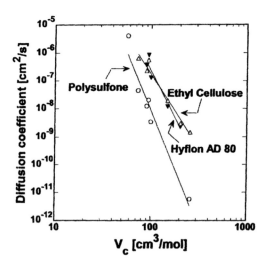

Figure 7. Infinite dilution diffusion coefficients in polysulfone (o) (30), ethyl cellulose (Δ) (23) and Hyflon AD 80 (▼) at 35°C as a function of penetrant critical volume. (Reproduced with permission from Elsevier Science, B.V.).

η in Eqs. 3 and 4. The best fit values are tabulated in Table V. Solubility and diffusivity values for higher hydrocarbons were obtained by extrapolation using these equations and the critical properties of the penetrants (cf. Table VI for critical properties).

For both composite membranes, the slope of the trendline of infinite dilution solubility coefficients with T_c (i.e. the value of b) of the fluoropolymer is much less than that of the hydrocarbon polymer, thus satisfying the first inequality (cf. Figure 6 and Table V). However, the size sieving ability of polysulfone (i.e. the value of η) is much greater than that of Hyflon AD 80, while ethyl cellulose is less strongly size sieving than Hyflon AD 80 (cf. Figure 7 and Table V). Thus, the ethyl cellulose/Hyflon AD 80 membrane satisfies both inequalities in Eq. 15 while the polysulfone/Hyflon AD 80 membrane does not.

The 3 ratios in Eq. 5 were calculated for 4 linear alkanes and for fluorocarbon coating-to-hydrocarbon membrane thickness ratios, $\left(l^{COAT} / l^{HP} \right)$, ranging from 0.05 to 5. This range was chosen to obtain a wide variation in values for $\left(p^* / p_{up} \right)_{HC}$ values. The results are shown in Figures 8 and 9 for polysulfone/Hyflon AD 80 and ethyl cellulose/Hyflon AD 80, respectively.

Table V. Parameter values for polysulfone, ethyl cellulose and Hyflon AD 80.

Polymer	a (cm³(STP)/(cm³·atm))	b (K⁻¹)	τ^a	η (-)
Polysulfone	0.0511	0.017	4.79×10^8	8.37
Ethyl cellulose	0.0148	0.017	1.48×10^3	5.03
Hyflon AD 80	0.1936	0.007	2.34×10^5	6.09

[a] τ has units of $[(cm^2/s)\cdot(cm^3/mol)^\eta]$

Results and Discussion

Figures 8 and 9 present the tradeoff between reducing hydrocarbon partial pressure at the interface and maintaining high CO_2 permeability and CO_2/CH_4 selectivity for polysulfone/Hyflon AD 80 and ethyl cellulose/Hyflon AD 80 composite membranes, respectively. The ordinate shows the ratio of CO_2 flux through the composite membrane to that through the original hydrocarbon

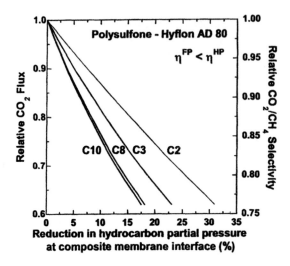

Figure 8. *Tradeoff between partial pressure reduction of C2, C3, C8 and C10 linear alkanes at the polysulfone/Hyflon AD 80 composite membrane interface and loss in CO2 flux and CO2/CH4 permselectivity. The two y-axes have been so plotted that each of the curves in the figure corresponds to values on both axes. (Reproduced with permission from Elsevier Science, B.V.).*

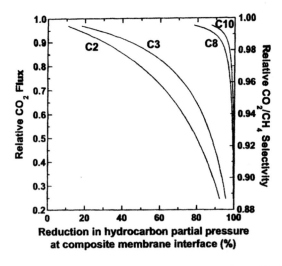

Figure 9. *Tradeoff between partial pressure reduction of C2, C3, C8 and C10 linear alkanes at the ethyl cellulose/Hyflon AD 80 composite membrane interface and loss in CO_2 flux and CO_2/CH_4 permselectivity. The two y-axes have been so plotted that each of the curves in the figure corresponds to values on both axes. (Reproduced with permission from Elsevier Science, B.V.).*

membrane and the ratio of CO_2/CH_4 selectivity of the composite relative to that of the original membrane.

Table VI. Critical properties of penetrants used in model calculations *(34)*.

Penetrant	Critical Temperature T_c (K)	Critical Volume V_c (cm^3/mol)
CO_2	304.21	93.9
CH_4	191.05	99.2
C_2H_6	305.35	148.3
C_3H_8	369.95	203.0
C_8H_{18}	568.80	492.0
$C_{10}H_{22}$	617.70	603.0

Figure 8 shows that the reduction of hydrocarbon partial pressure at the polysulfone/Hyflon AD 80 interface comes at the expense of a significant drop in CO_2 flux (throughput) and CO_2/CH_4 selectivity (purity). For example, a 15% reduction in C_3H_8 partial pressure at the interface is accompanied by a 25% loss in flux and a loss of more than 15% in CO_2/CH_4 selectivity. Also, as hydrocarbon penetrant size increases, the tradeoff becomes more unfavorable. The same loss in flux and CO_2/CH_4 selectivity mentioned above yields only an 11% reduction in $n\text{-}C_{10}H_{22}$ partial pressure at the hydrocarbon polymer-fluoropolymer interface.

In contrast, for the ethyl cellulose/Hyflon AD 80 membrane (*cf.* Figure 9), a large reduction in hydrocarbon penetrant partial pressure at the interface can be obtained with only moderate decreases in CO_2 flux and CO_2/CH_4 selectivity. For example, at 25% loss in CO_2 flux, the interface partial pressure of propane is reduced by 70%, which is a much greater reduction than the 15% reduction in interfacial partial pressure achieved in the polysulfone/Hyflon AD 80 membrane for this penetrant. Also, the associated loss in CO_2/CH_4 selectivity is only about 4% for the ethyl cellulose/Hyflon AD 80 membrane. Interestingly, the tradeoff between interfacial partial pressure reduction and flux and selectivity losses becomes more favorable with increasing penetrant size, which is opposite to the case of polysulfone/Hyflon AD 80. Thus, a coating that reduces CO_2 flux by 25% and CO_2/CH_4 selectivity by 4% provides over 95% reduction in $n\text{-}C_8H_{18}$ and $n\text{-}C_{10}H_{22}$ interfacial partial pressures.

The poor predicted performance of the Hyflon AD 80 coating on polysulfone relative to that on ethyl cellulose results from the unfavorable mismatch in size sieving ability for the polysulfone/Hyflon AD 80 composite membrane. Polysulfone is far more size sieving than Hyflon AD 80, so the critical volume term in Eq. 14 is less than unity for higher hydrocarbons, and its value decreases progressively with increasing hydrocarbon contaminant size. Figure 10 shows the value of the expression in Eq. 15 for the two composite membranes as a function of hydrocarbon penetrant critical volume. With increasing hydrocarbon penetrant size, the condition of Eq. 14 becomes progressively better satisfied for the ethyl cellulose containing membrane while it worsens for the polysulfone/Hyflon AD 80 composite.

In summary, a hydrocarbon-fluorocarbon composite polymer membrane satisfying the conditions of Eq. 15 could, in principle, achieve large reductions in interfacial partial pressure of higher hydrocarbon penetrants without a large sacrifice in flux and selectivity. Therefore, this approach might be useful for addressing the issue of plasticization of hydrocarbon membranes used in natural gas separations. However, it must be stressed that the model cases presented above are based on pure gas permeation properties determined under laboratory conditions and hence are, at best, only qualitatively suggestive of the potential benefits. The actual performance benefits can be analyzed only with the help of mixture permeation properties determined at process conditions, and such data are currently quite rare in the open literature. Also, if a fluoropolymer were available that was considerably more strongly size sieving than conventional hydrocarbon-based polymers, one might eliminate the hydrocarbon polymer membrane entirely.

Conclusions

A model is presented for using a lipophobic fluoropolymer coating on a hydrocarbon membrane to mitigate plasticization of the hydrocarbon membrane due to sorption of higher hydrocarbon contaminants in natural gas. Fluoropolymers can have much lower solubility values for higher hydrocarbons than hydrocarbon-based polymers, and the model calculations suggest that, under certain circumstances, this property may be exploited to reduce the exposure of the hydrocarbon polymer in the composite membrane to higher hydrocarbons. However, fluoropolymers reported to date in the open literature have only modest size-selectivities. Therefore, moderately size-sieving hydrocarbon polymers (*e.g.*, ethyl cellulose, cellulose acetate, *etc.*) might benefit more from this approach than more strongly size-sieving materials (*e.g.*, polysulfones, polyimides, *etc.*) To provide effective plasticization resistance to

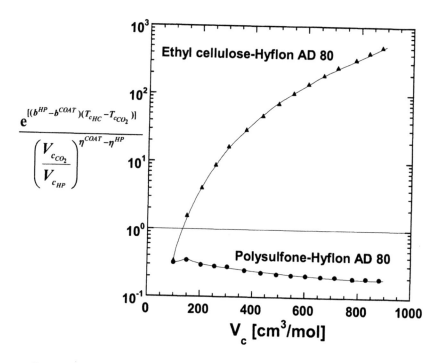

Figure 10. Comparison of the value of the expression in Eq. 15 for the two composite membranes as a function of critical volume of C1 to C15 linear alkanes.

the latter polymers, more strongly size-sieving fluoropolymers may need to be developed.

Appendix

The CO_2/CH_4 selectivity condition in Eq. 2 is:

$$\left(\frac{\alpha^C}{\alpha}\right)_{CO_2/CH_4} \geq 1 \tag{A1}$$

which can be written as follows:

$$\left(\frac{P_{CO_2}^C}{P_{CH_4}^C}\right)\left(\frac{P_{CH_4}^{HP}}{P_{CO_2}^{HP}}\right) \geq 1 \tag{A2}$$

Introducing the thicknesses of the hydrocarbon polymer membrane and the composite membrane into Eq. A2 converts the permeabilities into mass transfer resistances:

$$\frac{\left(\dfrac{l^C}{P_{CH_4}^C}\right)\left(\dfrac{l^{HP}}{P_{CO_2}^{HP}}\right)}{\left(\dfrac{l^C}{P_{CO_2}^C}\right)\left(\dfrac{l^{HP}}{P_{CH_4}^{HP}}\right)} \geq 1 \tag{A3}$$

Expressing the composite membrane resistances in terms of the resistances of the individual layers in the composite (*cf.* Eq. 8) and simplifying the resulting expression yields the following:

$$\frac{\left(1+\dfrac{l^{COAT}}{l^{HP}}\dfrac{P_{CH_4}^{HP}}{P_{CH_4}^{COAT}}\right)}{\left(1+\dfrac{l^{COAT}}{l^{HP}}\dfrac{P_{CO_2}^{HP}}{P_{CO_2}^{COAT}}\right)} \geq 1 \tag{A4}$$

This condition implies that

$$\left(\frac{P_{CO_2}}{P_{CH_4}}\right)^{COAT} \geq \left(\frac{P_{CO_2}}{P_{CH_4}}\right)^{HP} \quad (A5)$$

Using Eqs. 3 and 4 in the above expression gives the condition that ideal materials should meet to satisfy the selectivity constraint in Eq. A1:

$$\exp[(b^{COAT} - b^{HP})(T_{c_{CO_2}} - T_{c_{CH_4}})]\left(\frac{V_{c_{CH_4}}}{V_{c_{CO_2}}}\right)^{\eta^{COAT} - \eta^{HP}} \geq 1 \quad (A6)$$

References

1. Koros, W. J.; Mahajan, R. *J. Membrane Sci.* **2000**, *175*, 181-196.
2. Baker, R. W. *Membrane Technology and Applications* McGraw-Hill: New York, 2000.
3. Ratcliffe, C. T.; Diaz, A.; Nopasit, C.; Munoz, G. In *Application of Membranes in CO₂ Separation from Natural Gas: Pilot Plant Tests on Offshore Platforms*, Laurence Reid Gas Conditioning Conference, Norman, OK, 1999; pp 118-140.
4. Berens, A. R. *Makromol. Chem. Macromol. Symp.* **1989**, *29*, 95-108.
5. White, L. S.; Blinka, T. A.; Kloczewski, H. A.; Wang, I.-F. *J. Membrane Sci.* **1995**, *103*, 73-82.
6. Puleo, A. C.; Paul, D. R.; Kelley, S. S. *J. Membrane Sci.* **1989**, *47*, 301-322.
7. Thundyil, M. J.; Jois, Y. H.; Koros, W. J. *J. Membrane Sci.* **1999**, *152*, 29-40.
8. Dixon-Garrett, S. V.; Nagai, K.; Freeman, B. D. *J. Polym. Sci. Part B. Polym. Phys.* **2000**, *38*, 1461-1473.
9. Dixon-Garrett, S. V.; Nagai, K.; Freeman, B. D. *J. Polym. Sci. Part B. Polym. Phys.* **2000**, *38*, 1078-1089.
10. Ghosal, K.; Freeman, B. D. *Polym. Adv. Technol.* **1993**, *5*, 673-697.
11. Gee, G. *Quart. Rev. (London)* **1947**, *1*, 265-298.
12. Hildebrand, J. H.; Prausnitz, J. M.; Scott, R. L. *Regular and Related Solutions* Van Nostrand Reinhold Company: New York, 1970.

13. Wilhelm, E.; Battino, R. *J. Chem. Thermody.* **1971**, *3*, 761-768.
14. Prabhakar, R.; De Angelis, M.-G.; Sarti, G. C.; Freeman, B. D. *to be published* **2003**.
15. Jolley, J. E.; Hildebrand, J. H. *J. Amer. Chem. Soc.* **1958**, *80*, 1050-1054.
16. Pinnau, I.; He, Z.; Da Costa, A. R.; Amo, K. D.; Daniels, R. U.S. Patent 6,361,583 B1, 2002.
17. Pinnau, I.; He, Z.; Da Costa, A. R.; Amo, K. D.; Daniels, R. U.S. Patent 6,361,583 B1, 2002.
18. Michaels, A. S.; Bixler, H. J. *J. Polym.Sci.* **1961**, *50*, 393-412.
19. Kamiya, Y.; Naito, Y.; Terada, K.; Mizoguchi, K. *Macromolecules* **2000**, *33*, 3111-3119.
20. Ghosal, K.; Chern, R. Y.; Freeman, B. D.; Savariar, R. *J. Polym. Sci. Part B. Polym. Phys.* **1995**, *33*, 657-666.
21. Toi, K.; Paul, D. R. *J. Appl. Polym. Sci.* **1982**, *27*, 2997-3005.
22. Serad, G. E.; Freeman, B. D.; Stewart, M. E.; Hill, A. J. *Polymer* **2001**, *42*, 6929-6943.
23. Brandrup, J.; Immergut, E. H.; Grulke, E. A. *Polymer Handbook* 4th Ed.; Wiley Interscience: New York, 1999.
24. Pasternak, R. A.; Burns, G. L.; Heller, J. *Macromolecules* **1971**, *4*, 470-475.
25. Merkel, T. C.; Bondar, V.; Nagai, K.; Freeman, B. D.; Yampolskii, Y. P. *Macromolecules* **1999**, *32*, 8427-8440.
26. Alentiev, A. Y.; Shantarovich, V. P.; Merkel, T. C.; Bondar, V. I.; Freeman, B. D.; Yampolskii, Y. P. *Macromolecules* **2002**, *35*, 9513-9522.
27. Prabhakar, R.; Freeman, B. D. *Desalination* **2002**, *144*, 79-83.
28. Stern, S. A. *J. Membrane Sci.* **1994**, *94*, 1-65.
29. Wijmans, J. G.; Baker, R. W. *J. Membrane Sci.* **1995**, *107*, 1-21.
30. Freeman, B. D.; Pinnau, I. In *Polymer Membranes for Gas and Vapor Separation*; Freeman, B. D.; Pinnau, I., Eds.; American Chemical Society: Washington, DC, 1999; Vol. 733, pp 1-27.
31. Erb, A. J.; Paul, D. R. *J. Membrane Sci.* **1981**, *8*, 11-22.
32. Ghosal, K.; Chern, R. T.; Freeman, B. D. *Macromolecules* **1996**, *29*, 4360-4369.
33. McHattie, J. S.; Koros, W. J.; Paul, D. R. *Polymer* **1991**, *32*, 840-850.
34. Reid, R. C.; Prausnitz, J. M.; Poling, B. E. *The Properties of Gases and Liquids* 4th Ed.; McGraw-Hill: New York, 1987.

Chapter 8

Structure, Gas Sorption, and Gas Diffusion of a Low Density Liquid Crystalline Polyester

M. Tsukahara, Y. Tsujita, H. Yoshimizu, and T. Kinoshita

Polymeric Materials Course, Department of Material Science and Engineering, Nagoya Institute of Technology, Gokiso-cho, Showa-ku, Nagoya 466–8555, Japan

A low-density liquid crystalline polyester with an alkyl side chain, B-C14, was synthesized from the 1,4-di-(1-tetradecyl ester) of 1,2,4,5-benzenetetracarboxylic acid and 4,4'-biphenol, to clarify the relationship between the structure and gas sorption and diffusion properties of this polyester. B-C14 is a thermotropic liquid crystalline polymer and formed a layered structure composed of alternating rigid aromatic main chain layers and flexible alkyl side chain layers. CO_2 and Xe sorption of B-C14 was restricted to the side chain layer, which is almost a liquid-like environment, and these sorption isotherms obeyed Henry's law. This result also indicates that the main chain layer was very dense and could not sorb CO_2 and Xe. Therefore, the gas solubility of B-C14 was low, although the polymer density is relatively low.

Rigid aromatic polyesters consisting of dialkyl esters of 1,2,4,5-pyromellitic acid and 4,4'-biphenol, B-Cn (n is the number of carbon atoms in the alkyl side chain) form thermotropic liquid crystal polymers. The thermotropic transition behavior and mesophase structure of B-Cn have been examined by optical microscopy, DSC, NMR, and X-ray measurements (*1-5*). Recently, a highly oriented crystalline fiber of B-C6, prepared by spinning a nematic melt of this material, has been reported (*6*). The most interesting property of B-Cn is the ability to form layered structures in the crystalline and liquid crystalline states.

© 2004 American Chemical Society

130

That is, the aromatic main chains are packed into a layered structure, and the flexible alkyl side chains occupy the space between the layers (see Figure 1). In other words, the layered structure of B-Cn consists of a high-density layer containing aromatic main chains and a low-density layer containing alkyl side chains. Therefore, B-Cn polymers form low density liquid crystalline polyesters. Based on these structural features, the gas sorption and diffusion properties of B-Cn should be very unique relative to those of other dense and glassy polyesters such as bisphenol-A polycarbonate, PC. In this study, to clarify the relationship between the layered structure and the gas sorption and diffusion properties of B-Cn, B-C14 was synthesized from the 1,4-di-(1-tetradecyl ester) of 1,2,4,5-benzenetetracarboxylic acid and 4,4'-biphenol, and two kinds of film-like samples of B-C14 were prepared by annealing in the liquid crystalline state for a long time and cooling to room temperature from the isotropic liquid state. Structural characterization and gas sorption measurements were performed.

Experimental

Samples

The synthesis of B-C14 was performed according to the literature (*1-3*). The primary structure of B-C14 was determined by ^{1}H NMR. The molecular weight and degree of polymerization was determined to be Mw = 2.36 × 10^4 and Mw/Mn = 2.18, respectively, in THF at 40 °C by GPC. The GPC column was a TOSOH HLC-8020, and it was calibrated using polystyrene standards. An as-cast film-like sample of B-C14 was obtained by casting from 10 wt% THF solution and drying for 4 hrs at 80 °C *in vacuo*. From this as-cast sample, a melt-cooled sample was prepared by annealing at 150 °C for 3 hrs and then cooling to room temperature. The annealed sample was prepared by annealing the melt-cooled sample at 100 °C for 7 days. These samples were brittle at room temperature. CO$_2$ and Xe used in this study were at least 99.9 % pure and were used without further purification.

Methods

Densities of B-C14 samples at 25 °C were determined by the floatation method using aqueous solutions of NaBr. Differential scanning calorimetry, DSC, measurements were made using a DSC7 calorimeter (Perkin Elmer Co.). The wide-angle X-ray diffraction profiles at room temperature were obtained with a RAD-RC (Rigaku Denki Co., Inc.) apparatus using nickel-filtered copper Kα irradiation and power settings of 40 kV x 80 mA. Sorption isotherms of CO$_2$ and Xe at 25 °C were determined using a gravimetric sorption apparatus with an electromicrobalance 2000 (Cahn Instruments Inc.).

Results and Discussion

Characterization by DSC

Figure 2 shows DSC thermograms of an as-cast sample of B-C14. On heating, there are three endothermic peaks at about 50, 90, and 120 °C. These thermal events are assigned to the melting of partly crystallized side chains, the transition from the solid crystalline phase to the fluid liquid crystalline phase, and the transition to the isotropic phase, respectively, in accordance with previous literatures (*1-3*). The enthalpy of melting of the side chain crystals is very small (less than 5 J/g) relative to that of polyethylene (290 J/g (*7*)), indicating that the amount of side chain crystallinity in this sample was quite negligible. On cooling, exothermic peaks were observed at about 135 and 110 °C in the DSC thermogram. These peaks are assigned to the isotropic-liquid crystal and liquid crystal-liquid crystal transitions, respectively. These findings are consistent with results reported previously (*1-3*). Because B-C14 is a polymer, its thermal behavior shows considerable hysteresis as shown in Figure 3. In the DSC thermogram of melt-cooled B-C14, no peaks are observed, but some endothermic peaks appear after annealing at 100 °C. The enthalpies and temperatures of the endothermic DSC peaks, except for the liquid crystal-isotropic liquid transition, increase with annealing time until 7 days. Based on these results, B-C14 exhibits a liquid crystal phase at high temperature and is in a highly ordered state (crystal like) at room temperature after annealing at or near the crystal-liquid crystal transition temperature for a long time. Hereafter, an annealed sample of B-C14, prepared by annealing the melt-cooled sample at 100 °C for 7 days, is used in the characterization studies.

Characterization by X-ray diffraction

X-ray diffraction patterns of the annealed and melt-cooled samples are shown in Figure 4. The diffraction peak at about $2\theta = 4.5°$ corresponds to the layer spacing (*1-3*) and is clearly observed in both profiles. Since the overtone reflection appeared at about $2\theta = 9°$ and some diffraction peaks at higher angles are weakly detected only for the annealed sample, the regularity and content of the layered structure of B-C14 become higher after annealing. The layer spacing values of the annealed and melt-cooled samples of B-C14 determined by X-ray diffraction are tabulated in Table I, together with densities. Upon annealing, the layer spacing becomes slightly thinner and the density increases somewhat, but the differences are larger than the uncertainties in layer spacing ($\pm$ 0.05 Å) and density ($\pm$ 0.002 g/cm^3). These results support the hypothesis that annealing leads to the formation of highly ordered structure as mentioned above. For the melt-cooled sample, although a diffraction peak due to the layered structure was observed, there was no corresponding thermal event in the DSC

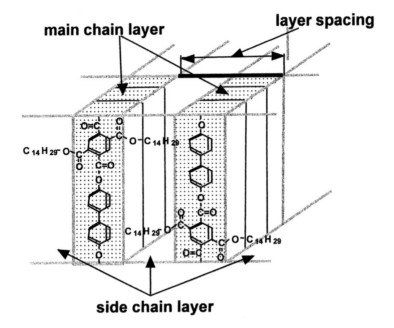

Figure 1. Schematic model of B-C14.

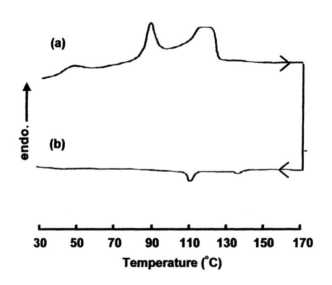

Figure 2. DSC thermograms of an as-cast sample of B-C14 obtained under a nitrogen atmosphere at a scanning rate of 10 °C/min; (a) heating curve, (b) cooling curve.

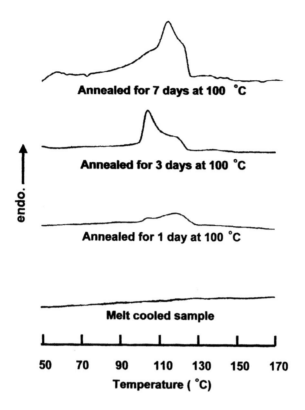

Figure 3. DSC thermograms of B-C14 obtained under a nitrogen atmosphere at a heating rate of 10 °C/min.

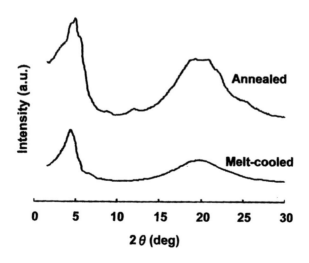

Figure 4. X-ray diffraction profiles of B-C14 at room temperature.

thermogram (see Figure 3). The reason for this discrepancy is not clear. However, the distribution of regularity of the layered structure in the melt-cooled B-C14 is wide, so the DSC peaks of some transitions become very broad and overlapped, and/or the enthalpies of these thermal events are appreciably reduced because of the small size of the highly ordered domains.

Table I. The Layer Spacing and Density of B-C14 at 25 °C

Sample	Layer Spacing (Å)	Density (g/cm³)
Melt-cooled	19.6	1.08
Annealed	18.7	1.09

Gas sorption and diffusion properties

Figures 5 and 6 show CO_2 and Xe sorption isotherms in annealed and melt-cooled B-C14 samples at 25 °C. CO_2 and Xe sorption amounts are almost identical to one another, which is consistent with the CO_2 and Xe sorption properties in PC (8, 9). This result is probably due to the similarity in critical temperatures and pressures of CO_2 and Xe, respectively (The values of T_c and p_c for CO_2 and Xe are 304.2 and 289.7 K, and 72.86 and 57.64 atm, respectively). If the gas sorption properties of B-C14 were similar to those of other polyesters in the glassy state (such as PC), the gas sorption isotherm should be concave to the pressure axis and interpreted by the dual-mode sorption model (10, 11). In fact, the CO_2 and Xe isotherms in PC can be explained by the dual-mode sorption model (8, 9). However, all sorption isotherms studied here obey Henry's law. Therefore, in the case of B-C14, gas sorption presumably occurs only in the alkyl side chain layer, which is almost liquid-like in character, because the gas sorption mechanism of rubbery polymers and liquids is, in general, described by Henry's law (12-14). The Henry's law constants of CO_2 and Xe for B-C14 samples are tabulated in Table II, together with those for PC at 760 cmHg. From these values, the amounts of CO_2 and Xe sorption in PC at 760 cmHg were about two (for the melt-cooled) or three (for the annealed sample) times larger than those of B-C14 samples, although the densities of the B-C14 samples are considerably lower than that of PC (ρ_{PC} is ca. 1.2 g/cm³). This result suggests that the aromatic main chain layer is very dense and cannot sorb any CO_2 and

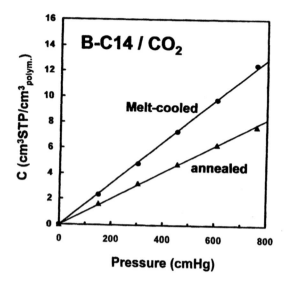

Figure 5. CO₂ sorption isotherms of melt-cooled and annealed samples of B-C14 at 25 °C.

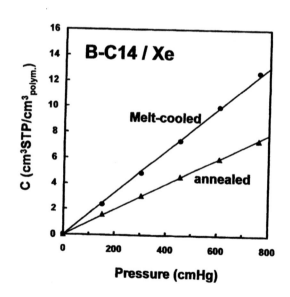

Figure 6. Xe sorption isotherms of melt-cooled and annealed samples of B-C14 at 25 °C.

Table II. CO₂ and Xe Henry's law constants in B-C14 and PC at 25 °C

| Sample | Henry's law constant $(cm^3 STP/cm^3_{polym.}\ cmHg)$ | |
	CO_2	Xe
Melt-cooled	1.61×10^{-2}	1.64×10^{-2}
Annealed	1.03×10^{-2}	0.99×10^{-2}
PC	3.39×10^{-2}	3.50×10^{-2}

NOTE: The polycarbonate (PC) values correspond to solubility coefficients of CO_2 and Xe at 760 cmHg, and they are calculated from sorption data in references (8) and (9).

Xe, similar to the behavior of crystals of PC (8), which is consistent with the structural features of B-C14.

On the other hand, CO_2 and Xe sorption levels in the melt-cooled B-C14 are larger than in the annealed sample. That is, the Henry's law constants of CO_2 and Xe are reduced by annealing. In spite of the small changes in density and layer spacing due to annealing (see Table I), the Henry's law constants decrease by about one-third, indicating that the change in gas sorption properties cannot be explained only by the volume change of the sorption sites. Based on these results, one might speculate that the microspaces around the aromatic main chain layer, whose size is the order of one gas molecule, are considerably reduced by annealing. The microspace size influences gas sorption properties strongly but hardly changes density. In other words, the effective surface area of the main chain layer in melt-cooled B-C14 is larger than that in annealed B-C14.

Since the gas sorption properties of B-C14 obey Henry's law, diffusion coefficients of CO_2 and Xe can be determined from the transient sorption curves (15). The diffusion coefficients of CO_2 and Xe for the annealed and melt-cooled B-C14 are listed in Table III. The diffusion coefficients of CO_2 for the both samples are larger than those of Xe because of the smaller size of CO_2. On the other hand, CO_2 and Xe diffusion coefficients of the melt-cooled sample are almost twice as large as those of the annealed sample. Because there were only small changes in density and layer spacing upon annealing, it cannot be expected that the mobility of gas diffusion sites, i.e., the alkyl side chain layers, was reduced significantly. Hence, changes in the diffusion coefficients mean that the diffusion pathways of gas molecules in the sample become longer as a result of annealing. That is, the hindrance of gas diffusion by the aromatic main chain layer is increased by the build-up of the layered structure during annealing. Consequently, the gas diffusion behavior of B-C14 is consistent with the structural differences between the annealed and melt-cooled B-C14 samples.

Table III. Diffusion coefficients of CO$_2$ and Xe for B-C14 at 25 °C

Sample	Diffusion coefficient (cm^2/s)	
	CO$_2$	Xe
Melt-cooled	3.58 x 10^{-7}	7.07 x 10^{-8}
Annealed	1.71 x 10^{-7}	3.21 x 10^{-8}

Concluding Remarks

The B-C14 sample synthesized in this study exhibits thermal characteristics of a thermotropic liquid crystal and forms a layered structure composed of rigid aromatic polyester main chain layers and flexible alkyl side chain layers. The sample of B-C14 prepared by cooling to room temperature from the isotropic liquid state (above 120 °C) forms a layered structure with many defects relative to the sample annealed in the liquid crystalline state (near 100 °C) for a long time (7 days). The CO$_2$ and Xe sorption and diffusion coefficients are dramatically reduced by annealing. This behavior is consistent with the structural characteristics of B-C14. Gas sorption in B-C14 occurs only in the flexible alkyl side chain layers, indicating that the aromatic main chain layers are very dense and can not sorb CO$_2$ and Xe to a significant extent. Based on these results, the gas solubility of B-C14 is reasonably lower than that of other glassy polyesters, such as PC, even though the density of B-C14 is relatively low.

Acknowledgements

The authors gratefully acknowledge partial financial support through a Grant-in-Aid for Scientific Research in Priority Area (B), "Novel Smart Membrane with Controlled Molecular Cavity", No. 13133202 (2001, 2002) from the Ministry of Education, Culture, Sports, Science, and Technology of Japan, and the Aichi Science and Technology Foundation of Japan.

References

1. Harkness, B.R.; Watanabe, J. *Macromolecules* **1991**, 24, 6759-6763.
2. Watanabe, J.; Harkness, B.R.; Sone, M. *Polym. J.* **1992**, 24, 1119-1127.
3. Sone, M.; Harkness, B.R.; Watanabe, J.; Torii, T.; Yamashita, T.; Horie, K. *Polym. J.* **1993**, 25, 997-1001.

138

4. Sone, M.; Harkness, B.R.; Kurosu, H.; Ando, I.; Watanabe, J. *Macromolecules* **1994**, 27, 2769-2777.
5. Watanabe, J.; Harkness, B.R.; Sone, M.; Ichimura, I. *Macromolecules* **1994**, 27, 507-512.
6. Fu, K.; Nematsu, T.; Sone, M.; Itoh, T.; Hayakawa, T.; Ueda, M.; Tokita, M.; Watanabe, J. *Macromolecules* **2000**, 33, 8367-8370.
7. Mandelkern, L.; Stack, G. M. *Macromolecules* **1984**, 17, 871-878.
8. Kato, S.; Tsujita, Y.; Yoshimizu, H.; Kinoshita, T.; Higgins, J.S. *Polymer* **1997**, 38, 2807-2811.
9. Suzuki, T.; Miyauchi, M.; Yoshimizu, H.; Tsujita, Y. *Polymer J.* **2001**, 33, 934-938.
10. Barrer, R. M.; Barrie, J. A.; Slater, J. *J. Polym. Sci.* **1958**, 27, 177-197.
11. Michaels, A. S.; Vieth, W. R.; Barrie, J. A. *J. Appl. Phys.* **1963**, 34, 1-12.
12. Hildebrand, J. H.; Scott, R. *The Solubility of Nonelectrolytes,* 3rd ed.; Reinhold Pub. Corp.: New York, 1958.
13. Van Amerongen, G. J. *J. Polym. Sci.* **1950**, 5, 307-332.
14. Van Krevelen, D. W.; Hoftyzer, P. J. *Properties of Polymers: Their Estimation and Correlation with Chemical Structure,* 2nd ed.; Elsevier Scientific Pub. Co.: New York, 1976; Chap. 18,
15. Crank, J. *The Mathematics of Diffusion, 2nd ed.;* Clarendon Press, Oxford, 1975.

Chapter 9

Gas Permeation Properties of Silane-Cross-Linked Poly(propylene oxide)

Kazukiyo Nagai and Tsutomu Nakagawa

Department of Industrial Chemistry, Meiji University, 1-1-1 Higashi-mita, Tama-ku, Kawasaki 214-8571, Japan

Gas permeation properties of rubbery silane-crosslinked poly(propylene oxide) films were evaluated and compared to those of polydimethylsiloxane (PDMS). The permeability, diffusivity, and solubility of poly(propylene oxide) [PPO] to various gases, except for carbon dioxide, obey common transport behavior in rubbery polymers, such as PDMS. The carbon dioxide diffusivity of PPO is lower than expected based on the trend of the data of other gases, whereas the carbon dioxide solubility is higher than expected. Because PPO is a very hydrophilic polymer, strong interactions exist between the propylene oxide units and carbon dioxide. For all gases, except for carbon dioxide, both permeability and diffusivity in PPO are lower than those of polydimethylsiloxane. On the other hand, the solubility in PPO is almost the same as that in PDMS.

© 2004 American Chemical Society

Carbon dioxide separation is one of the most important industrial gas separation applications, including the following: (i) carbon dioxide removal from flue gas and landfill gas, (ii) natural gas sweetening, and (iii) enhanced oil recovery. Because CO_2 is a greenhouse gas according to the 1997 Kyoto United Nations Protocol on Climate Change, most countries have tried to reduce carbon dioxide release into the atmosphere. Carbon dioxide separation from either air or methane are, therefore, target applications for the development of advanced membrane materials and novel processes.

Gas transport through dense polymer membranes is based on a solution-diffusion mechanism (1). Hence, the gas selectivity of the membrane is the product of the diffusivity selectivity and the solubility selectivity. To increase the gas selectivity, materials with improved diffusivity selectivity and/or solubility selectivity are desirable. Carbon dioxide is a polar gas relative to air (i.e., oxygen and nitrogen) and methane. For carbon dioxide/inert-gas separation applications, it is possible to enhance the overall selectivity by increasing the solubility selectivity of the membrane. In the past, several membranes have been developed that contain chemical structures which exhibit preferential interactions with carbon dioxide, especially, polyether-based materials (2-4).

Poly(propylene oxide)-based elastomers are used as sealing, coating, and adhesion materials (5). However, very few data on gas permeability, diffusivity, and solubility of this elastomer are available. In this study, gas permeation properties of silane-crosslinked poly(propylene oxide) are reported and compared to those of polydimethylsiloxane. The structures of the polymers are shown in Figure 1.

Experimental

Film Preparation

Linear poly(propylene oxide) having a trimethoxysilyl group at both polymer chain ends was cured (crosslinked) using a tin-catalyst (Scheme 1) according to previous work (6). The average number of the repeat units of poly(propylene oxide) was 51.

Characterization

The geometric density of the silane-crosslinked poly(propylene oxide) films was determined by measuring the film weight and volume at ambient conditions. The fractional free volume, FFV, of the films was estimated from

$$FFV = \frac{V - 1.3V_W}{V} \quad (1)$$

where V is the polymer specific volume (reciprocal of geometric density) and V_W is the specific van der Waals volume estimated from Bondi's group contribution

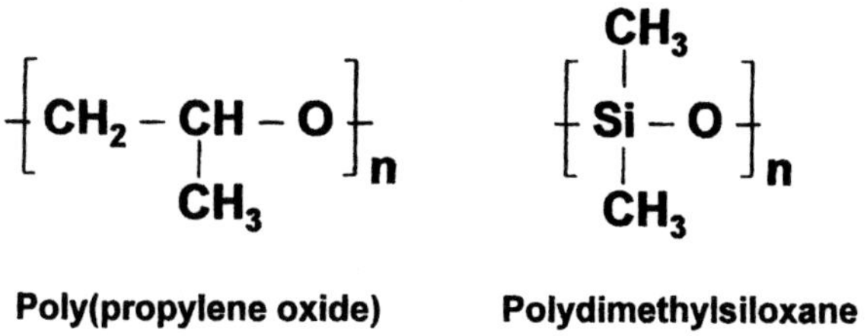

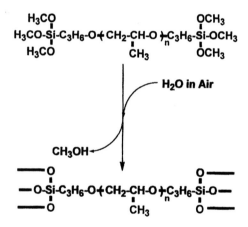

Figure 1. Repeat units of poly(propylene oxide) and polydimethylsiloxane.

Scheme 1. Preparation of silane-crosslinked poly(propylene oxide) films.

method (7). The polymer cohesive energy density, *CED*, and the solubility parameter, δ, of PPO were estimated using the group contribution method of Fedors (8). Thermal properties of the polymer were determined by differential scanning calorimetry using a Perkin-Elmer DSC7.

Gas Permeation Measurements

The gas permeability coefficient, *P*, of the films was determined with the constant volume/variable pressure method at 30°C, according to a previously reported procedure (9). The gas permeability is typically expressed in Barrer, where 1 Barrer = 1×10^{-10} cm^3(STP)·cm/(cm^2·s·cmHg). The feed pressure was 10-76 cmHg. The permeate pressure was about 0 cm Hg (i.e., vacuum). All film samples used in this study were rubbery under the conditions of measurement. The apparent diffusion coefficient, *D*, was calculated by the time-lag method using the following equation:

$$D = \frac{\ell^2}{6\theta} \qquad (2)$$

where θ is the time-lag (s) and ℓ is the film thickness (cm). In addition, the apparent solubility coefficient, *S* (cm^{3}gas(STP)/cm^{3}polymer·cmHg), was estimated from:

$$S = \frac{P}{D} \qquad (3)$$

The ideal selectivity, $\alpha_{A/B}$, of the polymer for gas A relative to gas B is the ratio of the permeability coefficients of the two gases:

$$\alpha_{A/B} = \frac{P_A}{P_B} \qquad (4)$$

When permeability is viewed as the product of solubility and diffusivity, this expression may be rewritten as the product of the two ratios:

$$\alpha_{A,B} = \left(\frac{S_A}{S_B} \right) x \left(\frac{D_A}{D_B} \right) \qquad (5)$$

where the first term is the solubility selectivity and the second is the diffusivity selectivity or mobility selectivity.

Results and Discussion

Characterization

The physical properties of PPO and PDMS are summarized in Table I. The glass transition temperature of the silane-crosslinked poly(propylene oxide) film is –67°C. The PPO film was amorphous on the basis of its wide-angle X-ray diffraction spectrum. The film density is 1.01 ± 0.01 g/cm^3 and its calculated fractional free volume is 0.19 ± 0.01. The solubility parameter and the cohesive energy density of the polymer are 17.6 MPa$^{0.5}$ and 310 MPa, respectively.

Replacing the dimethylsiloxane group with the hydrophilic propylene oxide group in the backbone of the polymer results in an increase of the glass transition temperature by 56°C. This replacement does not significantly influence the density and the fractional free volume of the films. The solubility parameter and the cohesive energy density are greater in poly(propylene oxide) than in polydimethylsiloxane. In this regard, even though poly(propylene oxide) has similar fractional free volume to polydimethylsiloxane, the chain mobility in poly(propylene oxide) is probably lower, because poly(propylene oxide) has a higher glass transition temperature and cohesive energy density than polydimethylsiloxane.

Table I. Properties of silane-crosslinked poly(propylene oxide) and polydimethylsiloxane films

Polymer	T_g (°C)	ρ (g/cm^3)	FFV	δ (MPa$^{0.5}$)	CED (MPa)
Poly(propylene oxide)	−67	1.01±0.01	0.19±0.01	17.6	310
Polydimethylsiloxane	−123	0.98±0.01	0.18±0.01	15.1	228

Gas Permeability

Figure 2 presents the permeability of various gases in silane-crosslinked poly(propylene oxide) at 30°C and polydimethylsiloxane at 35°C as a function of the critical volume of the gases, which is a relative measure of the penetrant size. The permeabilities in polydimethylsiloxane were calculated at a gas pressure of 1 atm using literature data at 35°C (*10*). As expected, polydimethylsiloxane is more permeable to all gases than poly(propylene oxide). For example, the N_2 permeability of polydimethylsiloxane is 400 Barrer, whereas that of poly(propylene oxide) is 33 Barrer. The CO_2 permeability is 3,800 Barrer in polydimethylsiloxane and 710 Barrer in poly(propylene oxide).

Poly(propylene oxide) and polydimethylsiloxane show the same general trend in permeability, as shown in Figure 2. The permeability of less condensable gases (i.e., hydrogen, oxygen, and nitrogen) decreases with increasing critical gas volume. On the other hand, the permeability of condensable hydrocarbons increases with increasing molecular size. For C_2- and C_3-hydrocarbon permeabilities in poly(propylene oxide), the olefin permeability is always higher than the paraffin permeability. It is interesting to note that the carbon dioxide permeability is much higher than that of gases with similar critical volume.

Figure 3 presents the ratio of the permeability in polydimethylsiloxane at 35°C relative to the permeability of poly(propylene oxide) at 30°C as a function of the critical volume of the gases. The permeability ratio of less condensable gases (i.e., hydrogen, oxygen, and nitrogen) increases with increasing critical volume. The permeability ratio is 7.0 for hydrogen, 9.6 for oxygen, and 12 for nitrogen. On the other hand, the permeability ratios of condensable hydrocarbons are about 10, regardless of the carbon number, considering the experimental uncertainties. The ratio of the carbon dioxide permeability is 5.4 and is much lower than that of gases with similar critical volume.

Gas Diffusivity

Figure 4 shows the diffusivity of various gases in silane-crosslinked poly(propylene oxide) at 30°C and polydimethylsiloxane at 35°C as a function of the critical volume (i.e., penetrant size) of the gases. The diffusivities in polydimethylsiloxane were calculated at a gas pressure of 1 atm using literature data at 35°C (10). The diffusivity of all gases is higher in polydimethylsiloxane than in poly(propylene oxide). For example, the nitrogen diffusivity is 3.4×10^{-5} cm^2/s in polydimethylsiloxane and 3.4×10^{-6} cm^2/s in poly(propylene oxide). Similar to their permeability properties, poly(propylene oxide) and polydimethylsiloxane show the same trend in gas diffusivity, except for carbon dioxide. The gas diffusivity decreases with increasing the critical gas volume (i.e., penetrant size), except for the carbon dioxide diffusivity in poly(propylene oxide). The carbon dioxide diffusivity in poly(propylene oxide) is much lower than that of gases with similar critical volume. This is probably due to strong interaction between the propylene oxide units and carbon dioxide molecules; this interaction reduces diffusion of carbon dioxide molecules in the poly(propylene oxide) matrix. The carbon dioxide diffusivity is 2.2×10^{-5} cm^2/s in polydimethylsiloxane and 5.2×10^{-7} cm^2/s in poly(propylene oxide).

The empirical relationship (11) between gas diffusivity, D, and the critical volume, V_c, is described as:

$$D = \frac{\tau}{V_c^{\eta}} \qquad (6)$$

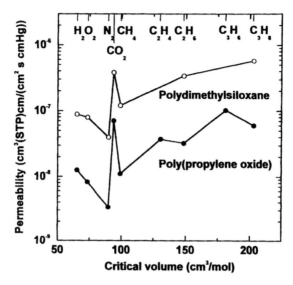

Figure 2. Permeability of various gases in silane-crosslinked poly(propylene oxide) (●) at 30°C and polydimethylsiloxane (O) at 35°C as a function of the critical volume of the gases. The permeabilities of PDMS at a gas pressure of 1 atm at 35°C were estimated from literature data.
(Reproduced with permission from reference 10. Copyright 2000 Wiley.)

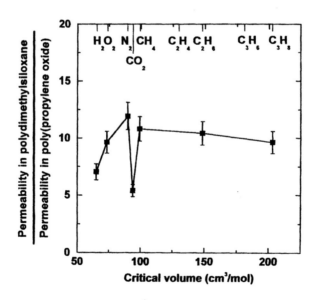

Figure 3. Ratio of the permeability in polydimethylsiloxane at 35° C relative to the permeability in poly(propylene oxide) at 30°C as a function of the critical volume of the gases. The permeability data are from Figure 2.

146

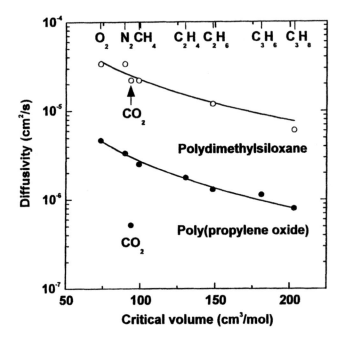

*Figure 4. Diffusivity of various gases in silane-crosslinked poly(propylene oxide)
(●) at 30°C and polydimethylsiloxane (○) at 35°C as a function of the critical
volume of the gases. The diffusivities of PDMS at a gas pressure of 1 atm at
34°C were estimated from literature data.*
(Reproduced with permission from reference 10. Copyright 2000 Wiley.)

where τ and η are adjustable constants. The exponent, η, is a crude measure of the size-sieving ability of the polymer. Polymers with larger η values will have greater size-sieving ability and higher diffusivity selectivity than those with smaller values.

These two adjustable constants of poly(propylene oxide), except for carbon dioxide, are $\tau = 8.3\pm3.1\times10^{-3}$ and $\eta = 1.7\pm0.1$. In Table II, this η value is compared to those of polymers with weak size-sieving ability. The η value is 1.7 for poly(propylene oxide) and 1.5 for polydimethylsiloxane. The size-sieving ability of poly(propylene oxide) is slightly larger than that of polydimethylsiloxane.

Figure 5 presents the ratio of the diffusivity in polydimethylsiloxane at 35°C relative to the diffusivity in poly(propylene oxide) at 30°C as a function of the critical volume of the gases. The ratio of the carbon dioxide diffusivity is much higher than that of the diffusivity of other gases. The diffusivity ratio of less condensable gases (i.e., oxygen and nitrogen) increases with increasing the critical volume. On the other hand, the diffusivity ratios of condensable hydrocarbons are almost constant regardless of the carbon number. This trend, except for the carbon dioxide diffusivity, is the same as that of the permeability, as shown in Figure 3.

Table II. Parameters τ and η in polymers

Polymer	τ	η	Data
Poly(propylene oxide)	8.3×10^{-3}	1.7	Figure 4
Polydimethylsiloxane	2.6×10^{-2}	1.5	Figure 4
Polydimethylsiloxane	4.3×10^{-1}	2.2	Reference *(11)*
Polysulfone	4.8×10^{8}	8.4	Reference *(11)*
Poly(vinyl chloride)	4.5×10^{11}	10	Reference *(11)*

Gas Solubility

Figure 6 shows the solubility of various gases in silane-crosslinked poly(propylene oxide) at 30°C and polydimethylsiloxane at 35°C as a function of the critical temperature of the gases, which is a relative measure of the penetrant condensability. The solubilities in polydimethylsiloxane were calculated at a gas pressure of 1 atm using the literature data at 35°C *(10)*. The solubility in both polymers varies from 10^{-3} to 10^{-1} cm^3(STP)/(cm^3·cmHg), depending on the gases tested. Poly(propylene oxide) and polydimethylsiloxane show the same general trend in solubility, except for carbon dioxide. In Figure 6, the logarithm of the gas solubility increases linearly with increasing the critical temperature of the gases (i.e., penetrant condensability), except for carbon

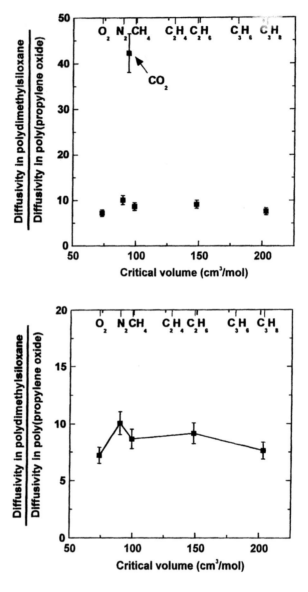

Figure 5. Ratio of the diffusivity in polydimethylsiloxane at 35 °C relative to the diffusivity in poly(propylene oxide) at 30 °C as a function of the critical volume of the gases. The diffusivity data are from Figure 4.

dioxide. The carbon dioxide solubility in poly(propylene oxide) is much higher than that of gases with similar critical temperatures, presumably due to strong interaction between the propylene oxide units and carbon dioxide molecules. This interaction probably enhances sorption of carbon dioxide molecules in the poly(propylene oxide) matrix.

The relationship between gas solubility, S, and the critical temperature, T_c, of the penetrant is defined as: (11)

$$LnS = M + 0.016T_c \qquad (7)$$

where M is a parameter that depends primarily on polymer-penetrant interactions and polymer free volume.

Table III. Parameter M in polymers

Polymer	M	Data
Poly(propylene oxide)	−8.8	Figure 6
Polydimethylsiloxane	−8.8	Figure 6
Polydimethylsiloxane	−8.7	Reference (11)
Polysulfone	−7.7	Reference (11)

The best fit line for poly(propylene oxide) solubility data in Figure 6, except for carbon dioxide, provides M = 8.8±0.9 and a slope of 0.017±0.003. The M value and the slope for poly(propylene oxide) are consistent with those of common polymers, as shown in Table III. As a result, the solubility selectivity of poly(propylene oxide) is the same as that of polydimethylsiloxane.

Figure 7 presents the ratio of the gas solubility in polydimethylsiloxane at 35°C relative to that in poly(propylene oxide) at 30°C as a function of the critical temperature of the gases. The ratio of the carbon dioxide solubility is much lower than that of the solubilities of other gases. On the other hand, the solubility ratios of other gases are almost constant.

Based on these results, differences in the permeability and permeability ratio between poly(propylene oxide) and polydimethylsiloxane are more dependent on those in the diffusivity and diffusivity ratio as compared to those in the solubility and solubility ratio. The differences in diffusivity probably result from the differences in chain mobility between these two polymers, as suggested earlier in Table I. In the case of carbon dioxide transport in poly(propylene oxide), the solubility and solubility ratio are the dominant factors of the permeability and permeability ratio rather than the diffusivity and diffusivity ratio. This is probably due to favorable interactions between the propylene oxide units and carbon dioxide molecules.

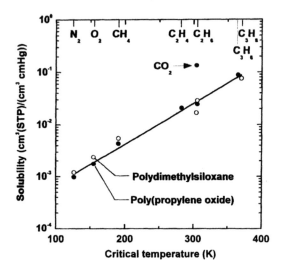

Figure 6. Solubility of various gases in silane-crosslinked poly(propylene oxide) (●) at 30°C and polydimethylsiloxane (○) at 35°C as a function of the critical temperature of the gases. The solubilities of PDMS at a gas pressure of 1 atm at 35 °C were estimated from literature data.
(Reproduced with permission from reference 10. Copyright 2000 Wiley.)

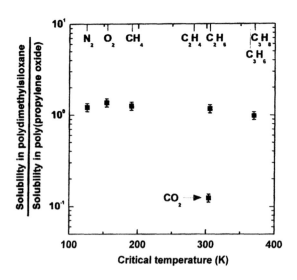

Figure 7. Ratio of the solubility in polydimethylsiloxane at 35°C relative to the solubility in poly(propylene oxide) at 30°C as a function of the critical temperature of the gases. The solubility data are from Figure 6.

Gas Selectivity

Tables IV-VI summarize the overall ideal selectivity, diffusivity selectivity, and solubility selectivity of gases relative to nitrogen in poly(propylene oxide) at 30°C and polydimethylsiloxane at 35°C, respectively.

In all tables, the experimental data are slightly different from the calculated values. This suggests the importance of effects of the shape of gas molecules for

Table IV. Overall ideal selectivity of gases relative to nitrogen in poly(propylene oxide) at 30°C and polydimethylsiloxane at 35°C

Gas	Poly(propylene oxide)		Polydimethylsiloxane	
	Cal.[a]	Exp.[b]	Cal.[a]	Exp.[b]
H_2	-	3.8	-	2.2
O_2	2.3	2.5	2.2	2.0
CO_2	19	21	19	9.6
CH_4	2.5	3.3	2.6	3.0
C_2H_6	9.0	9.7	9.9	8.5
C_3H_8	16	18	19	15

[a] The permeability of each gas was estimated as product of the diffusivity and the solubility using their calculated values from Tables V and VI, respectively.

[b] The permeabilities of polydimethylsiloxane at a gas pressure of 1 atm at 35°C were estimated from literature data *(10)*. Because the diffusivity in Table V and solubility in Table VI were calculated independently, their products were not simply equal to the permeability values.

Table V. Diffusivity selectivity of gases relative to nitrogen in poly(propylene oxide) at 30°C and polydimethylsiloxane at 35°C

Gas	Poly(propylene oxide)		Polydimethylsiloxane	
	Cal.[a]	Exp.[b]	Cal.[a]	Exp.[b]
O_2	1.40	1.40	1.40	1.00
CO_2	0.93	0.15	0.94	0.65
CH_4	0.84	0.75	0.86	0.65
C_2H_6	0.43	0.39	0.47	0.34
C_3H_8	0.25	0.24	0.29	0.18

[a] The diffusivity of each gas was calculated using equation (6) with the τ and η values given in Table II.

[b] The diffusivities of polydimethylsiloxane at a gas pressure of 1 atm at 35°C were estimated from literature data *(10)*.

gas diffusion in the polymer matrix and weak interactions (e.g., polarity of backbone chains) between gas molecules and polymer chains for gas solution. The overall selectivity is slightly greater in poly(propylene oxide) than in polydimethylsiloxane.

The carbon dioxide/nitrogen selectivity of the experimental values in poly(propylene oxide) is 2.2 times greater than that in polydimethylsiloxane. The carbon dioxide/nitrogen selectivity in poly(propylene oxide) is dependent on the solubility selectivity compared to the diffusivity selectivity, presumably due to strong interactions between the propylene oxide units and carbon dioxide molecules.

Table VI. Solubility selectivity of gases relative to nitrogen in poly(propylene oxide) at 30°C and polydimethylsiloxane at 35°C

Gas	Poly(propylene oxide)		Polydimethylsiloxane	
	Cal.[a]	Exp.[b]	Cal.[a]	Exp.[b]
O_2	1.6	1.8	1.6	2.0
CO_2	21	140	21	14
CH_4	3.0	4.4	3.0	4.7
C_2H_6	21	25	21	25
C_3H_8	63	78	63	64

[a] The solubility of each gas was calculated using equation (7) with the M values given in Table III.

[b] The solubilities of polydimethylsiloxane at a gas pressure of 1 atm at 35°C were estimated from literature data *(10)*.

Conclusions

Gas permeation properties of silane-crosslinked poly(propylene oxide) were studied and compared to those of polydimethylsiloxane. The permeability, diffusivity, and solubility of poly(propylene oxide) to various gases, except for carbon dioxide, obey common transport behavior in rubbery polymers. The carbon dioxide diffusivity is, however, lower than the value expected from the diffusion model, whereas the carbon dioxide solubility is higher than the value expected from the sorption model. It can be concluded that strong interactions occur between the propylene oxide units in PPO and carbon dioxide molecules. This interaction probably reduces the carbon dioxide diffusivity and enhances the carbon dioxide solubility. For all gases, except for carbon dioxide, the permeability and diffusivity in poly(propylene oxide) are lower than those of polydimethylsiloxane. On the other hand, the solubility in poly(propylene oxide) is almost the same as that in polydimethylsiloxane. Therefore, the difference in the permeability between poly(propylene oxide) and polydimethylsiloxane is more dependent on that in the diffusivity compared to that in the solubility. The difference in diffusivity probably results from the difference in chain mobility between these two polymers.

References

1. Baker, R. W. *Membrane Technology and Applications*; McGraw-Hill: New York, 2000.
2. Li, J.; Nagai, K.; Nakagawa, T.; Wang, S. *J. Appl. Polym. Sci.* **1995**, *58*, 1455-1463.
3. Okamoto, K.; Fujii, M.; Okamyo, S.; Suzuki, H.; Tanaka, K.; Kita, H. *Macromolecules* **1995**, *28*, 6950-6956.
4. Bondar, V. I.; Freeman, B. D.; Pinnau, I. *J. Polym. Sci.: Part B: Polym. Phys.* **2000**, *38*, 2051-2062.
5. Ito, K. *Silicone Handbook*; Nikkan Kogyo Shinbun: Tokyo, 1990.
6. Kimura, T.; Nagai, K.; Arai, M. *Jpn. Patent 04,359,018*, 1992.
7. Bondi, A. *J. Phys. Chem.* **1964**, *68*, 441-451.
8. Fedors, R. F. *Polym. Eng. Sci.* **1974**, *14*, 147-154.
9. Nagai, K.; Higuchi, A.; Nakagawa, T. *J. Appl. Polym. Sci.* **1994**, *54*, 1207-1217.
10. Merkel, T. C.; Bondar, V. I.; Nagai, K.; Freeman, B. D.; Pinnau, I. *J. Polym. Sci.: Part B: Polym. Phys.* **2000**, *38*, 415-434.
11. Freeman, B. D.; Pinnau, I. In *Polymer Membranes for Gas and Vapor Separation: Chemistry and Materials Science*; Freeman, B. D.; Pinnau, I., Eds.; American Chemical Society: Washington, DC, 1999; pp 1-27.

Chapter 10

Gas Transport Properties of Tetramethyl Polysulfones Containing Trimethylsilyl Group Side Substituents

Michael D. Guiver[1], Ying Dai[1], Gilles P. Robertson[1], Kwi Jong Lee[2], Jae Young Jho[2], and Yong Soo Kang[3]

[1]Institute for Chemical Process and Environmental Technology, National Research Council of Canada, Ottawa, Ontario K1A 0R6, Canada
[2]Hyperstructured Organic Materials Research Center and School of Chemical Engineering, Seoul National University, Seoul 151–744, Korea
[3]Center for Facilitated Transport Membranes and Polymer Physics Laboratory, Korea Institute of Science and Technology, P.O. Box 131, Cheongryang, Seoul 130–650, Korea

Polysulfones with rigid chain structures and trimethylsilyl (TMS) substituents were prepared as potential membrane materials having enhanced gas transport properties. Tetramethyl polysulfones were modified by bromination and lithiation methodology to introduce high levels of TMS groups. Direct lithiation resulted exclusively in ortho-sulfone TMS. Bromination followed by excess lithiation resulted in a high degree of substitution (DS) of TMS on both the bisphenol and phenylsulfone rings. Polymers with a high DS of TMS had high CO_2 and O_2 permeabilities and good permselectivities from N_2. Permeabilities were correlated with d-spacing and specific volume.

† NRCC No. 44385

Published 2004 American Chemical Society

155

Introduction

Bisphenol A Polysulfone (PSf) is a commercial thermoplastic widely used as a membrane material or support for liquid separation processes (*1,2*). It was used for fabricating the first commercial gas separation hollow fiber membranes about two decades ago because of its relatively high permselectivities and adequate permeabilities to various gases as well as its combination of good mechanical properties and fiber spinning qualities (*3*). Increased performance requirements provide incentives to develop new membrane materials since many studies have shown that an improvement in gas transport properties could be obtained by modifying or tailoring the polymer structure.

The groups of Koros and Paul studied parallel families of polysulfones in which systematic structural modifications were correlated with their gas transport properties (*4-8*). The groups of Koros and Paul (*4,6,9*) and Pilato et al. (*10*) showed that symmetrical substitution of methyl groups onto the bisphenol or biphenol segment of PSf and polyphenylsulfone (PPSf) to give tetramethylbisphenol-A polysulfone (TMPSf) and tetramethylbiphenol polysulfone (TMPPSf) respectively, resulted in a large increase in permeability without too great a loss in permselectivity because of increased chain stiffness and higher free volume. The objective of the present work was to modify these materials with TMS and other bulky substituents and correlate the structural effect on gas transport properties.

We reported the modification of polysulfones by direct lithiation (*11*) or by bromination-lithiation (*12*) to produce reactive intermediates that were converted by reaction with various electrophiles to yield TMS (*13*) and a number of other derivatives (*14,15*). Using this chemistry, we previously synthesized PSfs and PPSfs with sequentially increasing sizes of silicon substituents containing methyl and phenyl groups (*13*) and correlated their structural effect on gas transport properties (*16*). TMS substituents gave the most favorable properties with permeability coefficients for O_2 increasing with DS and with minimal loss in O_2/N_2 permselectivity. In the present work, we combine the good gas transport properties of TMPSf and TMPPSf with permeability-enhancing TMS groups.

Experimental

Polymer syntheses and characterization methods

Complete details of polymer syntheses and corresponding detailed structural characterization data are reported elsewhere (*17*). A DuPont 951 thermogravimetric analyzer was used for measuring glass transition temperature (T_g) by differential scanning calorimetry (DSC). Samples for DSC were heated initially to at least 30°C above T_g at 10°C/min, quenched with liquid nitrogen, held isothermally for 10 min and re-heated at 10°C/min for the T_g measurement. Dense polymer films (~100 μm thickness) were made from chloroform solutions as detailed in (17). The absence of residual solvent in the films was confirmed by observing T_g. Gas permeability coefficients were obtained by measuring downstream pressure change using the constant volume method at 35°C with an upstream pressure of 1 atmosphere. Steady-state pressure rate was chosen in time region above ten more than the time-lag and permeability was calculated using the following formula:

$$P = \frac{(\text{quantity of permeate})(\text{film thickness})}{(\text{film area})(\text{time})(\text{pressure drop across film})} \left[\frac{cm^3(STP)\,cm}{cm^2\,s\,cmHg} \right]$$

X-ray diffraction was used to determine d-spacing. A Macscience Model M18XHF22 was utilized with Cu K_α radiation of which wavelength (λ) was 1.54 Å and a scanning speed of 5°/min. The value of d-spacing was calculated by means of Bragg's law ($d = \lambda / 2\sin\theta$), using θ of the broad peak maximum. Densities of the films were measured by the displacement method using a Mettler density kit with anhydrous ethanol at 23°C.

Results and Discussion

Syntheses

The starting polymers TMPSf (**1a**) and TMPPSf (**1b**) were prepared by the potassium carbonate / dimethylacetamide McGrath polycondensation conditions (18-22) and gave polymers with inherent viscosities in NMP solvent of 0.84 and 0.66 respectively (17). Bromination reactions were performed with bromine at room temperature and resulted in fully dibrominated polymers as shown in Scheme 1. It was anticipated that when TMPSf-5-Br$_2$ (**2a**) and TMPPSf-5-Br$_2$ (**2b**) were lithiated with 2.2 mol equivalents of n-butyllithium, then quenched with TMSCl or TMSI electrophiles according to Scheme 1, simple metal-halogen exchange would occur giving TMS groups substituting only at the 5-bromine site. However, for both resulting "low DS" silylated polymers **3a** and **3b**, a DS of ~1.5 was obtained. Bromine analyses showed incomplete metal-halogen exchange, and NMR spectroscopy indicated that TMS groups were approximately equally distributed at both the bromine site (DS~0.75 at the 5-position) and *ortho* to the sulfone group (DS~0.75 at the 11-site) (2).

1a TMPSf, R is $-\overset{\displaystyle CH_3}{\underset{\displaystyle CH_3}{C}}-$ 1b TMPPSf, R is sigma bond

2a TMPSf-5-Br$_2$ 2b TMPPSf-5-Br$_2$

3a TMPSf-5,11-(SiMe$_3$) DS~1.5   3b TMPPSf-5,11,(SiMe$_3$) DS~1.5

4a TMPSf-5,11-(SiMe$_3$) DS~2.6   4b TMPPSf-5,11-(SiMe$_3$) DS~2.6

A: Br$_2$/CHCl$_3$; B: 2.2 mol eq. n-C$_4$H$_9$Li; C: 5 mol eq. n-C$_4$H$_9$Li

D: iodotrimethylsilane or chlorotrimethylsilane

eme 1 Preparation of TMS derivatives by bromination – lithiation

Both dibrominated polymers **2a** and **2b** were also lithiated with an excess of *n*-butyllithium (5 mol equivalents) for lithium-halogen exchange in addition to direct lithiation at the sulfone site in order to achieve the maximum amount of TMS groups on the polymer chain. For both resulting "high DS" silylated polymers **4a** and **4b**, a DS of ~2.6 was obtained and in a second synthesis, a DS 2.8 was obtained for **4b**. Bromine analyses of these products showed only trace residual amounts of bromine indicating almost complete metal-halogen exchange. NMR analysis (*17*) showed that the increase in DS for TMS groups occurred predominately at the 11-site, resulting in a polymer having the composition of TMS DS~0.75 at the 5-position and TMS DS~1.85 at the 11-position. Although almost all brominated sites underwent lithium-halogen exchange, the majority of sites were protonated during work-up since the steric hindrance was too high to allow any greater than DS~0.75 of TMS groups at the 5-site. As anticipated, direct lithiation of **1a** and **1b** resulted exclusively in 11-substituted TMS polymers **5a** and **5b**, each with a DS of 1.7 as shown in Scheme 2 (*17*).

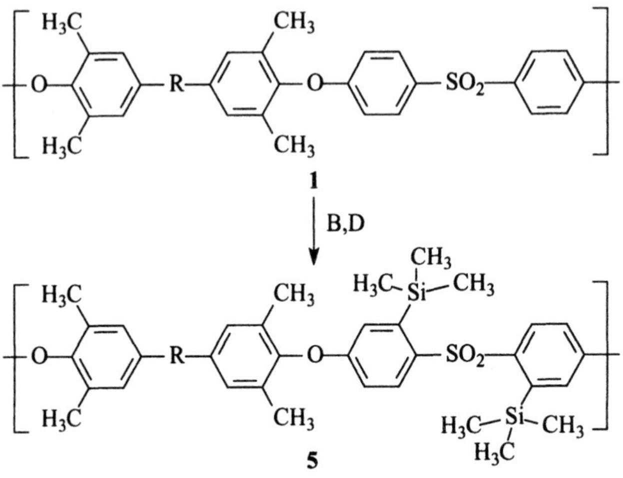

1a TMPSf, R is -C(CH₃)₂-

1a TMPSf, R is -$\overset{\text{CH}_3}{\underset{\text{CH}_3}{\text{C}}}$-   **1b** TMPPSf, R is sigma bond

5a TMPSf-11-(SiMe₃) DS~1.7   **5b** TMPSf-11-(SiMe₃) DS~1.7
B: 2.2 mol eq. *n*-C₄H₉Li;
D: iodotrimethylsilane or chlorotrimethylsilane

Scheme 2 Preparation of TMS derivatives by direct lithiation

Glass transition temperature

When compared with their non-tetramethylated counterparts (T_g (PSf) 188.1°C, T_g (PPSf) 225.6°C (*17*)) it is evident from Table I that the respective TM-polysulfones **1a** and **1b** have significantly higher T_gs due to the increased chain rigidity induced by the methyl groups. The T_gs of the TMS polymers were substantially reduced with increasing DS from those of **1a** and **1b** as found in our previous study of related polymers (*13*). This is possibly due to a disruption in symmetry of the phenylene rings (*4*). For the 5,11-TMS substituted polymers, increasing DS led to reduced T_g. The relative reduction in T_g was greater for the directly-lithiated 11-TMS polymer **5a** than for the 5,11-TMS polymer **4a**.

Table I Physical property data for polymers

Polymer	T_g °C	V_{sp} cm^3/g	*d*-space Å	[a]$P(CO_2)$	$P(O_2)$	$P(N_2)$	$P(CH_4)$
1a	231.3	0.86	5.2	[b]21	[b]5.6	[b]1.06	[b]0.95
[c]**3a**	212.4	0.90	5.3	32	6.9	1.5	2.5
[d]**4a**	193.0	0.94	6.0	66	14	3.1	4.5
[e]**5a**	189.1	–	–	51	12	2.6	–
1b	280.2	0.82	5.2	[b]32	[b]5.8	[b]1.2	[b]1.3
[c]**3b**	240.2	0.84	5.3	73	15	3.2	5.1
[d]**4b**	234.3	0.93	6.1	126	29	6.3	9.7
[e]**5b**	226.4	–	–	–	–	–	–

a Permeability coefficients measured at 35°C and at 1 atm upstream pressure. 1 Barrer = 10^{-10} [cm^3 (STP) · cm] / (cm^2 · sec · cm Hg). b Literature values (*23*); c DS~1.5; d DS~2.6; e DS~1.7

Gas transport

The permeability coefficients (P_{gas}) for four gases are also summarized in Table I, showing that they increase for all gases with the introduction of TMS. Permselectivity *versus* permeability data are plotted relative to the Robeson 'upper bound' line (*24*) for two representative gas pairs: O_2/N_2 in Figure 1 and CO_2/N_2 in Figure 2. Figure 1 shows there are considerable losses in permselectivity for the TMPSf derivatives **3a-5a**. Only the high DS polymer **4a** had a moderate overall improvement in gas transport properties since the drop in permselectivity was offset by a $P(O_2)$ increase more than double that of the low DS **3a** polymer. In contrast, the TMPPSf derivatives **3b** and **4b** show a greater increase in $P(O_2)$ with increasing DS, with only a moderate loss in selectivity.

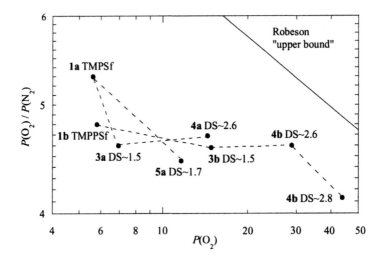

Figure 1. Permselectivity versus permeability for the O_2 / N_2 gas pair

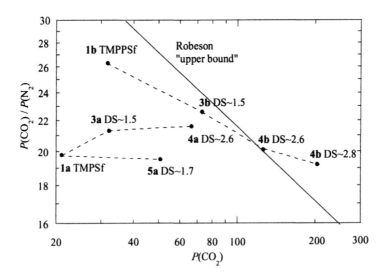

Figure 2. Permselectivity versus permeability for the CO_2 / N_2 gas pair

Here it is worth noting that the O_2/N_2 permselectivity of starting polymer **1b** is rather less than **1a**. Polymer **4b** with DS~2.6 maintained reasonable permselectivity with a five-fold increase in $P(O_2)$, but permselectivity unexpectedly dropped for the same polymer with DS~2.8.

The CO_2/N_2 gas pair data plotted in Figure 2 show different behavior; here the permselectivity for starting polymer **1b** is considerably higher than **1a** (*23*). The TMPSf series **3a-5a** show significant increases in $P(CO_2)$ as well as a surprising increase in CO_2/N_2 permselectivity. Both **3a** and **5a** have a similar DS, but **3a** having TMS substituents on both the 5-site and on the 11-site has higher permselectivity than **5a**, which is exclusively substituted on the 11-site. This supports our previous finding that the substitution site is an important parameter for gas transport properties (*16*). In this case, the higher permselectivity of **3a** over **5a** could be due to increased relative chain rigidity as shown by T_g. TMS substitution at the 11-site tends to give higher gas permeability while at the 5-site, higher permselectivity. Polymer **4b** (DS 2.6 and 2.8) was at or above the Robeson upper-bound limit, with a 4 to 6-fold permeability increase and with a moderate drop in permselectivity compared with TMPPSf. For the TMS polymers, the CO_2/CH_4 permselectivity was lower because $P(CH_4) > P(N_2)$. The relatively high permeability coefficients for both CO_2 and CH_4 condensable gases suggest that TMS increases the solubility coefficients.

Polymer chain packing

The bulky and spherical TMS group hinders interchain packing giving rise to increases in the d-spacing values as shown in Table I. Low DS~1.5 polymers **3a** and **3b**, where the TMS groups are distributed ~0.75 on the 5 and 11-sites, have only a small increase in the d-spacing values (see Table I) compared with starting polymers. The high DS~2.6 polymers **4a** and **4b** also have DS~0.75 on the 5-site, but the TMS on the *ortho*-sulfone 11-site is increased to ~1.85. The high DS combined with the distribution of TMS groups have a large effect on decreasing the packing ability of the polymers' chains. Figure 3 shows the extent of this effect of the increase in d-spacing with increasing DS for TMPSf derivatives. Comparative data (*17*) for PSfs TMS-substituted at the *ortho*-ether 6-site and *ortho*-sulfone 11-site are also included in the graph. Similar trends can be observed for the TMPPSf series (not shown). A single 11-TMS substituent per repeat unit on PSf gives a roughly equivalent d-spacing value compared with the tetramethyl substituted polymer **1a**, although the $P(CO_2)$ for the polymers are 10 and 21 Barrers respectively. Even polymer **3a**, where the TMS DS~1.5, there is only a small increase in the d-spacing value. As the TMS DS increases towards 2, the d-spacing values for the PSf derivatives substituted

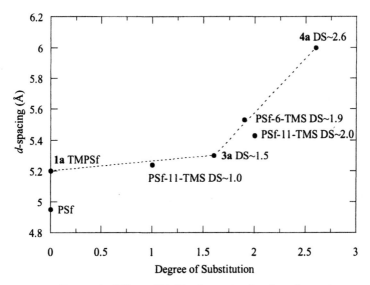

Figure 3. Effect of TMS substitution level on d-spacing.

ortho-sulfone (11-site) and *ortho*-ether (6-site) exhibit significant increases. It is noteworthy that TMS substituted at the more 'open' ether linkage site 6-TMS has a larger effect in reducing interchain packing and has higher permeability than the 11-TMS polymer. Although the PSf-11-TMS polymer has a larger *d*-spacing value than **3a**, it has lower gas permeability ($P(CO_2)$ 18). Both **3a** and **3b** have the same distribution and DS of TMS and *d*-spacing values, yet the gas permeability increase for **3b** is much greater than for **3a** as shown for CO_2 in Figure 4. This is possibly because TMS may induce a larger change in chain conformation of **3b** compared with **3a** through increasing the torsional angle on the biphenyl ring. At DS~2.6, much larger increases in *d*-spacing to 6.0 Å and above were observed for the tetramethylated polymers **4a** and **4b**

The specific volume (V_{sp}) increased with the introduction of TMS groups as shown in Table I, further supporting the assumption of reduced interchain packing. The effect of chain packing on gas permeability is often evaluated by correlating P with fractional free volume (FFV) (*25,26*). The FFV values calculated by group contribution date did not correlate well with P for the trimethylsilyl tetramethyl polysulfones or for modified PSfs investigated previously (*16*). The FFV calculation is very sensitive to small errors in density measurements, in this case measured by the displacement method.

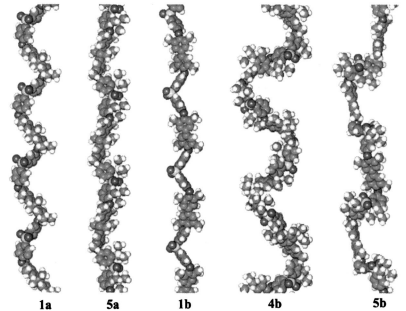

| 1a | 5a | 1b | 4b | 5b |

Figure 5. Chain conformation of various unmodified and TMS substituted tetramethylpolysulfones.

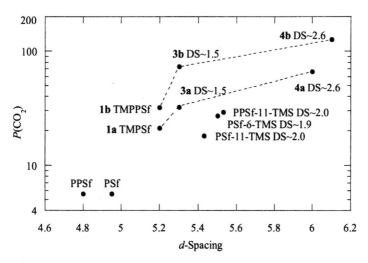

Figure 4. Correlation between P(CO₂) and d-spacing for various polysulfones.

Molecular modeling

Diphenylsulfone has a periplanar conformation allowing conjugative interaction between the phenyl π-electrons and the $3d$ sulfur orbitals (*27*). The high energy barrier to rotation about the Phenyl–Sulfur bond necessary to overcome the strong preference for this conjugative conformation is evident since disubstitution of methyl or chloro groups at the *ortho*-sulfone (H_{11} sites) did not result in significant conformational change in relative phenyl angles about the sulfone as shown by NMR studies (*27*). Our previous NMR data (*17*) for **4a** and **4b** suggest that conformational change by Phenyl–Sulfone–Phenyl rotation occurs to better accommodate the more bulky TMS substituents.

Conformational analysis of the polymers using HyperChem™ (Hypercube Inc. software, Florida) was performed on eight repeat unit chain lengths of the polymer structures 'as drawn' to study the effect of TMS substitution and distribution of TMS on chain geometry and steric interactions. Molecular modeling software gives a visual indication of major conformational changes in the polymers containing TMS groups. Figure 5 shows partial central views of eight repeat unit model lengths that were optimized for minimum energy. TMPSf **1a** is compared with **5a**, the 11-substituted TMS at the ortho-sulfone linkage. Surprisingly, although **5a** changes conformation compared with **1a**, both chains are relatively linear. However, in the case of the TMS-substituted polymers **4b** and **5b**, both polymers exhibit considerable visible conformational change compared with the linear TMPPSf **1b**. In particular, the chain of **4b** is non-linear and kinked with a somewhat corkscrew conformation. It was difficult to ascertain the relative pseudo-torsional angles between aromatic rings connected by linkages such as sulfone or ether (i.e. across five bonds), since both the linkage angle relative to the phenyl plane and the phenyl planar angle are combined. Polymer **1b** and **4b** were compared first by true torsional angles between the biphenyl rings. The unmodified polymer had torsional angles typically ~18°, whereas the TMS modified one had a range of angles from 45-56°, relieving the steric restriction of the bulky TMS group and supporting our previous NMR studies. The central repeat units of the eight repeat unit lengths were examined for pseudo-torsional angles between the planes of the diphenyl ether and the diphenylsulfone. Unexpectedly, a range of values was obtained, even for the unmodified **1b**. For **1b**, diphenylether was 56-63° and diphenylsulfone had several distinct angles of ~0, ~8, ~15, ~30 along the polymer chain repeat units. There are two types of diphenylether linkages in **4b**, those with and without TMS in the biphenol ring. The diphenyls around the ether linkage were typically twisted out of plane for the rings containing TMS and those around the sulfone linkage had increased relative angles between 58-83° for three repeat units calculated.

Conclusions

Tetramethyl polysulfone and polyphenylsulfone were modified with TMS substituents to investigate their potential as membrane materials for gas separation. The TMS substituents were introduced on the bisphenol and on the phenylsulfone segments of the polymer chains. The T_gs of the polymers were reduced by the addition of pendant TMS, possibly due to a disruption of symmetry in the phenylene rings. TMS Polymers with high DS had the largest increases in CO_2 and O_2 permeabilities and good permselectivities from N_2. The O_2/N_2 and CO_2/N_2 pairs showed considerable improvements in gas transport properties since performance was close to the Robeson upper-bound line. Polymer **4b** had the best overall gas transport properties, exhibited a 5-fold increase in $P(O_2)$ with a minor decrease in α (O_2/N_2). The $P(CO_2)$ increased ~4-fold with some decrease in α (CO_2/N_2). An investigation of d-spacing for TMS polymers showed that it increased sharply for DS>2, indicating increases in free volume. Molecular modeling of the polymer chains suggests that TMS groups introduced onto PPSf **1b** result in chain kinking and considerable conformational changes.

Acknowledgements

Partial financial support was provided from an international collaborative project between the National Research Council of Canada and the Korea Institute of Science and Technology. We are most grateful to Dr. Takashi Kowada of the Fine Chemicals Department, Mitsubishi Chemical Corporation, Tokyo, Japan, for kindly donating the monomer 3,3',5,5'-tetramethylbiphenol. We thank Fatima Haggar for assistance in molecular modeling.

References

1. Petersen, R. J. *J. Membr. Sci.* **1993**, *83*, 81-150.
2. Lundy, K. A.; Cabasso, I. *Ind. Eng. Chem. Res.* **1989**, *28*, 742-756.
3. Henis, J. M. S.; Tripodi, M. K. *Sep. Sci. Technol.* **1980**, *15*, 1059-1068.
4. Aitken, C. L; Koros, W. J.; Paul, D. R. *Macromolecules* **1992**, *25*, 3424-3434.
5. Aitken, C. L.; McHattie, J. S.; Paul, D. R. *Macromolecules* **1992**, *25*, 2910-2922.
6. McHattie, J. S.; Koros, W. J.; Paul, D. R. *Polymer* **1992**, *33*, 1701-1711.
7. McHattie, J. S.; Koros, W. J.; Paul, D. R. *Polymer* **1991**, *32*, 2618-2625.
8. McHattie, J. S.; Koros, W. J.; Paul, D. R. *Polymer* **1991**, *32*, 840-850.
9. Moe, M. B; Koros, W. J.; Paul, D. R. *J. Polym. Sci. Polym. Phys. Edn.* **1988**, *26*, 1931-1945.
10. Pilato, L. A; Litz, L. M.; Hargitay, B.; Osborne, R. C.; Farnham, A. G.; Kawakami, J. H.; Fritze, P. E.; McGrath, J. E. *Polym. Prepr. (Am Chem Soc, Div. Polym. Chem.)* **1975**, *16*, 41-47.
11. Guiver, M. D; ApSimon, J. W; Kutowy, O. *J. Polym. Sci. Polym. Lett. Ed.* **1988**, *26*, 123-127.
12. Guiver, M. D; Kutowy, O.; ApSimon, J. W. *Polymer,* **1989**, *30*, 1137-1142.
13. Guiver, M. D.; Robertson, G. P.; Rowe, S.; Foley, S.; Kang, Y. S.; Park, H. C.; Won, J.; Le Thi, H. N. *J. Polym. Sci.: Part A: Polym. Chem.* **2001**, *39*, 2103-2124.
14. Guiver, M. D.; ApSimon, J. W.; Kutowy, O. U.S. Patent 4,797,457 **1989** and U.S. Patent 4,833,219 **1989**.
15. Guiver, M. D.; Robertson, G. P.; Yoshikawa, M.; Tam, C. M. *ACS Symp. Ser.* 744 Membranes Formation and Modification, Pinnau I, Freeman B, Eds., American Chemical Society, Washington, DC, 1999, Chapter 10, pp. 137-161.
16. Kim, I.-W.; Lee, K. J.; Jho, J. Y.; Park, H. C.; Won, J.; Kang, Y. S.; Guiver, M. D.; Robertson, G. P.; Dai, Y. *Macromolecules,* **2001**, *34*, 2908-2913.
17. Dai, Y.; Guiver, M. D.; Robertson, G. P.; Bilodeau, F.; Kang, Y. S.; Lee, K. J.; Jho, J. Y.; Won, J. *Polymer,* **2002**, *43*, 5369-5378.

18. Mohanty, D. K.; Sachdeva, Y.; Hedrick, J. L.; Wolfe, J. F.; McGrath, J. E.; *Polym. Prepr. (Am. Chem. Soc. Div. Polym. Chem.)* **1984**, *25*, 19-22.
19. Mohanty, D. K. Ph.D. thesis; Virginia Polytechnic Institute and State University, 1983.
20. McHattie, J. S. Ph.D. thesis; University of Texas at Austin, 1990.
21. Aitken, C. L. Ph.D. thesis, University of Texas at Austin, 1992.
22. Viswanathan, R.; Johnson, B. C.; McGrath, J. E. *Polymer*, **1984**, *25*, 1827-1836.
23. Aitken, C.L.; Koros, W. J.; Paul, D. R. *Macromolecules*, **1992**, *25*, 3651-3658.
24. Robeson, L. M. *J. Membr. Sci.* **1991**, *62*, 165-185.
25. Park, J. Y.; Paul, D. R. *J. Membr. Sci.* **1997**, *125*, 3-39.
26. van Krevelen, Q. W. In *Properties of Polymers*; Elsevier: Amsterdam, 1990: pp 455-623.
27. Montaudo, G.; Finocciaro, P.; Trivellone, E.; Bottino, F.; Maravigna, P. *J. Molec. Struct.* **1973**, *16*, 299-306.

Chapter 11

Pure- and Mixed-Gas Permeation Properties of Poly(*p-tert*-butyl diphenylacetylene)

Ingo Pinnau[1], Zhenjie He[1], Toshio Masuda[2],
and Toshikazu Sakaguchi[2]

[1]Membrane Technology and Research, Inc., 1360 Willow Road, Suite 103,
Menlo Park, CA 94025
[2]Department of Polymer Chemistry, Kyoto University, Kyoto 606–8501,
Japan

Poly(*p-tert*-butyl diphenylacetylene) [P*pt*BDPA] is an amorphous, glassy, substituted acetylene-based polymer. P*pt*BDPA ranks amongst the most permeable polymers known. The oxygen permeability of P*pt*BDPA at 35°C is 1,930 x 10^{-10} $cm^3(STP) \cdot cm/cm^2 \cdot s \cdot cmHg$. As expected for a high permeability glassy polymer, the selectivity of P*pt*BDPA is low for the separation of supercritical gases; for example, its oxygen/nitrogen selectivity is only 1.9. On the other hand, P*pt*BDPA shows very high permeabilities for organic vapors and high organic-vapor/supercritical-gas selectivity. The permeability of supercritical gases, such as nitrogen or methane, was significantly reduced by co-permeation of a condensable gas. This behavior is very similar to that of other high-free-volume, glassy acetylene-based polymers and results from blocking of the excess free volume of the polymer by preferential sorption of the condensable gas mixture component.

© 2004 American Chemical Society

Introduction

The separation of organic vapors from supercritical gases with membranes has recently gained significant commercial importance in the chemical process industry. Examples of current large-scale industrial applications include (i) separation of olefins, such as ethylene and propylene, from nitrogen in polyolefin polymerization purge-gas streams *(1,2)*, (ii) recovery of vinyl chloride monomer from poly(vinyl chloride) off-gas streams *(3,4)*, and (iii) recovery of gasoline vapor from storage tank operations *(5)*. Future applications that could significantly expand the commercial use of membranes for vapor separations include the recovery of hydrocarbons from hydrogen in petrochemical processes and removal of C_{2+} hydrocarbons from natural gas.

Recently, the pure- and mixed-gas permeation properties of high-free-volume, glassy acetylene-based polymers, such as poly(1-trimethylsilyl-1-propyne) [PTMSP] *(6-10)* and poly(4-methyl-2-pentyne) [PMP] *(11,12)*, were reported for hydrocarbon/methane and hydrocarbon/hydrogen separations. These polymers are characterized by high glass transition temperatures, typically > 200°C, very high fractional free volume (> 0.25), and extremely high supercritical gas and organic vapor permeabilities. Specifically, poly(1-trimethylsilyl-1-propyne) [PTMSP], has the highest gas and vapor permeabilities of all known polymers *(13,14)*. The extraordinarily high permeability of PTMSP results from its very large amount of excess free volume and interconnectivity of free-volume-elements. PTMSP exhibits very unusual organic vapor permeation properties. In contrast to conventional low-free-volume glassy polymers, such as bispenol-A polycarbonate, PTMSP is significantly more permeable to large, organic vapors than to small, supercritical gases. In particular, for C_{2+} hydrocarbon/methane as well as C_{2+} hydrocarbon/hydrogen mixtures, PTMSP exhibits both the highest C_{2+} hydrocarbon permeability and the highest C_{2+}/methane and C_{2+}/hydrogen selectivity of any known polymer *(6,7,9)*.

In this study, the pure-gas permeation properties of poly(p-*tert*-butyl diphenylacetylene) [P*pt*BDPA], a glassy, diphenyl-substituted acetylene polymer, were determined for supercritical gases as well as for a series of hydrocarbon vapors. In addition, gas permeation experiments with P*pt*BDPA were carried out with ethylene/nitrogen, propylene/nitrogen, and *n*-butane/methane mixtures.

Experimental

Polymer Synthesis, Characterization, and Film Formation. Poly(p-*tert*-butyl diphenylacetylene) was synthesized as described previously by Masuda et al. *(15,16)*. The polymerization was carried out in toluene at 80°C in a Schlenk tube under dry nitrogen using a $TaCl_5$-*n*-Bu_4Sn co-catalyst system. The polymer yield was 84%. The chemical structure of P*pt*BDPA is shown in Figure 1. The molecular weight of the polymer, as determined by gel chromatography, was 3.6 x 10^6 g/mole.

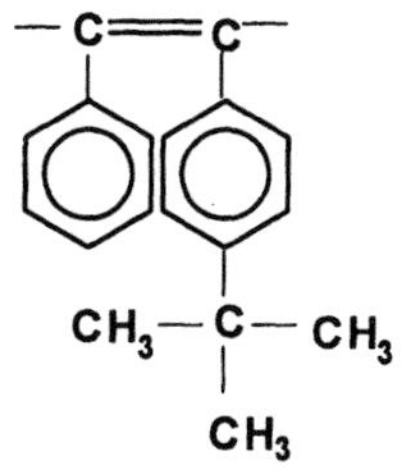

Figure 1. Repeat unit of poly(p-tert-butyl diphenylacetylene).

A dense, isotropic film of P*pt*BDPA was ring-cast from a polymer solution (0.5 wt% in toluene) onto a Teflon-coated glass plate. The film was dried gradually at ambient conditions for 72 hours and then under vacuum at 80°C for three days to completely remove the solvent. To ensure that the film was completely solvent-free, the film was removed periodically from the vacuum oven and weighed on an analytical balance. The P*pt*BDPA film used for the permeation measurements had a thickness of 63 μm (± 0.5 μm). The density of P*pt*BDPA was determined by a gravimetric method. Three film samples were weighed on an analytical balance and the density was determined from the known area and the thickness of the films, as determined with a precision micrometer. The density of P*pt*BDPA was 0.91 (±0.01) g/cm^3. The fractional free volume [FFV] (cm^3 free volume/cm^3 polymer), commonly used as a measure for the free volume available for chain packing can be determined from

$$FFV = \frac{v_{sp} - 1.3 v_w}{v_{sp}} \quad (1)$$

where v_{sp} is the specific volume (cm^3/g) of the polymer, as determined from density measurements, and v_w is the van der Waals volume, estimated from van Krevelen's group contribution method. The fractional free volume of P*pt*BDPA is 0.27, which ranks amongst the highest FFV values of glassy polymers reported to date.

Permeation Experiments. The pure-gas permeabilities of P*pt*BDPA to helium, hydrogen, nitrogen, oxygen, methane, ethane, ethylene, propane, propylene, and *n*-butane were determined using the constant pressure/variable volume method. The gas permeation experiments were carried out at 35°C. The feed pressure was 50 psig (except for *n*-butane: p_{feed} = 10 psig); the permeate-side pressure was atmospheric (0 psig). Volumetric permeate flow rates were determined with soap-bubble flowmeters. The steady-state flux, J (cm^3(STP)/cm^2·s), was calculated from:

$$J = \frac{dV \cdot 273 \cdot p_a}{dt \cdot A \cdot T \cdot 76} \qquad (2)$$

where (dV/dt) is the volumetric displacement rate of the soap film in the flowmeter (cm^3/s), A is the membrane area (cm^2), T is the gas temperature (K), and p_a is the atmospheric pressure. The permeability, P, (cm^3(STP)·cm/cm^2·s·cmHg), of the film was determined by:

$$P = \frac{J \cdot \ell}{(p_2 - p_1)} \qquad (3)$$

where ℓ is the film thickness and p_2 and p_1 are the feed and permeate pressure, respectively.

The selectivity, $\alpha_{a,b}$, was calculated by:

$$\alpha_{a,b} = \frac{P_a}{P_b} \qquad (4)$$

The mixed-gas permeation properties of PptBDPA were also determined using the constant pressure/variable volume method. The following gas mixtures were used: i) 20 vol% ethylene/80 vol% nitrogen, (ii) 10 vol% propylene/90 vol% nitrogen, and (iii) a series of n-butane/methane mixtures containing 1, 2, 4, and 6 vol% n-butane in methane, respectively. The feed pressure was 150 psig; the permeate pressure was atmospheric (0 psig). The mixed-gas permeability was calculated by:

$$P_{mix} = \frac{J_{perm} \cdot x_{perm} \cdot \ell}{(p_{feed} \cdot x_{feed}) - (p_{perm} \cdot x_{perm})} \qquad (5)$$

where p_{feed} and p_{perm} are the feed and permeate pressure (cmHg absolute), and x_{feed} and x_{perm} are the feed and permeate volume fractions, respectively. The gas mixture selectivity was then calculated from Eq. 4.

Results and Discussion

Pure-Gas Permeation Properties of PptBDPA

The gas permeabilities of PptBDPA as a function of the critical gas volume, a relative measure of penetrant size, are shown in Figure 2. In PptBDPA the permeability of larger gas molecules, specifically hydrocarbons, increases as the molecular size of the gases increases. This result is in qualitative agreement with the pure-gas permeation properties of other high-free-volume glassy disubstituted acetylene-based polymers, namely, poly(1-trimethylsilyl-1-propyne) [PTMSP], poly(4-methyl-2-pentyne) [PMP], poly[1-phenyl-2-[*p*-(trimethylsilyl) phenyl]acetylene], and poly[1-phenyl-2-[triisopropyl)phenyl]acetylene] (*6,8,17-18*). In these high-free-volume polymers, diffusion coefficients show a relatively weak dependence on gas size and, hence, permeability is affected significantly less by differences in diffusion coefficients but is more dependent on differences in gas solubility (*8*).

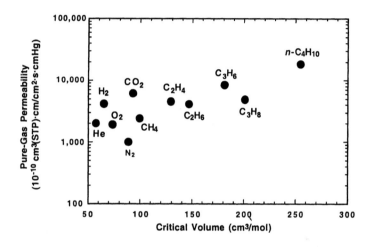

Figure 2. Pure-gas permeability of PptBDPA for a series of gases and hydrocarbon vapors as a function of critical gas volume.

The pure-gas permeation properties of P*pt*BDPA for various gases in terms of permeability and selectivity over nitrogen are summarized in Table I. As observed previously for other high-free-volume, disubstituted acetylene-based polymers, P*pt*BDPA exhibits very high gas permeabilities. For example, the oxygen permeability of P*pt*BDPA is $1,930 \times 10^{-10}$ cm^3(STP)·cm/cm^2·s·cmHg. The high gas permeabilities of P*pt*BDPA result from its very high fractional free volume (FFV = 0.27). Typical for a high-free-volume disubstituted acetylene polymer, the selectivity of P*pt*BDPA for supercritical gases is low; its oxygen/nitrogen selectivity is only 1.9. It is important to note, however, that P*pt*BDPA is more permeable to large, condensable C$_{2+}$ hydrocarbons than to supercritical gases, such as nitrogen or methane. For example, the pure-gas *n*-butane/nitrogen selectivity of P*pt*BDPA is 19. Hence, the hydrocarbon/nitrogen solubility selectivity of P*pt*BDPA is higher (>>1) than the hydrocarbon/nitrogen diffusivity selectivity (<1), similar to the behavior in high-free-volume PTMSP and PMP.

172

Table I. Pure-gas permeabilities and gas/nitrogen selectivities of poly(p-tert-butyl diphenylacetylene) [PptBDPA]. Feed pressure: 50 psig (except n-butane: 10 psig); permeate pressure: atmospheric (0 psig); T = 35°C

Gas	Lennard-Jones Diameter (Å)*	Permeability x10^10 (cm^3(STP)·cm/cm^2·s·cmHg)	Selectivity Gas/Nitrogen
He	3.16	2,010	2.0
H$_2$	3.27	4,150	4.1
O$_2$	3.43	1,930	1.9
N$_2$	3.66	1,000	-
CH$_4$	3.80	2,430	2.4
C$_2$H$_4$	4.19	4,540	4.5
C$_2$H$_6$	4.34	4,130	4.1
C$_3$H$_6$	4.69	8,440	8.4
C$_3$H$_8$	4.84	4,900	4.9
n-C$_4$H$_{10}$	5.25	18,600	19

*Determined from: $D_{LJ}=10^{-10}$ x (2.3551-0.087 x acentric factor)/(P$_c$/T$_c$)$^{1/3}$ (19)

Mixed-Gas Permeation Properties of PptBDPA

Recently, it was reported that high-free-volume acetylene-based polymers, such as PTMSP and PMP, exhibit very unusual gas permeation behavior with gas mixtures containing condensable feed components. Specifically, these materials show significantly higher organic-vapor/supercritical-gas selectivity in mixtures than expected based on pure-gas permeation data. In gas mixture experiments, the permeability of supercritical gases is reduced significantly by co-permeation of an organic vapor (6,20). The reduction in supercritical mixed-gas permeability is caused by preferential sorption of the organic vapor in the free volume elements of the polymers, which leads to a reduction in mixed-gas diffusion coefficient of the supercritical gas. This behavior is similar to that of nanoporous inorganic membrane with pore sizes in the range of 5 to 20 Å (21). In this study, we investigated the gas permeation properties of PptBDPA with olefin/nitrogen mixtures and n-butane/methane mixtures.

The olefin and nitrogen permeabilities and olefin/nitrogen selectivities of PptBDPA are shown in Table II. The ethylene/nitrogen and propylene/nitrogen selectivity of PptBDPA was 7.3 and 22, respectively. For comparison, the selectivity values for a commercial rubbery PDMS film, tested under the same conditions, were 7.5 and 18, respectively. It is noteworthy that the PptBDPA film exhibited reduced mixed-nitrogen permeability due to preferential olefin sorption in the excess free volume of the polymer. This behavior is similar to that of other high-free-volume acetylene-based polymers. As shown in earlier studies, blocking of the supercritical gas depends on the condensability, and hence, sorption capacity, of the co-permeating organic vapor. The more condensable the organic vapor, the higher is typically its solubility in the polymer matrix, which leads to increased blocking of the diffusion pathway of the supercritical gas.

Table II. Mixed-gas permeation properties of poly(*p-tert*-butyl diphenylacetylene) for olefin/nitrogen separation

Permeability $(10^{-10}$ cm^3(STP)·cm/cm^2·s·cmHg)		Selectivity Olefin / N_2	Permeability Ratio Mixed-Gas N_2 / Pure-Gas N_2
Olefin	N_2		
C$_2$H$_4$	3,580	491	7.3	0.65
C$_3$H$_6$	5,280	240	22	0.30

Note: Feed composition: (a) 20 vol% ethylene/80 vol% nitrogen and (b) 10 vol% propylene/90 vol% nitrogen; feed pressure: 150 psig; permeate pressure: atmospheric (0 psig); temperature: 25°C.

Additional gas permeation experiments were conducted with a series of *n*-butane/methane mixtures. The *n*-butane and methane permeability and *n*-butane/methane selectivity of P*pt*BDPA as a function of feed composition are shown in Figures 3a and 3b. P*pt*BDPA shows essentially the same qualitative mixed-gas permeation behavior as PTMSP and PMP. This result is expected, because the fractional free volume of P*pt*BDPA (FFV = 0.27) is very similar to that of PMP (FFV = 0.28) and PTMSP (FFV = 0.29). The *n*-butane mixed-gas permeability of P*pt*BDPA was about 6,000 x 10^{-10} cm^3(STP)·cm/cm^2·s·cmHg and essentially independent of *n*-butane feed concentration. On the other hand, the *n*-butane/methane selectivity increased from 8 to 13 by increasing the *n*-butane feed concentration from 1 to 6 vol%. This increase in selectivity resulted from a decrease in methane permeability due to increased sorption of *n*-butane in the free volume of the polymer at higher *n*-butane feed concentration.

Recently, Freeman and Pinnau noted an inverse permeability/selectivity relationship for organic-vapor/supercritical gas separation (*22*) as compared to that typically observed for supercritical-gas/supercritical-gas separation (*23*). In organic vapor/supercritical gas separations, membranes that exhibit the highest organic vapor permeability also have the highest organic-vapor/supercritical-gas selectivity. Our mixed-gas permeation results with P*tp*BDPA are in complete agreement with this trend.

An example for this behavior is shown in Figure 4 for *n*-butane/methane separation for a series of acetylene-based glassy polymers, rubbery PDMS, and a hypothetical liquid membrane (*n*-hexane). The permeabilities of methane and *n*-butane for the liquid membrane were calculated from their known CH$_4$ and *n*-C$_4$H$_{10}$ solubility values in *n*-hexane (*24*), and their diffusion coefficients were calculated based on the Wilke-Chang method (*25*).

P*tp*BDPA is one of the most permeable and most *n*-butane/methane selective polymers known. However, the mixed-gas permeation properties of glassy P*tp*BDPA are comparable to those of commercially used rubbery PDMS membranes. To provide more advanced glassy materials for organic-vapor/supercritical-gas separation applications future work should be directed towards the development of extremely high-free-volume polymers, such as PTMSP. However, advanced materials for hydrocarbon/methane and hydrocarbon/hydrogen separation require better chemical resistance than PTMSP as it is soluble in C$_{5+}$ hydrocarbons, which severely limits its industrial use for these applications.

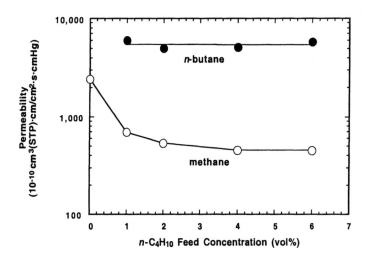

Figure 3a. Mixed-gas methane and n-butane permeability of PtpBDPA as a function of n-butane feed concentration.

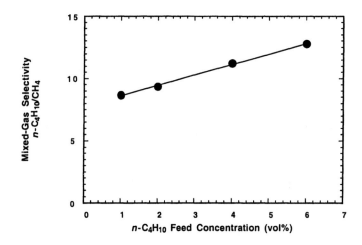

Figure 3b. Mixed-gas n-butane/methane selectivity of PtpBDPA as a function of n-butane feed concentration.

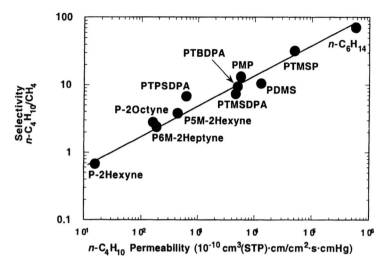

Figure 4. Relationship between n-butane permeability and n-butane/methane selectivity for a series of glassy acetylene-based polymers, rubbery polydimethylsiloxane (PDMS), and a hypothetical liquid membrane (n-hexane).

Conclusions

PtpBDPA is a rigid, amorphous, glassy, high-free-volume, acetylene-based polymer with excellent thermal and chemical stability. Because of its very high excess free volume (FFV = 0.27) it is one of the most permeable polymers known. Like other high-free-volume, glassy, acetylene-based polymers, PtpBDPA has very high supercritical-gas permeability coupled with low supercritical-gas/supercritical-gas selectivity. However, PtpBDPA has interesting properties for organic-vapor/supercritical-gas separation applications. Our studies demonstrated that PtpBDPA is one of the most selective polymeric materials for olefin/nitrogen and n-butane/methane separations. However, the mixed-gas permeation properties of glassy PtpBDPA are similar to those of PDMS, a conventional, commercial rubbery membrane material. Future work should be directed towards the development of ultra-high free-volume glassy, diphenyl acetylene-based polymers that exhibit the superior mixed-gas permeation properties of PTMSP, but have better chemical resistance and long-term permeation stability.

References

1.   Baker, R.W.; Wijmans, J.G.; Kaschemekat, J.H. *J. Membr. Sci.* **1998**, *151*, 55.

176

2. Baker, R.W.; Kaschemekat, J.; Wijmans, J.G. *Chemtech* **1996**, *26*, 37.
3. Baker, R.W.; Wijmans, J.G. In *Polymeric Gas Separation Membranes*, Paul, D.R.; Yampolskii, Y.P., Eds.; CRC Press: Boca Raton, FL, 1994, 353.
4. Lahiere, R.J.; Hellums, M.W.; Wijmans, J.G.; Kaschemekat, J. *Ind. Eng. Chem. Res.* **1993**, *32*, 2236.
5. Ohlrogge, K.; Peinemann, K.-V.; Wind, J; Behling, R.D. *Sep. Sci. Technol.* **1990**, *25*, 1375..
6. Pinnau, I.; Toy, L.G. J. Membr. Sci., **1996**, *116*, 199.
7. Schultz. J; Peinemann, K.-V. *J. Membr. Sci.* **1996**,*110*, 37.
8. Merkel, T.C.; Bondar, V.; Nagai, K.; Freeman, B.D. *J. Polym. Sci. Part B: Polym. Phys.* **2000**, *38*, 273.
9. Pinnau,I.; Casillas, C.G.; Morisato, A; Freeman, B.D. *J. Polym. Sci. Part B: Polym. Phys.* **1996**, *34*, 2613.
10. Pinnau I; Casillas, C.G; Morisato, A.; Freeman, B.D. *J. Polym. Sci. Part B: Polym. Phys.* **1997**, *35*, 1483.
11. Morisato; Pinnau, I. *J. Membr. Sci.* **1996**, *121*, 243.
12. Morisato, A.; He, Z.; Pinnau, I. In *Polymer Membranes for Gas and Vapor Separation: Chemistry and Materials Science*; Freeman, B.D.; Pinnau, I., Eds.; ACS Symposium Series 733; American Chemical Society: Washington, DC, 1999; pp 56-67.
13. Masuda, T.; Isobe, E.; Higashimura, T; Takada, K. *J. Am. Chem. Soc.* **1983**, *105*, 7473.
14. Ichiraku, Y.; Stern, S.A.; Nakagawa T. *J. Membr. Sci.* **1987**, *34*, 5.
15. Kouzai, H.; Masuda, T.; Higashimura, T. *J. Polym. Sci. Part A: Polym. Chem.* **1994**, *32*, 2523.
16. Masuda, T.; Teraguchi, M.; Nomura, R. In *Polymer Membranes for Gas and Vapor Separation: Chemistry and Materials Science*; Freeman, B.D.; Pinnau, I., Eds.; ACS Symposium Series 733; American Chemical Society: Washington, DC, 1999; pp. 28-37.
17. Toy, L.G.; Nagai, K.; Freeman, B.D.; Pinnau, I.; He, Z.; Masuda, T; Teraguchi, M.; Yampolskii, Y.P. Macromolecules **2000**, *33*, 2516.
18. Nagai, K.; Toy, L.G.; Freeman, B.D.; Teraguchi, M.; Masuda, T.; Pinnau, I. *J. Polym. Sci. Part B: Polym. Phys.* **2000**, *38*, 1474.
19. Broadkey, R.S.; Hershey, H.C. In *Transport Phenomena*; McGraw-Hill, New York, 1988: pp.722.
20. Srinivasan R,; Auvil, S.; Burban, P.M. *J. Membr. Sci.* **1994**, *86*, 67.
21. Ash, R.; Barrer, R.M.; Lowson, R.T. J. Chem. Soc. Faraday Trans. 1 **1973**, *69*, 2166.
22. Freeman, B.D.; Pinnau, I. *TRIP* **1997**, *5*, 167.
23. Robeson, L. *J. Membr. Sci.* **1991**, *62*, 165.
24. Fogg, P.G.T; Gerrard, W.; *Solubility of Gases in Liquids: A Critical Evaluation of Gas/Liquid Systems in Theory and Practice*; John Wiley & Sons: New York, 1991.
25. Reid, R.C.; Prausnitz, J.M.; Sherwood, T.K., *The Properties of Gases and Liquds*; 3rd Ed.; McGraw-Hill, 1977, p.567.

Chapter 12

Membrane Separations Using Functionalized Polyphosphazene Materials

Frederick F. Stewart[1], Christopher J. Orme[1], Mason K. Harrup[1], Robert P. Lash[1], Don H. Weinkauf[2], and John D. McCoy[3]

[1]Idaho National Engineering and Environmental Laboratory, P.O. Box 1625, Idaho Falls, ID 83415–2208
Departments of [2]Chemical Engineering and [3]Materials and Metallurgical Engineering, New Mexico Institute of Mining and Technology, 801 Leroy Place, Socorro, NM 87801

Polyphosphazenes are intriguing materials with potential application as membranes for selective mass transport. In this study, several polyphosphazenes, differing only in ratios of three different pendant groups, were synthesized and formed into supported thin dense film membranes. A balance of hydrophilic/hydrophobic behaviors was achieved through varying the amount of hydrophilic 2-(2-methoxyethoxy)ethanol (MEE) attached to the polymer. The remaining sites were substituted with hydrophobic 4-methoxyphenol with a small amount of 2-allylphenol added to provide a cross-linking moiety. Resulting polymers were studied using swelling determinations and pervaporation.

© 2004 American Chemical Society

178

Solubility control in membrane processes offers the ability to tailor a membrane to provide a high affinity for a specific permeate while rejecting other chemical species. Thus, separations are determined by the intermolecular interactions between specific permeates and the polymer membrane substrate. Maximizing these interactions then becomes the goal. Addition of functional groups to polymers that interact with specific permeates can be performed through chemical synthesis. In this study, polyphosphazenes have been employed to serve as a stable "platform" for chemical synthesis through attachment of pendant groups with specific functionality.

Background

Polyphosphazenes can be thought of as hybrid organic-inorganic polymers with a backbone comprised of alternating phosphorus and nitrogen atoms. Each phosphorus is pentavalent resulting in two pendant groups per mer (Figure 1). Typical pendant groups consist of organic nucleophiles *(1)*. Selection of the pendant groups is critical in determining the physical and chemical properties of the resulting polymer. In this work, several polyphosphazenes with differing chemical properties are discussed. Homopolymers, phosphazenes with only one type of pendant group, vary widely. For example, poly[bis-(2-(2-methoxyethoxy)phosphazene] (MEEP) is a hydrophilic elastomer, whereas poly[bis-4-methoxyphenoxyphosphazene] is a semi-crystalline solid material.

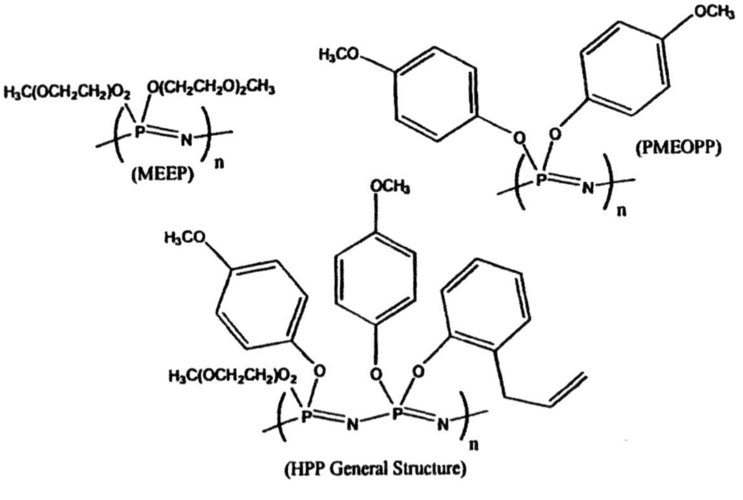

Figure 1. Structures for three polyorganophosphazenes studied in this work The general structure for HPP is a representation of both HPP1 and HPP2.

To provide an enhanced level of control over the properties of phosphazenes, the use of blended pendant group mixtures has been employed to yield HPP polymers (Figure 1). For example, hydrophobic 4-methoxyphenol was added to a polymer containing hydrophilic MEE with the resultant polymer properties dependent on the relative loading of each pendant group *(2)*. Higher levels of MEE in the polymer matrix yielded membranes with high affinities for polar permeates. Recent data for gas transport show a clear linear correlation between the amount of MEE on the polymer and the permeability of CO_2, suggesting a strong solubility interaction between the polymer and the gas *(3)*.

Synthesis of polymers *a priori* is inherently inefficient. An accurate description of solubility behavior is necessary such that polymers may be designed and synthesized for transport of specific permeates. Tools that recently have been explored to describe the solubility behavior of blended pendant group phosphazenes are Hanson solubility parameters *(2)*. Hanson parameters provide a numerical method to describe solubility such that estimations of mutual solubility can be made using the chemist's "rule of thumb" that like materials dissolve like materials. Thus, polymer membranes and permeates with similar Hanson parameters should exhibit mutual solubility resulting in increased solubility driven transport.

Three Hanson parameters are used to characterize the types of interactions that are possible for solvents in terms of molecular forces: 1) hydrogen bonding (δ_h), 2) polarity (δ_p), and 3) dispersion (δ_d) *(4)*. For solvents, determination of solubility parameters can be accomplished using the molar heat of vaporization (ΔH_{vap}). However, for polymers, this is generally not possible. Polymers can be characterized for their solubility behavior through simple immersion experiments and described in terms of the Hanson solubility parameters for each solvent showing degrees of solubility behavior. An additional method to estimate Hanson parameters for polymers uses group contributions. Using this method, the solubility parameters are derived as a sum of individual atomic parts that comprise the polymer where each part contributes its own individual characteristics. In this work, both of these methods have been applied to phosphazenes with the goal of characterizing the effect of varying the pendant groups upon membrane transport.

Experimental Section

Methods and Materials

Hexachlorocyclotriphosphazene was obtained from Esprit Chemicals and purified by sublimation prior to use. Other chemicals were obtained from Aldrich and were used without further purification. NMR analyses were performed using a Bruker Instruments DMX300WB spectrometer operating at 7.04 T. ^{31}P NMR spectra were collected at 131 MHz and referenced to external H_3PO_4. ^1H NMR data were collected in $CDCl_3$ solvent using external tetramethylsilane (TMS) as a reference (0 ppm). Thermal analyses were determined using a TA Instruments model 2910 Differential Scanning Calorimeter and TA Instruments Model 2950 Thermogravimetric Analyzer. Laser Light Scattering was employed to measure polymer molecular weights using a Wyatt Technologies Dawn-DSP system that uses polarized light having a wavelength of 633 nm and measures scattered light intensities at 18 angles ranging from 22.5° to 147°.

Polymer Synthesis

Four polymers were employed in this study. MEEP*(5)* and poly[bis(4-methoxyphenoxy)phosphazene] (PMEOPP)*(2)* were synthesized according to literature procedures.

Poly[(4-methoxyphenoxy)$_{0.96}$(MEE)$_{0.96}$(2-allylphenoxy)$_{0.08}$ phosphazene] (HPP1) was synthesized using a sequential pendant group addition method where 2-allylphenol (2AP) was added first, followed by a mixture of 4-methoxyphenol (MEOP) and MEE.*(2)* Characterization data for the resulting polymer: ^1H NMR (CDCl$_3$) δ (ppm) 7.3 (brs), 7.0 (brs), 6.9 (brs), 6.5 (brs), 6.1 (brs), 4.9 (brs), 3.9 (brs), 3.5 (brs), 3.3 (brs). Integrated ^1H NMR: 2-(2-methoxyethoxy)ethanol (MEE) 48%, 4-methoxyphenol 48%, and 2-allylphenol 4%. ^{31}P NMR (CDCl$_3$) δ (ppm) -8, -12, -13, -18. DSC T$_g$ -43 °C. TGA T$_d$ 288 °C. M$_w$ = (3.1±0.2) x 10^6, Polydispersity Index (M$_w$/M$_n$) 3.09 ± 0.49.

Poly[(4-methoxyphenoxy)$_{1.44}$(2-(2-methoxyethoxy)ethoxy)$_{0.46}$(2-allylphenoxy)$_{0.10}$] (HPP2) was also synthesized using a sequential addition route.*(6)* HPP2 characterization data: ^1H NMR (CDCl$_3$) δ (ppm) 7.3 (brs), 7.0 (brs), 6.9 (brs), 6.5 (brs), 6.1 (brs), 4.9 (brs), 3.9 (brs), 3.5 (brs), 3.3 (brs). Integrated ^1H NMR: 2-(2-methoxyethoxy)ethanol (MEE) 23%, 4-methoxyphenol 72%, and 2-allylphenol 5%. ^{31}P NMR (CDCl$_3$) δ (ppm) -8, -12, -18. (M$_w$) = (6.7±0.6) x 10^5, RMS Radius = 74.1±9.1 nm, 2nd Virial Coeff. = (-4.54±2.0) x 10^{-4}. DSC T$_g$ –10 °C. TGA T$_d$ 300 °C. Density 1.294 g/mL.

Sorption Experiments and Hanson Parameter Determinations

Hanson parameters for the four polyphosphazenes were estimated using group contributions. Chemical affinities for each polymer were determined through solvent immersion. MEEP and PMEOPP determinations are simple solvent/non-solvent behavior. Samples of HPP1 and HPP2 were cross-linked using 20 Mrad electron beam irradiation to yield cross-linked elastomer samples. Determinations of chemical affinities for these two polymers were made through measurement of swelling. From these observations, judgments of swelling behavior were delineated as "good" (swelling greater than 50%), "moderate" (10-50%), and "poor" (<10%). All sorption determination were made using bulk polymer samples. Samples were weighed prior to analysis, after one week immersion, and finally upon desorption of solvent. Desorption was performed under reduced pressure until a constant mass was obtained. Degree of swelling was calculated from:

$$\text{Degree of Swelling} = \frac{(W_s - W_d)}{W_d} \times 100 \qquad (1)$$

where W_s is the swollen mass and W_d is the dry weight after desorption of solvent. In all experiments, insignificant differences were noted between the initial dry mass and the final desorbed mass suggesting little soluble entrained material within the samples and little induced chain scission.

Estimation of Hanson parameters using group contribution elements (^7F_p, polar element, $-U_h$, hydrogen bonding element, ^7F_d, dispersion element) was made using data from literature sources for each "group" within the mer (7). This was done for HPP1 and HPP2 by determination of the parameters for each homogenous mer [(PN(MEE)$_2$), (PN(MEOP)$_2$, and (PN(2AP)$_2$] and then applying these with respect to the loading of each pendant group on the particular polymer.

Group contribution elements for phosphazene nitrogen and phosphorus were not found in the literature, thus they had to be estimated. Nitrogen in the backbone is analogous to carbon-bound azo-nitrogens, so for this work, parameters associated with an azo-nitrogen were used, except the parameter for the dispersion element, ^7F_p, was reduced to give better agreement with the experimental data. Phosphorus in a phosphazene backbone has no lone pairs of electrons, thus the hydrogen bonding ($-U_h$) and polar (^7F_p) elements were set to zero. Likewise, it was anticipated that the phosphorus would have similar dispersion characteristics to the nitrogen so ^7F_p was set to that of nitrogen.

Membrane Formation and Pervaporation Experiments

Membranes were formed as thin dense films supported by porous ceramic disks (Whatman Anopore®, 0.2 μm pore size, 47 mm diameter). Application of the polyphosphazene film was performed by solution casting. Casting solutions were made by dissolution of the polymer in tetrahydrofuran. Membrane thickness was governed by the concentration of the solution where more highly concentrated solutions yielded thicker films. Into HPP1 and HPP2 casting solutions was added 2% (polymer weight percent) benzoyl peroxide as a free radical initiator. Films of HPP1 and HPP2 were cured in an oven at 130 °C for a minimum of 10 minutes prior to use to effect cross-linking. Pervaporation experiments were conducted using an apparatus described elsewhere (2).

Results and Discussion

Estimation of Hanson Parameters

Application of group contribution parameters to PMEOPP, MEEP, HPP1, and HPP2 was accomplished using literature values *(7)* for the organic portions of each mer in addition to the estimated parameters for phosphorus and nitrogen, as shown in Table I.

Table I. Group contribution parameters for nitrogen and phosphorus

	V (cm³/mol)	^zF_d (MPa)$^{0.5}$	^zF_p (MPa)$^{0.5}$	^zU_h (MPa)$^{0.5}$
Phosphorus	8.8	164	0	0
Nitrogen	4.0	164	820	1759

To account for the small portion of 2-allylphenol contained in HPP1 and HPP2, the parameters and molar volume of $PN(2AP)_2$ was calculated from the group contribution elements. Calculation of the parameters for the two HPP polymers was accomplished through a fractional application of each set of data from the symmetrically substituted mers with respect to their abundance on the backbone. Previous data has demonstrated that the pendant groups are randomly substituted *(8)*. For example, each HPP polymer has only one T_g, indicating that there are no homogeneously substituted blocks within the polymer. In fact, application of the Fox relation to phosphazenes suggests that random heteropolymers act very much like organic co-polymers where the T_g is influenced by all component polymers *(9)*.

Table II. Hansen parameters and molar volumes calculated from group contributions

	V (cm³/mol)	δ_d (MPa)$^{0.5}$	δ_p (MPa)$^{0.5}$	δ_h (MPa)$^{0.5}$
MEEP	230.6	20.6	11.2	8.3
PMEOPP	204.6	19.6	5.8	5.0
$PN(2AP)_2$	263.4	18.8	4.6	4.6
HPP1	219.4	20.0	8.3	6.6
HPP2	219.8	19.9	7.7	6.2

SOURCE: Adapted from Reference 2. Copyright 2002 Elsevier

Hanson parameters derived for the MEEP and PMEOPP are clearly different with respect to δ_p, the polar component, as shown in Table II. This difference is reflected in their differences in chemical properties where MEEP is water-soluble and PMEOPP is hydrophobic. However, neither polymer has a strong hydrogen bonding component. Values for δ_p and δ_h for the two HPP polymers expectedly fall in between the extremes defined by MEEP and PMEOPP. HPP1 has higher values of δ_p and δ_h due to the higher MEE content.

Sorption Behavior

Sorption experiments on MEEP and PMEOPP were conducted using 15 different solvents of varying character to define the solubility behavior induced by the pendant groups. In this study, behaviors were defined only as "solvent" or "non-solvent" where a determination of "solvent" indicated dissolution of the sample. Data derived from these experiments are shown in Figure 2. In these plots, a new parameter, δ_v, is defined as:

$$\delta_v = (\delta_d^2 + \delta_p^2)^{0.5} \qquad (2)$$

where δ_v is plotted against δ_h (10). Clearly, the solubility behavior of PMEOPP is more narrowly defined than that for MEEP. Instead of the expected shift in solubility behavior, the solvents that dissolve PMEOPP also dissolve MEEP, however, many more solvents dissolve MEEP. This is potentially due to the high level of backbone and pendant group flexibility that also leads to MEEP having a low T_g (-84°C) (11).

Of the fifteen solvents examined, they can be classified into three general categories:

- Non-polar hydrocarbons - pentane, octane, cyclohexane
- Polar aprotic solvents - toluene, chloroform, methylene chloride, tetrahydrofuran, diethyl ether, 1,4-dioxane
- Polar protic solvents - water, ethylene glycol, tetraethylene glycol, methanol, 2-propanol, MEE

Non-polar hydrocarbons had little interaction with either polymer. An examination using group contributions show that the backbone nitrogen has a polar component, thus the phosphazene backbone itself can also dictate aspects of solubility behavior and must be accounted for in a solubility analysis.

Polar protic solvents, on the other hand, interact far more with MEEP than they do with PMEOPP. The critical characteristic of these solvents is the ability to hydrogen bond, as reflected in the high values of δ_h. Both MEEP and PMEOPP contain no free hydroxyls that can participate in both hydrogen bond donation and acceptance. However, aliphatic etherial oxygens, such as found in MEE, can accept through electron lone pair interactions with hydroxy-containing species. Furthermore, a high degree of backbone and pendant group

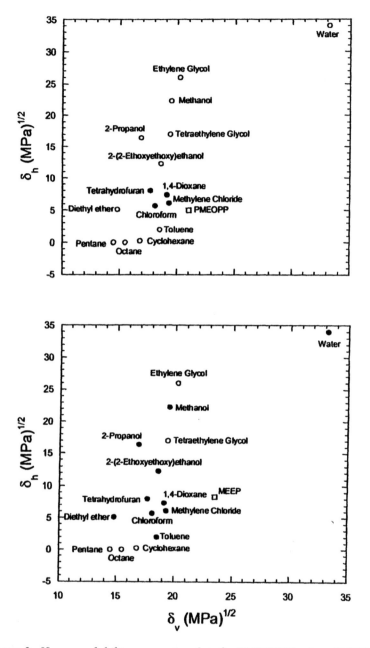

Figure 2. Hansen solubility parameter plots for PMEOPP(top) and MEEP (bottom)(● - solvent, ○ - non-solvent, □ - group contribution prediction for each polymer.) (Reproduced from Reference 2. Copyright 2002 Elsevier.)

flexibility encourages these interactions, flexibility that is not enjoyed by phenoxy-phosphazenes such as PMEOPP.

Thus, MEE containing polymers can take advantage of the hydrogen bonding capability of polar protic solvents resulting in higher solubilities. Polar aprotic solvents exhibit the highest level of interaction with both MEEP and PMEOPP. Good solubility behavior of these solvents can be visualized through interpretation of the solubility parameter data. Values for all three Hanson parameters for this solvent category are numerically most similar to that of MEEP and PMEOPP, as compared the two aforementioned solvent classes.

Solubility characteristics of HPP1 and HPP2 are best described as a combination of those of the two homopolymers, MEEP and PMEOPP. Neither heteropolymer displays the range of solubilities that MEEP exhibits, however they show solubility in a larger number of solvents than does PMEOPP. For example, the solubility behavior of HPP1 is shown in Figure 3 in addition to the predicted values for HPP1 and HPP2. Polar protic solvents exhibit a moderate interaction with both polymers where they swell without yielding a gel. Thus, these solvents are excellent targets for membrane separations since they sorb into the polymer without significant mechanical degradation of the polymer. Solvents that have the highest level of interaction with both HPP1 and HPP2, the polar aprotic solvents, tend to form gels, regardless of the cross-link density. Typically, these gels are far less dimensionally stable than simple swollen polymers and exhibit poor membrane durability.

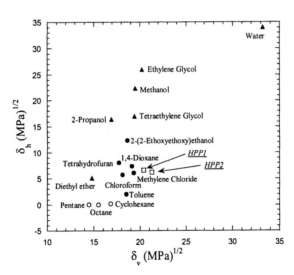

Figure 3. Hansen solubility parameter plot for polymer HPP1 (●-"good" solvent, ▲-"moderate" solvent, ○-"poor solvent", □ - group contribution prediction for HPP1 and HPP2.) (Adapted from Reference 2. Copyright 2002 Elsevier.)

Pervaporation Experiments

Pervaporation of single component water and short-chain aliphatic alcohols exhibit a typical thermal response behavior where increased fluxes are observed with increased operating temperatures. For example, permeation of water with respect to temperature is shown in Figure 4 for both HPP1 and HPP2. HPP1, the more polar polymer containing a higher amount of MEE, exhibits higher fluxes. Likewise, alcohols such as 2-propanol, methanol, and ethanol exhibited the same general trend (12). Additionally, fluxes of the alcohols were generally greater in magnitude than that of water. Using these pure component experiments, separation of the alcohols from water should be possible.

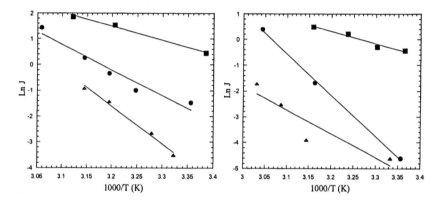

Figure 4. Variable temperature pervaporation through membranes formed from HPP1 (Left) and HPP2 (Right). Pure solvents include water (▲), 2-propanol (●), and methanol (■).

Pervaporation of alcohol-water solutions were conducted using two differing alcohols, methanol and 2-propanol. In general, fluxes of solvent compare favorably with other membrane systems, however the permselectivity was modest, at best, suggesting strong interactions of all three solvents with the MEE pendant groups.

Mixtures of methanol and water were characterized for flux in terms of the weight fraction of methanol in the feed. At low concentrations of methanol, fluxes of less than 0.5 kg/m^{2}h were obtained with separation factors greater than 4, as shown in Figure 5. As the weight fraction of methanol was increased, fluxes gradually increased to greater than 2 kg/m^{2}h. However, this increase in flux occurred with a loss of separation factor to less than two. The temperature for these experiments was 49°C.

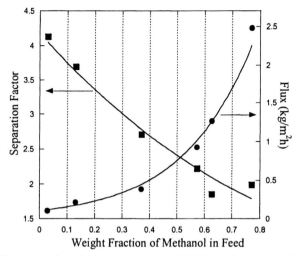

Figure 5. Water/methanol separation using HPP1 performed at 49 ℃. ■ - separation factor; ● - flux (kg/m²h). (Reproduced from Reference 2. Copyright 2002 Elsevier.)

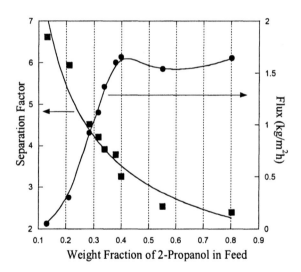

Figure 6. Water/2-propanol separation using HPP1 performed at 49 ℃. ■ - separation factor; ● - flux (kg/m²h). (Reproduced from Reference 2. Copyright 2002 Elsevier.)

Similar experiments performed using a 2-propanol/water feed revealed the same general behavior where flux increased and separation factor decreased with increasing weight fraction of 2-propanol in the feed, as shown in Figure 6. However, instead of the gradual increase in flux observed for the methanol-water system, fluxes increased sharply from 0.1 to over 1.5 kg/m^{2}h as the percentage was increased to 40% 2-propanol. At this point, the flux remained approximately unchanged, although the separation factor continued to decrease as the 2-propanol weight fraction was increased allowing these two characteristics to be effectively decoupled.

Conclusions and Future Work

Characterization of two mixed pendant group phosphazenes is discussed in this work. Using sorption measurements a data for solubility parameter maps, group contributions for Hansen Parameter assignment to the polymers, and pervaporation, two polymers with differing polarity were studied. Increased MEE on the polymer resulted in increased permeability for both water and short-chained alcohols. Mixing of water with the methanol and 2-propanol feeds allows for an assessment of separation behavior. Methanol/water pervaporation gave the largest fluxes and the poorest separation factors as compared to 2-propanol/water. The difference in performance was expected considering the smaller molecular size of methanol as compared to 2-propanol.

Hanson solubility parameters applied to functionalized polyphosphazenes allow for the prediction and design of separation systems. Implied by the predictive and experimental data is that certain useful separations may be possible and are the objective of future work. For example, separations of solvents with high or moderate affinity, such as toluene or methanol, may be separated efficiently from simple hydrocarbons. Thus, the Hanson parameter analysis using group contributions can aid in the characterization of novel synthetic membranes for selected separations.

Membranes formed from functionalized polyphosphazenes have been shown to permeate water and short-chained aliphatic alcohols through intimate intermolecular interactions with the polymer's pendant groups. Thus, it has been demonstrated that specific solubility character can be designed into phosphazene polymers through chemical synthesis. Specific interactions between the MEE pendant group and three differing classes of solvents have been identified. Little interaction with nonpolar organics observed in this work suggest that aliphatic hydrocarbons would be poorly transported by MEE containing phosphazenes. In order to overcome the polar influence of the backbone, large pendant groups with a high affinity for hydrocarbons would need to be attached.

Polar aprotic and polar protic solvents each showed affinity for the MEE substituted polymers, with the polar aprotic solvents showing a higher affinity level. This behavior was attributed to the poor hydrogen bond donating ability of MEE, thus there was a better match of affinities by the non-hydrogen bonding polar solvents. Consequently, an increase in the affinity of phosphazene polymers for polar protic solvents would be accomplished through attachment of hydroxyl functionalities onto the polymer.

Acknowledgements

The work described in this paper was supported by the United States Department of Energy through Contract DE-AC07-99ID13727. The authors also thank Dr. Daniel Goodman and Dr. Catherine Byrne (Science Research Laboratory, Somerville, MA) for their assistance with the electron beam crosslinking.

References

1. Allcock, H. R. *Chem. Rev.* **1972**, *72*, 315.
2. Orme, C. J.; Harrup, M. K.; McCoy, J. D.; Weinkauf, D. H.; Stewart, F. F. *J. Membr. Sci.* **2002**, *197*, 89-101.
3. Orme, C. J.; Harrup, M. K.; Luther, T. A.; Lash, R. P.; Houston, K. S.; Weinkauf, D. H.; Stewart, F. F. *J. Membr. Sci.* **2001**, *186*, 249-256.
4. Hansen, C. M.; Beerbower, A. In *Kirk-Othmer Encyclopedia of Chemical Technology, Supplemental Volume*; Standen, A., Ed.; Interscience: New York, 1971; p 889.
5. Harrup, M. K.; Stewart, F. F. *J. Appl. Polym. Sci.* **2000**, *78*, 1092-1099.
6. Stewart, F. F.; Harrup, M. K.; Luther, T. A.; Orme, C. J.; Lash, R. P. *J. Appl. Polym. Sci.* **2001**, *80*, 422-431.
7. Barton, A. F. M. *Handbook of Solubility Parameters and Other Cohesion Parameters*; 2nd ed.; CRC Press: Boca Raton, FL, 1991.
8. Stewart, F. F.; Peterson, E. S.; Stone, M. L.; Singler, R. E. Am. Chem. Soc. *Polym Prepr.* **1997**, *38*, 836.
9. Stewart, F. F.; Harrup, M. K.; Lash, R. P.; Tsang, M. N. *Polym. Int.* **2000**, *49*, 57-62.
10. Jou, J. D.; Yoshida, W.; Cohen, Y. *J. Membr. Sci.* **1999**, *162*, 269-284.
11. Blonsky, P. M.; Shriver, D. F.; Austin, P.; Allcock, H. R. *J. Am. Chem. Soc.* **1984**, *106*, 6854-6855.
12. Orme, C. J.; Klaehn, J. R.; Harrup, M. K.; Lash, R. P. *J. Appl. Polym. Sci.* Submitted for Publication.

Chapter 13

Novel Carbon–Silica Membranes for Improved Gas Separation

Young Moo Lee and Ho Bum Park

National Research Laboratory for Membrane, School of Chemical Engineering, College of Engineering, Hanyang University, Seungdong-ku, Seoul 133-791, Korea

Novel carbon-silica ($C\text{-}SiO_2$) membranes were prepared by pyrolyzing microphase-separated block or random copolymers consisting of two different domains – carbon-rich or silicon-rich domains. The size and shape of the ultra-micropores were dominated by the initial morphology of the $C\text{-}SiO_2$ precursor - poly(imide siloxane) (PIS). The morphological changes in polymeric nanomaterials, such as block or random copolymers consisting of two phases, affected the gas permeation properties to a large extent. In a molecular probe study using small molecules (He, CO_2, O_2, and N_2) having sizes from 2.60 to 3.64 Å, the $C\text{-}SiO_2$ membranes exhibited outstanding molecular sieving capability, together with a high gas permeability. The present study demonstrates that the main geometry of the $C\text{-}SiO_2$ precursor determines the microstructure and the gas separation capability of the final pyrolytic $C\text{-}SiO_2$ membrane.

© 2004 American Chemical Society

Gas separation processes using membranes have been developed since the mid-1960s following the discovery of asymmetric membranes prepared by the phase-inversion technique. Thereafter, most studies on membrane-based gas separation have focused on polymeric materials. However, new membrane materials are needed to compete with conventional gas separation processes, such as distillation and pressure swing adsorption. In particular, inorganic membranes such as zeolites (*1*), sol-gels (*2*), and pyrolytic carbon materials (*3*) offer significant potential advantages over conventional polymeric membranes. A main advantage of the inorganic membranes is that they can effectively perform the separation of small molecules such as H_2/N_2, He/N_2, O_2/N_2, and CO_2/N_2 under harsh conditions at elevated temperature and high pressure. To attain a desirable flux and selectivity, ultramicroporous inorganic membranes with pore sizes below 4Å need to have a proper molecular sieving capability performed by micropores with dimensions near the sizes of the penetrants.

The performance of carbon membranes is mainly controlled by the pyrolysis conditions, such as heating rate, atmosphere (inert or vacuum), temperature as well as the choice of the precursor. Up to now, many researchers have focused only on these pyrolysis conditions because in most cases they used homopolymeric precursors such poly(furfuryl alcohol) (*4*), phenolic resin (*5*) and polyimides (*6-8*). Furthermore, carbons obtained from these precursors are probably the most thermo-chemically stable materials but are very susceptible to oxidation, which is an obvious drawback to easy and reproducible fabrication of a carbon membrane.

In this paper, we report the properties of membranes consisting of carbon with silicon or silica via an inert pyrolysis by using microphase-separated copolymers such as poly(imide siloxane)s (PISs). PISs are polymeric nanostructured materials having block or random segmented domains. The imide and siloxane domains in PISs are thermally stable so that the spatial properties of PISs are expected to be kept even after heat treatment.

Experimental

Precursor Film Preparation

Five poly(imide siloxane)s (PISs) of two classes using building blocks consiting of imide and siloxane domains were designed and prepared by using pyromellitic dianhydride (PMDA), benzophenone-3,3',4,4'-tetracarboxylic dianhydride (BTDA), 4,4'-oxydianiline (ODA) (Tokyo Kasei Co., Inc., Tokyo, Japan), and α,ω-aminopropyl poly(dimethylsiloxane) (PDMS) (Shinetsu Chemical Co., Inc., Kyoto, Japan). The average number (*n*) of siloxane repeat units in the PDMS oligomer was 10 and M_n (g/mol) of PDMS determined by gel permeation chromatography (GPC) was about 900. The chemical compositions of precursors prepared in this work are summarized in Table I. All PIS films were prepared by thermal imidization at 100°C, 200°C, and 300°C under a

Table I. Sample designation and composition of precursors

Sample code	Class I			Volume fraction of siloxane moiety*
	PMDA (mmol)	ODA (mmol)	PDMS (mmol)	
PIS I	10	9.8	0.2	0.06
PIS II	10	9.0	1.0	0.27
PIS III	10	8.0	2.0	0.46
	Class II			Volume fraction of siloxane moiety
	BTDA (mmol)	ODA (mmol)	PDMS (mmol)	
r-PIS b-PIS	10	9.0	1.0	0.34

*Calculated from a group contribution method (9).

vacuum after casting siloxane-containing poly(amic acid) solutions onto glass plates.

In class I (-PMDA-ODA-PDMS-), three copolymers having different volume fractions of the siloxane moiety (-Si-O-) were prepared in order to study the effect of the siloxane content on the gas permeation properties of C-SiO$_2$ membranes obtained by pyrolysis of the precursors. In class II (-BTDA-ODA-PDMS-), two copolymers having different morphology, that is, random and block segmented copolymers, at a fixed siloxane content were prepared via one-step and two-step polymerization in order to study the gas permeation properties of C-SiO$_2$ membranes related to the initial morphology of precursors. Further information on the sample prepapation method can be found in our previous study (10).

C-SiO$_2$ Membrane Preparation

The pyrolysis protocol was followed thoroughly to obtain reproducible gas permeation properties of the C-SiO$_2$ membranes. Before each pyrolysis step, the free-standing PIS films were rinsed with deionized water and stored at 120°C in a vacuum oven until any organic contaminants and dust were completely removed. Prior to heating the furnace, the PIS films (thickness: 20-30 μm) were kept for 1 hour under an argon purge to stabilize the atmosphere and to remove any oxidant in a quartz tube. The heating rate used at the initial stage was 10°C/min from room temperature to 400°C. The heating rate was decreased to 3°C/min until the temperature reached 600°C. Then, the PIS films were kept at 600°C for 2 hours (denoted as C-SiO$_2$ at 600°C). From 600°C, the heating rate was again ramped to 3°C/min up to 800°C. Pyrolyzed films were held at 800°C for 2 hours (denoted as C-SiO$_2$ at 800°C). Finally, from 800°C, the heating rate was increased to 3°C/min until the temperature reached 1000°C and the resultant films were held at this temperature for 2 hours (denoted as C-SiO$_2$ at 1000°C). At each stage, the furnace was allowed to slowly cool down to room temperature. The final pyrolytic C-SiO$_2$ membranes were taken from the quartz tube and then stored in a dessicator filled with dry silica gel to eliminate any effects of humidity on membrane properties.

Gas Permeation Experiments

In the present work, the gas permeation properties of the polymeric precursors and C-SiO$_2$ membranes were measured. A constant volume, variable pressure method (*11,12*) was used to determine the steady-state permeability of the pure penetrants at 25°C over a range of upstream pressures up to 76 cmHg. A membrane was loaded into the permeation cell and degassed by exposing both sides of the membrane to vacuum. After degassing, the upstream side of the membrane was exposed to a fixed pressure of the penetrant. The steady-state rate of pressure rise on the downstream side was used to determine the permeability coefficient. At all times during the experiments, the downstream pressure was kept below 10^{-5} torr. At steady-state, when the rate of downstream pressure rise was constant, the following expression was used to determine the permeability coefficient, *P*:

$$P = \frac{22,414Vl}{RTAp_1} \cdot \frac{dp_2}{dt} \tag{1}$$

where *V* is the downstream reservoir volume, *R* the universal gas constant, *T* the absolute temperature, *A* the cross-sectional area of the membrane, *l* the thickness of membrane, p_1 the upstream pressure, and dp_2/dt is the rate of change in downstream pressure. The diffusion coefficient *D* was obtained from the time-lag θ as follows:

$$D = \frac{l^2}{6\theta} \tag{2}$$

Results and Discussion

Strategy to Develop Membranes with High Selectivity

Separation of small gases, such as He/N$_2$, O$_2$/N$_2$, and CO$_2$/N$_2$, presents the greatest challenge for membrane materials because their kinetic diameters differ by only a few tenths of an Angstrom. In particular, to obtain a high selectivity for O$_2$/N$_2$ and CO$_2$/N$_2$, the following strategy can be considered. Generally, the gas permeability, P (1 Barrer = 10^{-10} cm^3(STP)·cm/cm^2·s·cmHg), is expressed by the following equation:

$$P_A = D_A \times S_A \qquad (3)$$

The permeability (P_A) of component A can be expressed as the product of a kinetic factor, the diffusion coefficient (D_A), and a thermodynamic factor, the sorption coefficient (S_A). The selectivity, $\alpha_{A/B}$, is defined as the ratio of the permeabilities P_A and P_B:

$$\alpha_{A/B} = \frac{P_A}{P_B} \qquad (4)$$

Typical target gas pairs:	O_2/N_2	CO_2/N_2
Range of gas selectivity:	5-20	40-100

Main contribution: $\qquad \alpha_{A/B} = \dfrac{P_A}{P_B} = \left(\dfrac{D_A}{D_B}\right) \cdot \left(\dfrac{S_A}{S_B} \leq 2\right) \qquad \alpha_{A/B} = \left(\dfrac{D_A}{D_B} \leq 4\right) \cdot \left(\dfrac{S_A}{S_B}\right)$

Diffusion selectivity   Sorption selectivity

D_A/D_B is the ratio of the diffusion coefficients of the two gases and is referred to as the diffusivity selectivity, reflecting the different molecular sizes of the gases. S_A/S_B is the ratio of the solubility coefficients of the gases and can be viewed as the solubility (sorption) selectivity, reflecting the relative condensabilities of the gas molecules interacting with the membrane matrix. Commonly, the sorption-selectivity for the O_2/N_2 pair is in the range of 1-2 for almost all glassy and rubbery polymers and in the range of 0.7-2 for molecular sieving materials. Therefore, diffusivity selectivity is the most significant factor to obtain highly selective membrane materials for the separation of small gases.

Structural Analysis of C-SiO$_2$ Membranes

We carried out a structural analysis of C-SiO$_2$ membranes by using ^{29}Si-NMR, FE-SEM, ESCA, AFM, and TEM. Before analyzing the structure of the C-SiO$_2$ membranes, it was assumed that they were composed of a SiO$_2$-rich domain having sparse carbon clusters within a continuous carbon-rich matrix which serves as a molecular sieve. To confirm the existence of SiO$_2$ in a C-SiO$_2$

membrane, solid-state ^{29}Si-NMR was used to examine the local environment of Si atoms in the bulk state of C-SiO$_2$ pyrolyzed at 600 and 800°C. In the solid-state ^{29}Si-NMR spectra (Figure 1a), the transition of D unit [(CH$_3$)$_2$SiO] (-20.4 ppm) in the PIS to Q units [Q^3: Si(OR)$_1$(OSi)$_3$ with R=H or CH$_3$, Q^4: Si(OSi)$_4$] of SiO$_2$ derivatives was found after pyrolysis.

In microphase-separated polymer systems, to minimize the total energy of the system as well as the bulk composition, the block length and block sequence distribution lead to preferential surface aggregation of the low surface energy constituent (13-15). Usually, siloxane-containing polymers are used as a surface modifier through blending or co-polymerization with other polymers due to free rotatability and polarizability of the Si-O bond (16). Thus, the Si-O-Si chain is able to align itself, resulting in a rich in-depth distribution of the surface in copolymers or blends. In the present study, our interest was focused on the perculiar feature of siloxane-containing copolymers, assuming that the SiO$_2$ domains derived from siloxane domains would accumulate on the surface of the C-SiO$_2$ membranes. To confirm this assumption, hydrofluoric acid (HF) etching was carried out to remove the SiO$_2$ domains on the C-SiO$_2$ membrane surface, and then electron spectroscopy was used for chemical anylysis (ESCA) to examine the surface composition before and after the acid treatment (Figure 1b). The Si$_{2p}$ spectrum can be fitted with two components in Si compounds: Si(I) at 101.5 eV and Si(II) at about 103.0 eV; Si(I) and Si(II) are attributed to PDMS and SiO$_2$, respectively (17). A Si(I) spectrum was not observed after pyrolysis, indicating conversion of PDMS to a SiO$_2$ phase.

Tapping mode-atomic force microscopy (TM-AFM) was used to study the surface morphology of C-SiO$_2$ membranes before and after acid etching. Note that the surface topographies of C-SiO$_2$ membranes before and after acid treatment were very different because of removal of the SiO$_2$-rich phase on the surface, as shown in Figure 1c. The surface morphology of the C-SiO$_2$ membranes was rougher, indicating a selective removal of abundant SiO$_2$ domains on the surface after acid treatment. The topmost surface morphology of the C-SiO$_2$ membrane after acid treatment is further illustrated by the field emission-scanning electron microscopy (FE-SEM) image, shown in Figure 1d. While observing the surface morphology on both sides of the C-SiO$_2$ membranes after HF treatment via FE-SEM, we found very interesting features on the top surface. The surface of only one side (air-polymer interface when the PIS film was cast) in the C-SiO$_2$ membranes was covered with well-distributed carbon hemispheres with grained shapes in the form of discrete bright spots. The spheres had nearly a constant size of about 200-220 nm in diameter.

A separate energy dispersive X-ray spectrometry (EDX) analysis revealed that these carbon hemispheres were only composed of elemental carbon. This result indicates that the carbon phase derived from aromatic blocks was embedded in the continuous SiO$_2$ matrix on the surface. On the other hand, the SiO$_2$ domain was embedded in the continuous carbon matrix in the bulk solid

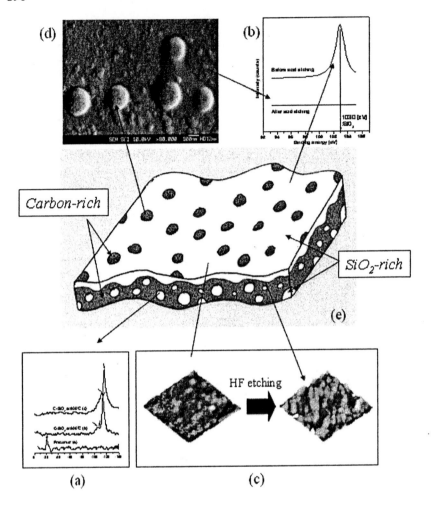

Figure 1. Structural analysis of a C-SiO₂ membrane (a-d) and its proposed morphology (e).

state. Accordingly, the present C-SiO$_2$ membrane may have an asymmetric structure, as shown in Figure 1e.

Gas Permeation Properties of Class I Membranes

To examine the molecular sieving capability of the C-SiO$_2$ membranes, molecular probe studies were performed to measure the permeability of pure-component gases (He[2.60 Å], CO_2[3.30 Å], O_2[3.46 Å], N_2[3.64 Å]) through C-SiO_2 membranes. The values in brackets correspond to the kinetic diameter of each gas. Class I membranes were made to investigate the effect of SiO_2 content within C-SiO_2 membranes derived from pyrolysis of PISs consisting of PMDA, ODA and PDMS. The gas permeabilities are in the order of He > CO_2 > O_2 > N_2. These relative gas permeabilities are common in microporous membranes, such as carbon molecular sieve and silica membranes. Therefore, gas transport occurring in these membranes depends strongly on diffusion, which results in a size-sieving mechanism. The correlation of gas permeabilities of C-SiO_2 membranes with their kinetic diameters indicates that the permeability is primarily dependent on diffusion. Hence, diffusion-controlled selectivity dominates the separation properties of the present C-SiO_2 membranes.

Table II. Gas permeation results of C-SiO_2 membranes (Class I) at 25°C

Precursor	Pyrolysis Temp.(°C)	Permeability (Barrer*)				Selectivity		
		He	CO_2	O_2	N_2	He/ N_2	CO_2/ N_2	O_2/ N_2
	600	315	21	8	0.3	955	62	23
PIS I	800	442	89	27	1.5	304	61	18
	1000	121	12	5	0.2	643	64	124
	600	1258	84	30	1.4	1033	62	22
PIS II	800	1393	204	68	3.7	341	56	19
	1000	133	7	2	0.1	1330	74	20
	600	610	386	111	18.0	34	21	6
PIS III	800	981	765	168	16.8	58	46	10
	1000	207	36	10	0.6	56	57	15

* 1 Barrer = 10^{-10} cm^3(STP)·cm/cm^2·sec·cmHg

Table II summarizes the gas permeation data of He, CO_2, O_2, and N_2 through C-SiO_2 membranes (class I) derived from PIS I, PIS II and PIS III at different pyrolysis temperatures (600°C, 800°C, 1000°C). As shown in Table II, the gas separation properties of C-SiO_2 membranes were influenced considerably by the siloxane content of the PISs. The gas permeabilities increased but the gas selectivities decreased with an increase of siloxane content in PISs. This result might indicate that the SiO_2 domains play an important role in the overall gas transport of C-SiO_2 membranes.

In this study, it has been observed that the pyrolysis temperature has a marked influence on the permeation characteristics of C-SiO_2 membranes. All gas permeabilities increased with an increase of the pyrolysis temperature up to 800°C. As summarized in Table II, the gas permeabilities of C-SiO_2 membranes show a maximum value at 800°C. However, as the pyrolysis temperature is further increased to 1000°C, the resulting C-SiO_2 membranes are less permeable than the ones pyrolyzed below 800°C. In the case of $C-SiO_2$ membranes derived from PIS II, the pyrolysis of the PIS at 1000°C leads to an O_2 permeability of only 2 Barrer in comparison with 68 Barrer at 800°C.

In addition, the gas permeabilities of $C-SiO_2$ membranes tended to increase and the gas selectivities decreased with the volume fraction of the initial siloxane content in the original poly(imide siloxane) membranes. It is worth investigating the gas permeation properties of the polymeric precursor used before an inert pyrolysis, because the microstructure of the polymer is closely related to its gas transport behavior. Our previous work on the gas permeation behavior of precursors reported the transport behavior of rigid-flexible block copolymer membranes, such as poly(amide imide siloxane) and poly(imide siloxane) (*18,19*). The gas permeation properties of these polymers were influenced by the volume fraction of the siloxane moiety. The same results were obtained in this study for $C-SiO_2$ membranes, that is, the SiO_2 domains in $C-SiO_2$ play a significant role in the overall gas permeation properties.

Gas Permeation Properties of Class II Membranes

In the previous section, we focused on the effect of the siloxane content in PISs on the gas permeation properties of the $C-SiO_2$ membranes. In additional experiments, the gas permeation properties of $C-SiO_2$ membranes (class II), derived from PISs, by varying the geometry of random and block sequences (see Table I), were investigated in with respect to their morphology. PISs consisting of BTDA, ODA and PDMS were prepared by using different synthetic methods: (i) one-step (*r*-PIS: random sequence-favorable PIS) and (ii) two-step (*b*-PIS: block sequence-favorable PIS) polymerization. At the same PDMS content, the microphase-separated siloxane domain size in *r*-PIS was expected to be smaller than that in *b*-PIS.

The microphase-separated structures of these PISs were confirmed by differential scanning calorimetry (DSC). The two PISs showed two T_gs of a siloxane and an imide domain, indicating microphase separation of PISs (T_gs of *r*-PIS: -112°C and 221°C; T_gs of *b*-PIS: -121°C and 241°C). Thus, we assumed that this difference might lead to changes in the gas permeation properties of PIS and $C-SiO_2$ membranes.

The gas permeation properties of PISs (*r*-PIS and *b*-PIS) and the $C-SiO_2$ membranes (*r*-$C-SiO_2$ and *b*-$C-SiO_2$) are summarized in Table III. The gas permeabilities (He, CO_2, O_2, and N_2) of the *b*-PIS membrane were higher than those of the *r*-PIS membrane. On the other hand, the gas selectivities (He/N_2, CO_2/N_2, and O_2/N_2) of the *b*-PIS membrane were lower than those of the *r*-PIS membrane. This result indicates that the domain size in multiphase polymers could have caused these differences in their gas permeation properties. The formation of larger siloxane domains in the *b*-PIS membrane leads to higher gas permeability but lower gas selectivity in comparison with the *r*-PIS membrane.

It is important to note that the two PISs contained the same siloxane content (0.34 vol%).

Table II. Gas permeabilities of precursors and their pyrolytic C-SiO$_2$ membranes at 25°C

	Permeability (Barrer*)				Selectivity		
	He	CO$_2$	O$_2$	N$_2$	He/N$_2$	CO$_2$/N$_2$	O$_2$/N$_2$
b-PIS	9.70	3.29	0.94	0.22	44.1	15.0	4.3
r-PIS	6.75	0.93	0.27	0.06	112.5	15.5	4.5
b-C-SiO$_2$	1449	155	55.8	3.9	371.5	39.7	14.3
r-C-SiO$_2$	1107	26	19.3	0.6	1845.0	43.3	32.1

* 1 Barrer = 10^{-10} cm^3(STP)·cm/cm^2·sec·cmHg

Interestingly, the gas permeation properties of C-SiO$_2$ membranes showed a similar tendency to those of their PIS precursors. The gas permeabilities of the b-C-SiO$_2$ membrane were higher and the gas selectivities were lower than those of the r-C-SiO$_2$ membrane. The gas permeation results of these C-SiO$_2$ membranes implied that the peculiar microstructure of phase-separated PISs influenced the gas separation performance even after pyrolysis. That is, the initial skeleton of the PISs was essentially maintained after pyrolysis despite its thermal deformation during the pyrolysis. During the heat treatment of the thermo-stable phases, a siloxane domain can be transformed into a SiO$_2$-rich domain, whereas the aromatic imide domain can be converted into a carbon-rich domain.

To confirm this assumption, the morphologies of two C-SiO$_2$ membranes were observed by transmission electron microscopy (TEM, JEOL JSF-2000FX, Inc., MA, U.S.A.). TEM images were taken at 200 kV to observe the microstructure of the pyrolytic C-SiO$_2$ membranes. Figure 2 shows two C-SiO$_2$ samples derived from r-PIS and b-PIS membranes. In these TEM images, the black spots are the amorphous, carbon-rich domains, whereas the white spots are the SiO$_2$-rich domains containing a small carbon-poor domain. The formation of larger siloxane domains in a b-PIS membrane results in a larger SiO$_2$-rich phase compared to that in r-PIS.

On the other hand, the r-PIS membrane shows a well-distributed microstructure with small siloxane domains in a continuous imide matrix when inferred from TEM images after pyrolysis. Consequently, these structural differences of C-SiO$_2$ membranes resulting from the initial microstructure of their precursors led to changes of the gas transport behavior, as mentioned above. Thus, it can be concluded that proper control of the microstructure in a two-phase polymeric precursor, particularly coupled with thermo-stable phases, is the most important factor that determines the gas separation performance of the resulting pyrolytic membranes.

(a) *r*-C-SiO₂   (b) *b*-C-SiO₂

Figure 2. TEM images of C-SiO₂ membranes derived from r-PIS and b-PIS.

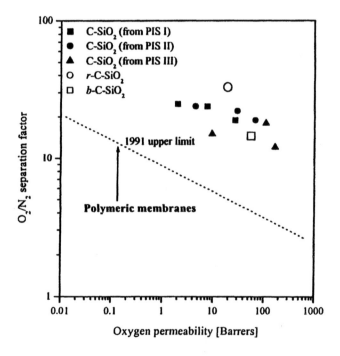

Figure 3. O_2 permeability vs. O_2/N_2 selectivity of C-SiO₂ membranes prepared in this study compared to the upper bound for polymer membranes (20).

Conclusions

In this work, novel carbon-silica ($C-SiO_2$) membranes were prepared by pyrolyzing microphase-separated block or random copolymers consisting of two different domains: carbon-rich or siloxane-rich domains. The resulting membranes had pores with different size or shape dictated by the initial morphology of the precursors. Moreover, $C-SiO_2$ membranes based on microphase-separated copolymers were prepared and characterized. The change of morphology in a polymeric material (block or random copolymers consisting of two phases on a nanoscale) was found to affect the gas permeation properties to a large extent. In the case of air separation, the O_2 permeability and O_2/N_2 selectivity of the $C-SiO_2$ membranes was, to the best of our knowledge, higher than the values obtained with other gas separation membranes (see Figure 3).

The combination of two nanoscale building blocks with different carbon densities can provide a new type of membrane material for gas separation. Furthermore, the stage has clearly been set for the widespread utilization of a variety of novel $C-SiO_2$ membranes in a range of important applications.

Acknowledgements

This work was supported by the Korea Institute of Science and Technology for Evaluation and Planning (KISTEP) under National Research Laboratory Program (NRL). PDMS samples from Shinetsu Chemical Co., Inc., are greatly appreciated.

References

1. Boudreau, L.C.; Kuck, J.A.; Tsapatsis, M. *J. Membr. Sci.* **1999**, *152*, 41.
2. Nair, B.N. *J. Membr. Sci.* **1997**, *135*, 237.
3. Koresh, J.; Soffer, A. *Sep. Sci. Technol.* **1983**, *18*, 723.
4. Cheng, Y.D.; Yang, R.T. *Ind. Eng. Chem. Res.* **1994**, *22*, 3146.
5. Wang, S.; Zeng, M.; Wang, Z. *Sep. Sci. Technol.* **1996**, *31*, 2299.
6. Hatori, H.; Yamada, Y.; Shiraishi, M.; Nakata, H.; Yoshitomi, S. *Carbon* **1992**, *30*, 305.
7. Jones, C.W.; Koros, W.J. *Carbon* **1994**, *32*, 1419.
8. Suda, H; Haraya, K. J. *J. Phys. Chem. B* **1997**, *101*, 3988.
9. Van Krevelen, D.W. *Properties of Polymers*, Elsevier, Amsterdam, 1990.
10. Park, H.B.; Suh, I.Y.; Lee, Y.M. *Chem. Mater.* **2002**, *14*, 3034.
11. Barrer, R.M. *Trans. Faraday. Soc.* **1939**, *35*, 628.
12. Felder, R.M.; Huvard, G.S. In *Methods of Experimental Physics;* Marton, L.; Marton, C., Eds.; Academic Press: New York, NY, 1980, p.315.

13. Zhao, J.; Rojstaczer, S.R.; Chen, J.; Xu, M.; Gardellar Jr., J.A. *Macromolecules* **1999**, *32*, 455.
14. Chen, X.; Gardellar Jr., J.A.; Ho, T.; Wynne, K.J. *Macromolecules* **1995**, *28*, 1635.
15. Li, L.; Chan, C.M.; Weng, L.T. *Macromolecules* **1997**, *30*, 3698.
16. Yilgör, I.; McGrath, J.E. *Adv. Polym. Sci.* **1988**, *86*, 1.
17. Feng, J.; Weng, L.-T.; Chan, C.-M.; Xhie, J.; Li, L. *Polymer* **2001**, *42*, 2259.
18. Park, H.B.; Ha, S.Y.; Lee, Y.M. *J. Membr. Sci.* **2000**, *177*, 143.
19. Ha, S.Y.; Park, H.B.; Lee, Y.M. *Macromolecules* **1999**, *32*, 2394.
20. Robeson, L.M. *J. Membr. Sci.* **1991**, *62*, 165.

Chapter 14

Gas Separation and Pervaporation through Microporous Carbon Membranes Derived from Phenolic Resin

Hidetoshi Kita, Koji Nanbu, Hiroshi Maeda, and Ken-ichi Okamoto

Department of Advanced Materials Science and Engineering, Faculty of Engineering, Yamaguchi University, Tokiwadai, Ube, Yamaguchi 755–8611, Japan

Carbon membranes were prepared by coating thin layers of a phenolic resin on the outer surface of a porous alumina substrate and then carbonizing the polymer under nitrogen atmosphere at 500-800°C for 1 hour. With increasing carbonization temperature, the gas permeance and selectivity decreased, because the membranes became denser and their pore size decreased. The membrane carbonized at 500°C showed high benzene selectivity for pervaporation and vapor permeation of benzene/cyclohexane and benzene/n-hexane mixtures. The carbon membrane behaves like a molecular sieve in the benzene/cyclohexane system. On the other hand, preferential sorption of benzene in the micropores determines the overall permeation performance for the benzene/n-hexane system.

Conventional separation processes, such as distillation, absorption, and adsorption, play a critical role in manufacturing and their optimum performance can significantly reduce overall process costs. Alternative, more energy-efficient separation processes are expected to be implemented in the chemical process industry. Membrane separation appears to be a promising new technology because of (i) low energy consumption, (ii) compact system size, (iii) simple operation, and (iv) low environmental impact. Therefore, strong interest exists in the development of novel materials that exhibit both higher permeability and higher selectivity than presently available membranes (*1*). In this report, gas separation and pervaporation properties of microporous carbon membranes derived from a phenolic resin precursor are described. Recently, carbon membranes have been prepared from numerous polymeric precursors (*2-4*). We have explored the use of polyimides and phenolic resins as polymeric precursors to obtain hollow fibers or supported carbon membranes (*5-8*). This report includes part of this ongoing work. A phenolic resin was selected as the precursor polymer because of its high carbon yield and excellent thermosetting properties without deforming the deposited layer on a substrate during heating and pyrolysis.

Experimental

Materials and Membrane Characterization

A phenolic resin (Bellpearl S-895, Kanebo Ltd., Japan) was used as starting material for the formation of microporous carbon membranes. Figure 1 shows the schematic structure of the phenolic resin. Porous α-alumina tubes were supplied by Mitsui Grinding Wheel Co. The substrate had a mean pore diameter of 1 μm with a porosity of about 50%. The tubes had an outer diameter of 1 cm, a thickness of 0.2 cm, and a length of 10 cm. Colloidal silica (EG-ST, supplied by Nissan Chemical Ind.) was dip-coated onto the α−alumina tubes to reduce the pore size of the substrate. Membranes made from phenolic resin were formed by a dip-coating process in which the porous substrate capped with silicone rubber was dipped into a 40 wt% phenolic resin solution in methanol for 20 min at room temperature. The coated substrate was then removed from the solution at a rate of 1 cm/min. After air-drying at 50°C for one day and then vacuum-drying at 50°C for one day, the membranes were carbonized at 500-800°C for 1 h under nitrogen atmosphere at a heating rate of 5°C/min and then allowed to cool to ambient temperature. The coating and carbonization cycle was repeated 3 to 4 times. After the first cycle, a 20 wt% phenolic resin solution was used for the coating process.

Thermogravimetric-mass spectrometry was performed using a Rigaku TG-8120-Shimadzu GCMS-QP5050 under a He atmosphere at a heating rate of 10°C/min. N_2 adsorption and desorption measurements at 77 K were carried out using a BEL-18SP automated adsorption apparatus. Each sample was degassed at 390°C for 24 h before performing a N_2 adsorption experiment. Micropore volume and pore size distribution of the samples were determined from the Dubinin-Radushkevich equation and Dollimore-Heal method, respectively.

Gas permeation experiments were carried out using a vacuum time-lag method at a feed pressure of 1 atm. Gas was fed to the outer side of the membrane in a permeation cell. The membrane was sealed in the cell module with a fluoro-rubber o-ring as described elsewhere (9). The effective membrane area was 15.2 cm^2. Pervaporation experiments were carried out using the apparatus described previously (9).

Results and Discussion

Characteristics of Carbonized Phenolic Resin

Thermogravimetric results for the phenolic resin are shown in Figure 2. A significant weight loss occurred around 350°C and the total weight loss at 800°C was about 45%. Figure 3 shows the pyrolysis mass spectrum of gaseous products evolved from the phenolic resin up to 800°C. Two decomposition stages were observed during the pyrolysis process: (a) in the first stage, 20% weight loss of phenolic resin occurred up to 500°C. The most abundant pyrolysis fragments were due to phenol, cresols, and xylene together with carbon monoxide, carbon dioxide and water; (b) in the second stage, about 25% weight loss occurred between 500 and 800°C. This weight loss was also attributed to evolution of carbon dioxide, and carbon monoxide together with higher molecular weight fragments, such as phenol, cresols and xylene. However, the peak intensities of higher molecular weight fragments during pyrolysis up to 500°C were 1.5 times higher than those above 500°C. These evolved pyrolysis fragments seem to contribute effectively to the micropore formation of the carbonized phenolic resin.

Figure 4 shows the sorption isotherm of N_2 for the phenolic resin carbonized at 600°C, which belongs to a typical Type I sorption isotherm, indicating the existence of a microporous structure. The micropore volume and the average pore diameter determined from nitrogen adsorption were 0.23 cm^3/g and 0.64 nm, respectively, which are both similar to values reported for carbonized polyimides (7).

Gas Separation

Figure 5 shows the effect of the carbonizing temperature on the gas permeances through the carbon membranes at 35°C. Gas permeances of the membranes carbonized at 500 and 600°C decreased approximately in the order of increasing the kinetic molecular diameter of the penetrant gas. By increasing the carbonizing temperature to 700°C, permeances of each gas decreased. Furthermore, an increase of the carbonizing temperature to 800°C resulted in a significant decrease of the gas permeance accompanied with a decrease in

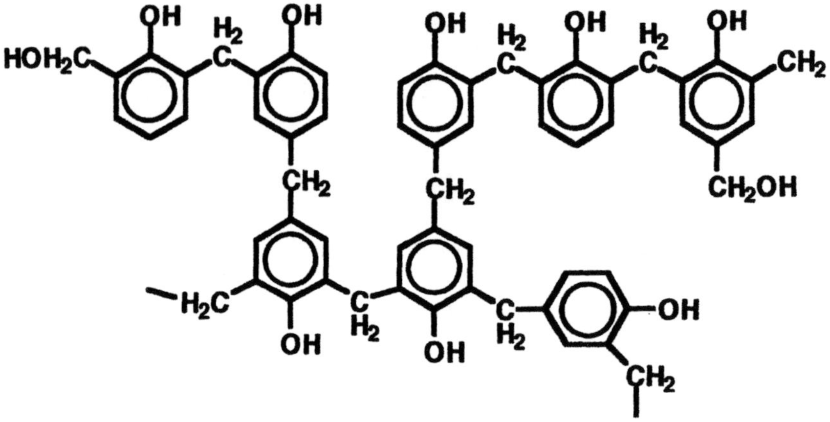

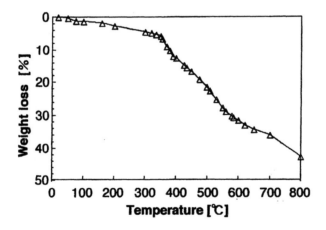

Figure 1. Chemical structure of a phenolic resin.

Figure 2. Thermograms of the phenolic resin at a heating rate of 5°C/min in N₂.

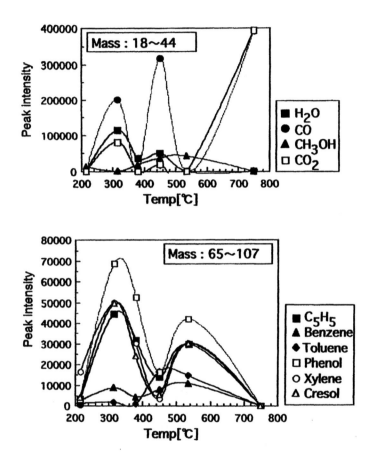

Figure 3. Mass spectra of gases evolved during pyrolysis of phenolic resin.

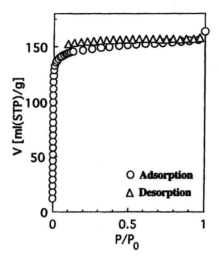

Figure 4. Sorption isotherm of N_2 at 77K for a phenolic resin carbonized at 873K for 1h.

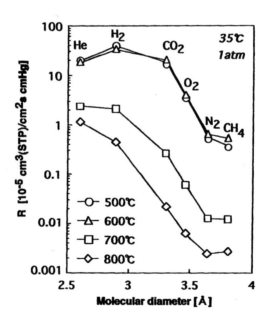

Figure 5. Gas permeances (R) vs. molecular diameter of gases through carbon membranes derived from phenolic resin at 500-800°C for 1 h in N_2.

selectivity. This may be caused by densification of the polymer matrix by cleavage of phenolic resin, leading to a decrease in the pore size of the microporous membrane. Thus, the carbon membrane derived from phenolic resin at 500°C showed the best performance. Gas selectivities of this membrane type for H_2/CH_4, CO_2/CH_4, and O_2/N_2 at 35°C were 110, 48, and 7.0, respectively. The activation energies of permeation were 5.1(He), 5.5(H_2), 3.8(CO_2), 11(O_2), 20(N_2), and 21(CH_4) kJ/mol, respectively. The activation energy increased with increasing the molecular gas diameter, except for CO_2. The activation energy of permeation for CO_2 is smaller than those of He and H_2, which could result from adsorption and surface flow of CO_2.

Pervaporation

Pervaporation has gained widespread acceptance in the chemical industry as an effective process for separation of azeotropic and close boiling point mixtures. In addition, it has been applied to dehydration of organic liquids. Application of pervaporation for separation of organic liquid mixtures, such as aromatic/aliphatic hydrocarbons, however, is still very limited because of low selectivity due to severe swelling of polymer membranes. In the design of optimized selective membranes for organic liquid separation it is important to suppress membrane swelling, yet providing enhanced penetrant solubility. Microporous inorganic materials with excellent thermal, chemical, and mechanical stability, are, therefore, of particular interest for preparing membranes for the separation of organic liquid mixtures.

Table I shows pervaporation and vapor permeation properties of carbon membranes made from phenolic resin. The carbon membranes preferentially permeated water from water/ethanol mixtures and methanol from methanol/benzene mixtures. Carbon membranes also showed high benzene selectivity for benzene/cyclohexane and benzene/n-hexane mixtures, similar to the properties of FAU zeolite membranes, as reported previously (9,10). Both the separation factor and the flux of the membrane carbonized at 500°C were larger than those of the membrane carbonized at 700°C, probably due to a less permeable structure of the 700°C-carbonized membrane, as mentioned above.

The effects of benzene feed concentration on benzene and cyclohexane flux at 75°C (pervaporation) and 150°C (vapor permeation) are shown in Figures 6 and 7, respectively. In both cases, the benzene flux was more than one order of magnitude larger than the cyclohexane flux and the benzene concentration in the permeate was higher than 97 wt% over the whole range of feed compositions. As shown in Figure 8, carbon membranes showed a very good performance, especially in the vapor permeation mode. The flux increased from 0.07 kg/m^{2}h at 75°C to 0.16 kg/m^{2}h at 150°C and the benzene concentration in the permeate was higher than 99 wt% in the vapor permeation mode. Figure 9 shows the effect of benzene feed concentration on benzene/n-hexane separation at 150°C. With increasing the benzene feed concentration, n-hexane flux decreased

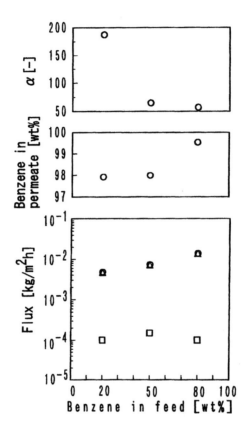

Figure 6. Effect of benzene feed concentration on pervaporation performance (75°C) for benzene/cyclohexane separation through a carbon membrane derived from phenolic resin at 500°C: (o) total flux, (Δ) benzene flux, (□) cyclohexane flux.

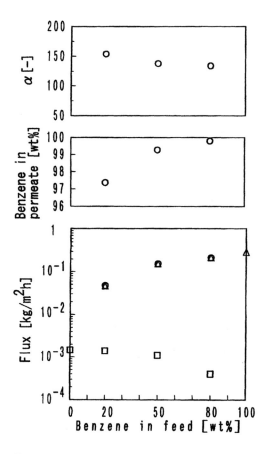

Figure 7. Effect of benzene feed concentration on vapor permeation performance (150°C) for benzene/cyclohexane separation through a carbon membrane derived from phenolic resin at 500°C: (o) total flux, (Δ) benzene flux, (□) cyclohexane flux.

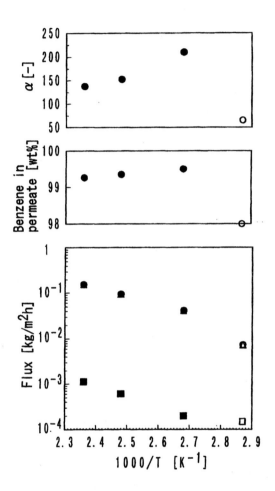

Figure 8. Membrane performance for pervaporation (open sign) and vapor permeation (closed sign) of benzene/cyclohexane (50/50 wt%) through a carbon membrane derived from phenolic resin at 500°C: (o) total flux, (Δ) benzene flux, (□) cyclohexane flux.

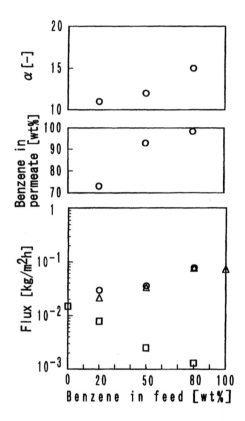

Figure 9. Effect of benzene feed concentration on vapor permeation performance (150°C) for benzene/n-hexane separation through a carbon membrane derived from phenolic resin at 500°C: (o) total flux, (∆) benzene flux, (▢) n-hexane flux.

linearly and became much smaller than the benzene flux, whereas the benzene flux gradually increased. Compared with the benzene/cyclohexane system, the pure n-hexane flux was only about half as large as the pure benzene flux, probably due to the molecular size difference between n-hexane and cyclohexane, as shown in Table II. Benzene and n-hexane molecules with smaller molecular cross-section can pass through the membrane, but cyclohexane having a larger molecular size can hardly pass through the membrane. Therefore, the carbon membrane behaves more like a "molecular sieve" in the benzene/cyclohexane system.

Table I. Pervaporation (50°C and 75°C) and vapor permeation (100°C and 150°C) performances of carbon membranes made from phenolic resin

Feed solution (A/B)	Pyrolysis condition		Temp	Flux	Separation Factor(A/B)
(wt% of A)	Temp (°C)	No.	(°C)	(g/m²h)	
Water/ethanol (50)	500	3	75	94	100
	700	3	75	110	570
Methanol/benzene (50)	500	3	50	45	51
	500	4	50	80	46
	500	4	100	370	280
	700	3	50	20	7
	700	4	50	13	22
Benzene/cyclohexane (50)	500	3	75	7	28
	500	4	75	7	65
			100	41	210
			150	160	140
	700	3	75	9	6
	700	4	75	<5	-
Benzene/n-hexane (50)	500	4	75	<5	-
			100	15	12
			150	110	11

Note: Separation factor was determined as $(Y_A/Y_B)/(X_A/X_B)$, where X_A, X_B, Y_A, and Y_B denote the weight fraction of component A and B in the feed and the permeate, respectively, and A is the preferentially permeating component.

Figure 10 shows the vapor permeation behavior of a benzene/cyclohexane mixture through a NaY zeolite membrane (10). The pore size of a Y-type zeolite membrane (0.74 nm) is larger than the molecular dimensions of benzene, cyclohexane, and n-hexane, and is also larger than the size of the graphitic slit pore of the carbonized membrane (0.64 nm). Thus, both benzene and cyclohexane can pass through the NaY zeolite membrane. In a binary system, however, selective sorption of benzene determines the separation performance. Adsorbed benzene molecules obstruct the transport of cyclohexane or n-hexane

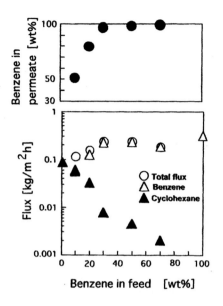

Figure 10. Effect of benzene feed concentration on vapor permeation performance (150°C) for benzene/cyclohexane separation through Y-type zeolite membrane: (o) total flux, (Δ) benzene flux, (□) cyclohexane flux.

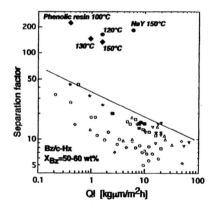

Figure 11. Comparison of membrane performance of carbon membranes (this study), zeolite membranes (10) and various polymer membranes (11) for pervaporation and vapor permeation of benzene/cyclohexane mixtures.

216

molecules, as reported previously (10). In the case of benzene/n-hexane separation through the carbon membrane, benzene molecules obstruct transport of n-hexane molecules. In this system preferential sorption of benzene presumably determines the permeation performance. Benzene sorption in the membrane increases with increasing benzene feed concentration and this preferential sorption causes a decrease in the flux of n-hexane.

Table II. Physical properties of penetrants

Penetrant	$D_{LJ}(nm)$	$a_D(nm^2)$
Benzene	0.527	0.21
Cyclohexane	0.609	0.33
n-hexane	0.591	0.18

Note: D_{LJ} is the Lennard-Jones collision diameter; a_D is the diffusional cross-section area determined as the product of the smallest and the next smallest dimensions measured by using the Stuard type molecular model (CPK model) (11).

Figure 11 shows the membrane performance of a carbonized phenolic resin and zeolite Y membrane (10) for the benzene-cyclohexane system together with those of polymer membranes (11). The performance of the microporous inorganic membranes is far superior than any polymeric membrane previously reported for both pervaporation and vapor permeation. Separation of these systems is difficult by conventional distillation because the components form close boiling point mixtures. Although azeotropic distillation and extractive distillation are presently used in industry, these processes suffer from complexity and high energy consumption. It may, therefore, be concluded that carbon membranes provide a promising new class of materials for the separation of organic liquid mixtures.

Acknowledgements

This work was supported by Grand-in-Aid for Scientific Research from the Ministry of Education, Culture, Sports, Science and Technology of Japan.

References

1. Stern, S. A.; Koros, W. J. *Chimie Nouvelle* **2000**, *18*, 3201-3215.
2. Hsieh, H. P. *Inorganic Membranes for Separation and Reaction*; Membrane Science and Technology Series 3; Elsevier: Amsterdam, 1996; p.65.
3. *Fundamentals of Inorganic Membranes Science and Technology*; Burggraaf, A. J.; Cot, L. (Eds.); Membrane Science and Technology Series 4; Elsevier: Amsterdam, 1996; p.312, 354, 546.

4. *Recent Advances in Gas Separation by Microporous Ceramic Membranes*; Kanellopoulos, N. K., Ed.; Membrane Science and Technology Series 6; Elsevier: Amsterdam, 2000; p.323, 473.
5. Kita, H.; Maeda, H.; Tanaka, K.; Okamoto, K. *Chem. Lett.* **1997**, *2*, 179-180.
6. Okamoto, K.; Kawamura, S.; Yoshino, M.; Kita, H.; Hirayama, Y.; Tanihara, N.; Kusuki, Y. *Ind. Eng. Chem. Res.* **1999**, *38*, 4424-4432.
7. Okamoto, K.; Yoshino, M.; Noborio, K.; Maeda, H.; Tanaka, K.; Kita, H. In *Membrane Formation and Modification*; Pinnau, I.; Freeman, B. D. (Eds.); ACS Symposium Series 744, American Chemical Society: Washington, DC, 2000; p.314-329.
8. Zhou, W.; Yoshino, M.; Kita, H.; Okamoto, K. *Ind. Eng. Chem. Res.* **2001**, *40*, 4801-4807.
9. Kita, H.; Asamura, H.; Tanaka, K.; Okamoto, K. In *Membrane Formation and Modification*; Pinnau, I.; Freeman, B. D. (Eds); ACS Symposium Series 744, American Chemical Society: Washington, DC, 2000; p.330-341.
10. Kita, H.; Fuchida, K.; Horita, T.; Asamura, H.; Okamoto, K. *Separation and Purification Technology*, **2001**, *25*, 261-268.
11. Okamoto, K.; Wang, H.; Ijyuin, T.; Fujiwara, S.; Tanaka, K.; Kita, H. *J. Membr. Sci.* **1999**, *157*, 97-105.

Chapter 15

Novel Nanostructured Polymer–Inorganic Hybrid Membranes for Vapor–Gas Separation

Zhenjie He, Ingo Pinnau, and Atsushi Morisato

Membrane Technology and Research, Inc., 1360 Willow Road, Suite 103, Menlo Park, CA 94025

Novel polymer/inorganic hybrid membranes exhibit extraordinary gas separation properties for vapor/gas separation applications. In contrast to conventional polymer/filler systems, the incorporation of nano-sized, nonporous inorganic fillers in rigid, high-free-volume, glassy polymers simultaneously increased vapor permeability and vapor/gas selectivity significantly. For example, a hybrid membrane of poly(4-methyl-2-pentyne) [PMP] containing 22 vol% TS-530 fumed silica had a mixed-gas n-butane permeability of 25,000 Barrer, four times higher than that of pure PMP. In addition, the n-butane/methane selectivity increased from 13 for pure PMP to 21 for the silica-filled PMP membrane. The enhanced gas separation properties of polymer/filler hybrid membranes result from an increase of the free volume in the polymer matrix due to disruption of polymer chain packing induced by the filler. The general applicability of this approach was demonstrated with other nano-sized inorganic filler/polymer hybrid systems, such as PMP/carbon black and Teflon AF 2400/fumed silica.

© 2004 American Chemical Society

During the past 20 years, most membrane-based gas separation applications have involved the separation of permanent gases, such as nitrogen from air; carbon dioxide from methane; and hydrogen from nitrogen, argon, or methane. However, a significant potential market lies in separating mixtures containing condensable gases. The separation of organic monomers from nitrogen, C_{2+} hydrocarbons from methane, and C_{2+} hydrocarbons from hydrogen are all processes of considerable industrial importance. For example, approximately 1% of the 30 billion lb/year of monomer used in polyethylene and polypropylene production is lost in the nitrogen vent streams from resin purge operations (*1*). Recovery of these monomers would save U.S. producers about $100 million/year. Similarly, in the production of natural gas, raw gas must often be treated to separate propane and higher hydrocarbons from the methane to bring the heating value and dew point to pipeline specification, and to recover the valuable higher hydrocarbons as chemical feedstock. Finally, most refineries produce off-gas streams that contain 20-30% hydrogen and 70-80% hydrocarbons. Currently, these streams are used as fuel or simply flared. It has been estimated that about 800 MMscfd of hydrogen in off-gas streams, with a potential value of $300 million, are lost annually in refineries in the United States (*2*).

Poly(1-trimethylsilyl-1-propyne) [PTMSP], a glassy, high-free-volume acetylene-based polymer, exhibits both the highest C_{2+} hydrocarbon/methane and C_{2+} hydrocarbon/hydrogen selectivity and the highest C_{2+} hydrocarbon permeability of any known polymer (*3,4*). However, a major disadvantage of PTMSP as a membrane material for organic-vapor/permanent-gas separation applications is its very limited solvent resistance. PTMSP is soluble in essentially all liquid hydrocarbons, which prohibits its use in most industrial applications. An alternative acetylene-based polymer, poly(4-methyl-2-pentyne) [PMP] is stable against condensed hydrocarbon vapors, but its organic-vapor/permanent-gas selectivities and permeabilities are lower than those of PTMSP (*5,6*).

In this study, the gas permeation properties of PMP films containing nonporous, nano-sized, fumed silica fillers were evaluated. Incorporation of these ultrafine, dense fillers in the glassy polymer matrix leads to an unexpected increase in permeability as well as vapor/gas selectivity (*7,8*).

Background

Gas and vapor permeation through nonporous polymer membranes is usually rationalized by the solution/diffusion model. This model assumes that the gas phases on either side of the membrane are in thermodynamic equilibrium with their respective interfaces, and that the interfacial sorption process is rapid compared with the rate of diffusion through the membrane. Thus, the rate-limiting step is diffusion through the polymer membrane, which is governed by Fick's law of diffusion. In simple cases, Fick's law leads to the equation:

$$J = \frac{D \cdot S \cdot \Delta p}{l} \tag{1}$$

where J is the volumetric gas flux (cm³(STP)/cm²·s), D is the diffusion coefficient of the gas or vapor in the membrane (cm²/s), l is the membrane thickness (cm), S is the solubility coefficient (cm³(STP)/cm³·cmHg), and Δp is the pressure difference across the membrane (cmHg).

The gas permeability, P, is usually expressed in Barrers (1 Barrer = 1 x 10⁻¹⁰ cm³(STP)·cm/cm²·s·cmHg), and is given by:

$$P = D \times S \qquad (2)$$

For a given membrane material and two gases, A and B, the ideal selectivity, $\alpha_{A/B}$ is defined as the ratio of the permeability coefficients of the gases:

$$\alpha_{A/B} = \frac{P_A}{P_B} = \frac{D_A}{D_B} \times \frac{S_A}{S_B} \qquad (3)$$

The membrane selectivity is the product of two terms, the ratio of the permeant diffusion coefficients and the ratio of the permeant solubility coefficients. The ratio of diffusion coefficients (D_A/D_B) is referred to as the diffusivity selectivity, reflects the ability of the polymer to differentiate molecules of different sizes. The solubility selectivity, (S_A/S_B), is the ratio of the solubility coefficients of the gases, reflecting the relative condensibilities of the gases. Solubility coefficients typically increase with molecular size, because large molecules are normally more condensable, and hence more soluble in polymers than small molecules. The balance between the diffusivity selectivity and the solubility selectivity determines whether a membrane material is permanent-gas-selective or organic-vapor-selective. For conventional glassy polymers, a tradeoff relationship exists between the gas selectivity and permeability, that is, the higher the selectivity, the lower the gas permeability and vice versa (9). Conventional glassy polymers are permanent-gas/organic-vapor selective, because the diffusivity selectivity is much larger than the solubility selectivity. On the other hand, rubbery materials, such as polydimethylsiloxane [PDMS], and rigid, high-free-volume, glassy PTMSP and PMP are more permeable to large organic vapors. In the separation of organic vapors from permanent gases, this permeability/selectivity tradeoff relationship does not exist (10). An example of this behavior, the relationship between n-butane permeability and n-butane/methane selectivity for several polyacetylenes and PDMS, is shown in Figure 1. In this case, polymers having higher n-butane permeability also exhibit higher n-butane/methane selectivity. This behavior is typically for any membrane process that involves the separation of large, organic vapors from small, permanent gases in which the larger penetrant is the more permeable mixture component. Because the diffusion coefficient of the large molecule will always be lower than that of the smaller penetrant, the diffusivity selectivity, D_{vapor}/D_{gas}, will always be less than one. Therefore, advanced polymers with improved selectivity for organic-vapor/permanent-gas separation should have a diffusivity selectivity as close to one as possible. For rubbery polymers this can be achieved by increasing the chain mobility, whereas glassy polymers must have high free volume coupled with large inter- and intrachain spacing. PTMSP, a high-free-volume, glassy polymer exhibits the highest gas permeability and highest n-butane/methane and n-butane/hydrogen selectivities of any known polymer. These properties derive directly from its high free volume, interconnectivity of free volume elements, and large interchain spacing. However, PTMSP has very limited solvent resistance in hydrocarbon environments. An alternative, high-free-volume, glassy polymer, PMP, has much better solvent resistance than PTMSP, but its organic-vapor/permanent-gas selectivity and organic vapor permeability are lower than those of PTMSP. In this work, we investigated the possibility of enhancing the gas separation performance of high-free-volume PMP and Teflon AF2400 (Du Pont, Wilmington, DE) by incorporating nano-sized filler particles, such as fumed silica or carbon black, into the polymer matrix.

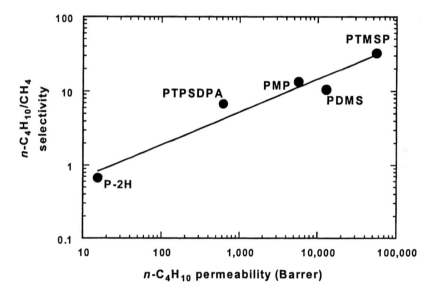

Figure 1.Mixed-gas n-butane/methane selectivity as a function of n-butane permeability. PMP: poly(4-methyl-2-pentyne); PTMSP: poly(1-trimethylsilyl-1-propyne); PDMS: polydimethylsiloxane; PTPSDPA: poly(1-phenyl-2-[p-triisopropylsilyl)phenyl]acetylene); P-2H: poly(2-hexyne).Feed: 2 vol% n-C_4H_{10}/98 vol% CH_4; feed pressure: 150 psig; permeate pressure: 0 psig; temperature: 25°C

Experimental

Polymer Synthesis

Poly(4-methyl-2-pentyne) [PMP] was synthesized following the procedure of Pinnau and Morisato (*11*). The monomer, 4-methyl-2-pentyne, was purchased from Lancaster (Windham, NH). The monomer was dried over calcium hydride for 24 hours. The catalysts, niobium pentachloride (NbCl$_5$) and triphenyl bismuth (Ph$_3$Bi) were purchased from Aldrich Chemicals and used without further purification. A solution of 1.32 g of NbCl$_5$ and 2.14 g Ph$_3$Bi in 188 ml cyclohexane was stirred at 80°C for 10 minutes under dry nitrogen. Thereafter, a monomer solution of 20 g 4-methyl-2-pentyne in 28 ml cyclohexane was added dropwise to the catalyst solution, and the mixture was reacted for 4 hours at 80°C. The viscosity of the solution increased very rapidly. The highly viscous polymer solution was poured into a large amount of methanol, filtered to recover the polymer, and dried under vacuum. The polymer was dissolved in cyclohexane and reprecipitated twice from methanol to remove excess monomer, oligomers, and catalysts. The yield was 87%.

Polymer Film Preparation

Isotropic, dense films of PMP were prepared by casting from a cyclohexane solution (2 wt% polymer) onto a glass plate. The films were dried at ambient conditions for 24 hours and then under vacuum at 80°C for three days. To ensure that the films were completely solvent-free, the films were removed from the oven and periodically weighed on an analytical balance. The thicknesses of the films were determined with a precision micrometer. Film samples with thicknesses of 20 to 50 μm were used for the permeation measurements.

Several different silica types (CAB-O-SIL, Cabot Corp., Tuscola, IL) were used in this study; the properties of these materials are listed in Table I. The L-90, M-5 and EH-5 fumed silica which have an -OH surface group, are hydrophilic. The TS-530 type fumed silica is modified with a -Si(CH$_3$)$_3$ surface group and is hydrophobic. As well as differences in the surface group chemistry, the various fumed silica types have different particle diameters. The size of the silica particles is an important parameter in the functionality of polymer/filler hybrid membranes, as shown later. Silica-filled PMP films were made by dispersing 10-45 wt% (based on polymer) ultrafine fumed silica particles in the PMP solution. The dispersion was mixed in a high-speed blender for 10 minutes and then cast immediately onto a glass plate. The silica-filled PMP membranes were dried as described above. The film thicknesses of the silica-filled PMP membranes were about 50 to 200 μm.

Carbon black-filled PMP films were made by the same method as the silica-filled PMP films.

Another high-free-volume glassy polymer Teflon AF 2400, a copolymer containing 87 mol% 2,2-bistrifluoromethyl-4,5-difluoro-1,3-dioxole and 13 mol% tetrafluoroethylene, obtained from Du Pont (Wilmington, DE), was also blended with fumed silica to form nanostructured hybrid membranes.

Table I. Physical properties of filler types used to prepare nanostructured PMP and Teflon AF 2400 membranes

Filler Type	Surface Groups	Surface Area (m^2/g)	Particle Size (nm)
L-90	-OH	90	30
M-5	-OH	200	14
EH-5	-OH	380	7
TS-530	$-Si(CH_3)_3$	210	13
Carbon black		1500	12

Pure-Gas Permeation Measurements

The pure-gas permeation properties of the blend membranes were measured at 25°C with nitrogen, oxygen, methane, hydrogen, and carbon dioxide at a feed pressure of 50 psig and atmospheric permeate pressure (0 psig). The experiments were performed using the constant pressure/variable volume method. For comparison, a pure, unfilled PMP film was tested under the same conditions.

Mixed-Gas Permeation Measurements

The mixed-gas permeation properties of pure polymer and polymer/filler hybrid films were determined with a gas mixture comprising 2 vol% n-butane/98 vol% methane. The experiments were also performed with the constant pressure/variable volume method. The feed pressure in all experiments was 150 psig; the permeate pressure was atmospheric (0 psig). The compositions of feed, residue, and permeate were determined with a gas chromatograph equipped with a thermal conductivity detector. The stage-cut, that is, the permeate to feed flow rate was always less than 1%. Under these conditions, the residue composition was essentially equal to the feed composition, and concentration polarization effects were negligible. PMP and silica-filled PMP films were also tested with a 50 vol% propane/50 vol% hydrogen mixture.

Results and Discussion

Effect of Filler on Pure-Gas Permeation Properties of PMP/Silica Hybrid Membranes

Nonporous fillers usually function as barriers to gas transport in polymers. The obstructive behavior of nonporous fillers to gas transport in polymers has been used to increase the barrier properties of polymers in packaging applications (12,13). Adding impermeable fillers to conventional polymers leads to a reduction in gas permeability by reducing the fraction of the membrane through which diffusion can occur and by increasing the average diffusion path length required for a gas to permeate the membrane (14-16). The reduction in gas permeability by addition of nonporous, impermeable fillers can be predicted by the well-known Maxwell model (14). This model predicts that for membranes containing dispersed, impermeable, spherical fillers, the permeability of the membrane, P_b, can be expressed as:

$$P_b = P_c \left(\frac{2 - 2\phi}{2 + \phi} \right) \tag{4}$$

where P_c is the permeability of the pure polymer and ϕ is the volume fraction of the filler. The Maxwell model predicts that the permeability of the polymer/filler hybrid membrane decreases by increasing the filler content.

Filled PMP membranes were prepared using a surface-modified, hydrophobic silica type, TS-530, with silica contents of 5.7, 10, 19, 22, and 25 vol%. Figure 2 shows the pure-gas nitrogen permeability of PMP and TS-530 filled PMP as a function of the filler content. The prediction by the Maxwell model for the permeability of the hybrid system is also shown for comparison. The data in Figure 2 show surprisingly that the nitrogen permeability of the silica-filled PMP film increases significantly as the filler content increases. The nitrogen permeability of PMP with 25 vol% TS-530 is three times that of pure PMP. Contrary to this unexpected increase, the calculated permeability of 25 vol% TS-530 filled PMP, using equation 4, is only 67% of that of pure PMP. This unusual increase in gas permeability is in singular contrast to the behavior of conventional polymer/filler hybrid membranes (14-16).

The permeabilities of silica-filled PMP for oxygen, methane, hydrogen, and carbon dioxide as a function of filler content are shown in Figure 3. The data show the same trend for all gases tested, that is, the gas permeabilities increase dramatically by increasing the filler content in the hybrid PMP membranes.

The decrease in gas permeability in conventional polymer/nonporous filler systems is due to a decrease in the fraction of membrane through which permeation

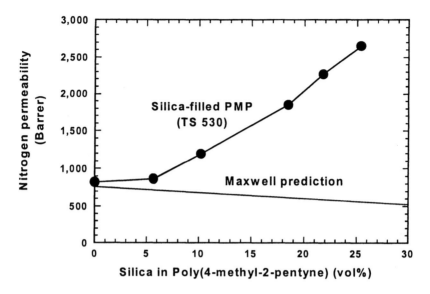

Figure 2. Pure-gas nitrogen permeability of filled PMP films as a function of nano-sized filler content. Feed pressure: 50 psig; permeate pressure: 0 psig; temperature: 25°C.

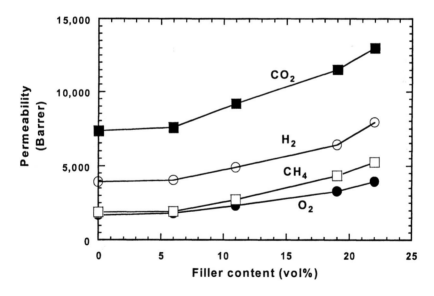

Figure 3. Pure-gas permeability of filled PMP films as a function of filler content. Feed pressure: 50 psig; permeate pressure: 0 psig; temperature: 25°C.

can occur and an increase in the tortuous diffusion path around the filler particles. Flexible rubbery materials, such as polydimethylsiloxane or natural rubber, can pack efficiently around the filler particles, which provide an impermeable barrier to gas transport. On the other hand, PMP is an extremely rigid, high-free-volume glassy polymer, in which the polymer chains pack very inefficiently. The nano-sized silica filler particles inhibit the ability of the rigid PMP polymer chains to pack in their equilibrium state. As a result, the addition of nano-sized fumed silica to PMP resulted in a disruption of polymer chain packing, leading to an increase in free volume and permeability in the hybrid system. This hypothesis is illustrated schematically in Figure 4.

Unfilled PMP   **Silica-particle-filled PMP**

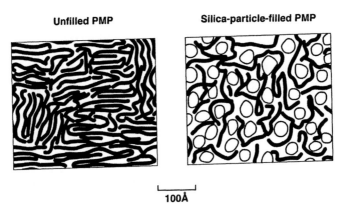

100Å

Figure 4. Schematic representation of PMP and silica-filled PMP. The strings represent PMP polymer chains, and the circles represent silica particles.

The permeability increase in PMP/silica hybrid membranes is coupled with a decrease in gas selectivity. Figure 5 shows the pure-gas oxygen/nitrogen and carbon dioxide/methane selectivities of PMP and PMP/TS-530 silica membranes as a function of the filler content. As the filler content in PMP increases, the oxygen/nitrogen and carbon dioxide/methane selectivities of the membranes decrease. The addition of impermeable, nano-sized fumed silica to PMP increases the free volume in the polymer matrix, which leads to increased gas diffusion coefficients. However, the diffusion coefficients for large molecules, such as nitrogen and methane, increase more relative to those of small molecules, such as oxygen and carbon dioxide. Consequently, oxygen/nitrogen and carbon dioxide/methane selectivities decrease with increased filler content. The decrease in selectivity seems to imply the existence of macroscopic defects in the hybrid membrane. However, the results of our mixed-gas experiments show that the selectivity decrease is not a result of gross defects and/or contribution of Knudsen diffusion, as discussed below.

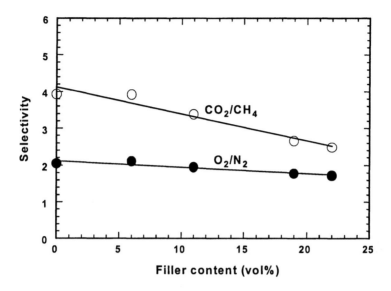

Figure 5. Pure-gas selectivity of filled PMP films as a function of silica filler content (TS-530). Feed pressure: 50 psig; permeate pressure: 0 psig; temperature: 25°C.

Effect of Filler Type on Mixed-gas Permeation Properties of PMP/Silica Hybrid Membranes

Mixed-gas permeation experiments provide a simple, efficient method to prove that incorporation of nano-sized filler particles to rigid, high-free-volume, glassy polymers can lead to molecular disruption of polymer chain packing. The mixed-gas permeation properties of silica-filled (13 vol%) PMP films are compared to those of a pure PMP film for a 2 vol% *n*-butane/98 vol% methane mixture in Figure 6. The *n*-butane/methane selectivity of pure PMP was 13 coupled with a *n*-butane permeability of about 6,000 Barrer. The addition of hydrophilic (L-90, M-5, and EH-5) fillers to PMP increased both the *n*-butane/methane selectivity and *n*-butane permeability significantly.

The type and fractional loading of filler particles in the polymer matrix affect the membrane permeability and selectivity. For example, at equivalent filler loading, the *n*-butane permeability and *n*-butane/methane selectivities increase significantly as the average filler particle size decreases. Results illustrating this effect are shown in Figure 6 for PMP films containing silica particles of different sizes. At the same filler loading, the number of particles present is inversely proportional to the cube of the particle diameter. Consequently, a large number of small particles disrupt polymer chain packing more efficiently than a few large particles. The largest increases in *n*-butane permeability and *n*-butane/methane selectivity were observed for the PMP film containing 7 nm EH-5 filler. The *n*-butane/methane selectivity for a PMP film containing 13 vol% hydrophilic silica EH-5 was 26 coupled with a *n*-butane permeability of 19,000 Barrer. Thus, the *n*-butane/methane selectivity of the EH-5 filled PMP film increased two-fold and the *n*-butane permeability increased more than three-fold over that of a pure PMP membrane. The PMP/silica hybrid membranes clearly show a significantly better performance for *n*-butane/methane separation than the pure PMP membrane.

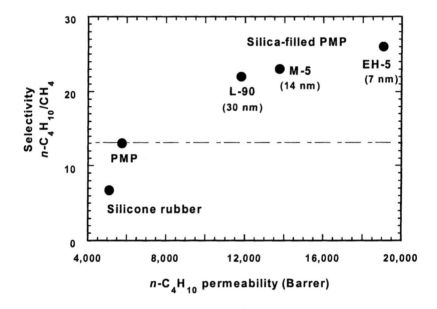

Figure 6. Mixed-gas permeation properties of PMP and PMP/silica (13 vol%) hybrid films determined with a 2 vol% n-butane/98 vol% methane mixture. Feed pressure: 150 psig; permeate pressure: 0 psig; temperature: 25°C.

In additional mixed-gas permeation experiments, the properties of PMP and silica-filled PMP films were evaluated with a 50 vol% propane/50 vol% hydrogen mixture. Figure 7 compares the mixed-gas propane/hydrogen selectivity and propane permeability of 13 vol% silica-filled PMP with those of pure PMP and rubbery PDMS. Again, with the addition of filler, propane/hydrogen selectivity as well as propane permeability of silica-filled PMP increase significantly. The largest increases in propane/hydrogen selectivity and propane permeability occurred with

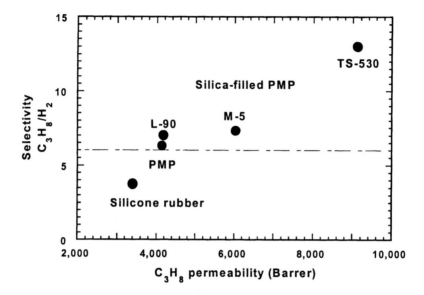

Figure 7. Mixed-gas permeation properties of PMP and PMP/silica (13 vol%) hybrid films determined with a 50 vol% propane/50 vol% hydrogen mixture. Feed pressure: 100 psig; permeate pressure: 0 psig; temperature: 25°C.

the hydrophobic filler TS-530. The PMP/TS-530 hybrid membrane showed a two-fold increase in selectivity and permeability over that of a pure PMP film.

Effect of Filler Content on Mixed-Gas Permeation Properties of PMP/Silica Hybrid Membranes.

The effect of silica content on the permeation properties of filled PMP films was determined at 25°C with a 2 vol% n-butane/98 vol% methane gas mixture. The filler content was varied between 6 and 22 vol%, using a hydrophobic fumed silica type (TS-530). The effect of increasing the silica filler content in the PMP film on the permeation properties is shown in Figure 8.

The data show that the n-butane permeability increased four-fold, from about 6,000 Barrer for the pure PMP film to about 25,000 Barrer for a PMP film containing 22 vol% fumed silica TS-530. More importantly, the mixed-gas n-butane/methane selectivity increased from 13 for a pure PMP film to 21 for the filled PMP film.

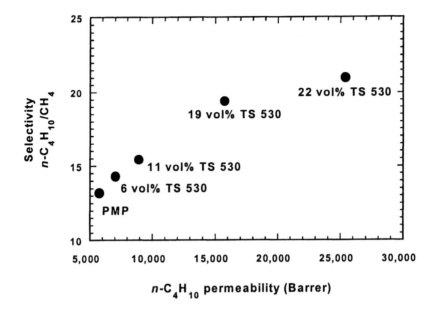

Figure 8. Mixed-gas permeation properties of PMP and PMP/silica hybrid films as a function of filler content. Feed composition: 2 vol% n-butane/98 vol% methane; feed pressure: 150 psig; permeate pressure: 0 psig; temperature: 25°C.

To investigate the mechanism and demonstrate the general applicability of nano-sized filler/polymer hybrid membranes, we evaluated the permeation properties of different polymer/filler types. Additional blend membranes were made from (i) PMP and carbon black (individual particle size of 12 nm) and (ii) Teflon AF 2400 (a rigid, high-free-volume, perfluorinated glassy polymer) and fumed silica (TS-530). Both types of nanostructured hybrid membranes were tested with a 2 vol% n-butane/98 vol% methane mixture under the same test conditions as described above. The effect of various filler types on different, high-free-volume glassy polymers, illustrated as the ratio of permeability and selectivity of filled membranes to the permeation properties of the pure polymer membranes, is shown in Figure 9.

The data in Figure 9 show clearly that for different high-free-volume, glassy polymer/nano-sized filler hybrid membranes, the effect of fillers on the gas

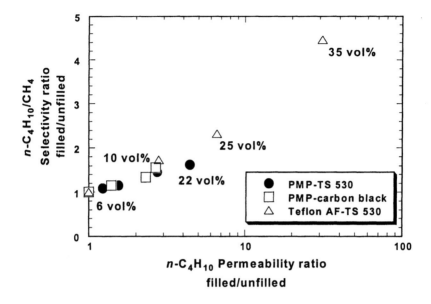

Figure 9. Mixed-gas n-butane/methane selectivity ratio as a function of the n-butane permeability ratio of filled/unfilled membrane for different polymer/filler hybrid membranes. Feed composition: 2 vol% n-butane/98 vol% methane; feed pressure: 150 psig; permeate pressure: 0 psig; temperature: 25°C.

permeation properties follows the same trend. As the filler content in the hybrid membranes increases, the n-butane permeability increases and is accompanied by a dramatic increase in n-butane/methane selectivity. It is noteworthy that a pure Teflon AF 2400 membrane is methane/n-butane selective. However, the addition of fumed silica reverses the selectivity of Teflon AF 2400, that is, the hybrid membrane becomes n-butane/methane selective. The n-butane/methane selectivity of a Teflon AF 2400/35 vol% TS-530 membrane is 4.5 times higher than that of the pure Teflon AF 2400 membrane. More dramatically, the n-butane permeability of the filled Teflon membrane increases 30-fold. Our results indicate that any nano-sized, nonporous, inorganic filler incorporated into high-free-volume, glassy polymers can function as a spacer material that disrupts polymer chain packing. To achieve enhanced vapor permeability and increased organic-vapor/permanent-gas selectivity, the size of the filler particles should be similar or smaller than the chain dimensions of the polymer (<50 nm).

The accomplishment of our study is summarized in Figure 10. The results demonstrate clearly that silica-filled PMP membranes show significantly improved mixed-gas separation properties compared to those of unfilled PMP. The addition of nano-sized, nonporous fumed silica in PMP pushes the selectivity and the permeability of the hybrid membranes towards that of PTMSP.

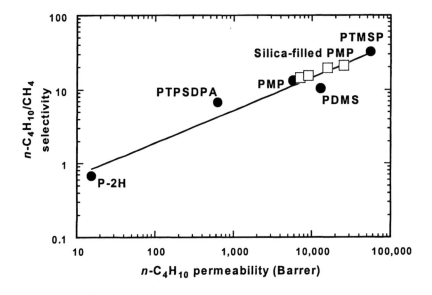

Figure 10. Mixed-gas n-butane/methane selectivity as a function of n-butane permeability for various polymers and TS-530-filled PMP. (□) 6, 11, 19, 22 vol% TS-530 filled PMP.

Conclusions

In previous studies, addition of nonporous, inorganic fillers to rubbery or low-free-volume, glassy polymers resulted in a decrease in the permeability of the hybrid membrane (*14-16*). By choosing rigid, high-free-volume, glassy polymers, we observed completely opposite behavior. In our study, incorporation of nonporous, nano-sized fillers to the matrix of a high-free-volume glassy polymer, such as PMP, resulted in unexpected increases in the organic-vapor/permanent-gas selectivity and organic-vapor permeability. It is suggested that the permeability increase is due to an increase of the free volume in the polymer matrix by disruption of the polymer chain packing. The general applicability of this approach was demonstrated using different polymer and filler types. Our study shows that the organic vapor/permanent-gas separation properties of rigid, high-free-volume, glassy polymers can be improved significantly by molecular manipulation of polymer chain packing by incorporation of nano-sized fillers.

233

Acknowledgement

t number DMI-9760767.

References

1. EPA Guideline Series, "Control of Volatile Compounds from Manufacture of High Density Polyethylene, Polypropylene, and Polystyrene Resins," PB 84 134600 (November 1984).
2. Anand, M. "Novel Selective Surface Flow Membranes," Proceedings of the Industrial Waste Program, Department of Energy, Albuquerque, NM, May 17-19, 1994.
3. Pinnau, I; Toy, L.G. *J. Membrane Sci.* **1996**, *116*, 199.
4. Pinnau, I.; Casillas, C.G.; Morisato, A.; Freeman, B.D. *J. Polym. Sci. Part B: Polym. Phys.* **1996**, *34*, 2613.
5. Morisato, A.; Pinnau, I. *J. Membrane Sci.* **1996**, *121*, 243.
6. Morisato, A.; He, Z.; Pinnau, I. in *Polymer Membranes for Gas and Vapor Separation: Chemistry and Materials Science*; Freeman, B.D.; Pinnau, I., Eds.; ACS Symposium Series 733; American Chemical Society: Washington, DC, 1999; pp 56-67.
7. Pinnau, I; He, Z. *U.S. Patent 6,316,684*, 2001.
8. Merkel, T. C.; Freeman, B. D.; Spontak, R. J.; He, Z.; Pinnau, I.; Meakin, P.; Hill, A. J. *Science*, **2002**, 296, 519.
9. Robeson, L. M. *J. Membrane Sci.* **1991**, 62, 165.
10. Freeman, B. D.; Pinnau, I. *TRIP*, 1997, 5, 167.
11. Pinnau, I.; Morisato, A. *U.S. Patent 5,707,423,* 1998.
12. Koros, W. J. *Barrier Polymers and Structures*; Koros, W. J., Eds.; ACS Symposium Series 423; American Chemical Society: Washington, DC, 1990; pp 1.
13. Bissot, T. C. *Barrier Polymers and Structures*; Koros, W. J., Eds; ACS Symposium Series 423; American Chemical Society: Washington, DC, 1990; pp 225.
14. Barrer, R.M. in *Diffusion of Polymers*, Crank, J; Park, G.S. (Eds.), Academic Press, New York, 1968, pp 165-217.
15. Van Amerongen, G.J., *Rubber Chem. Tech.*, **1964**, 37, pp 1065-1152.
16. Barrer, R. M.; Barrie, J. A.; Raman, N. K. *Polymer*, **1962**, 3, 605.

Chapter 16

Molecular Chemistry Control of Hybrid Morphology and Transport Properties in Polyimide–Silica Membranes

Chris J. Cornelius[1], Eva Marand[2], Pavla Meakin[3], and Anita J. Hill[3,4]

[1]Sandia National Laboratories, P.O. Box 5800, Albuquerque, NM 87185–0710
[2]Department of Chemical Engineering, Virginia Polytechnic and State University, Blacksburg, VA 24061–0211
[3]CSIRO Manufacturing and Infrastructure Technology, Private Bag 33, South Clayton, MDC Victoria 3169, Australia
[4]School of Chemistry, Monash University, Wellington Road, Clayton, Victoria 3800, Australia

Polyimide-silica hybrid membranes were characterized by dynamic mechanical thermal analysis, differential scanning calorimetry, positron annihilation lifetime spectroscopy, and gas permeability measurements in order to investigate the role of inorganic-organic interactions on chain mobility, free volume, and transport properties. Current composite theories neglect inorganic-organic interactions (or interfacial phenomena); however, the present work illustrates that this interaction is critical to the transport properties of these hybrid materials. Transmission electron microscopy of samples with low levels of alkoxysilane shows well-dispersed inorganic domains of sizes ranging from 25 nm to 100 nm. These nanodomains are shown to increase the glass transition temperature of the polymer, the free volume, and the permeability.

© 2004 American Chemical Society

Introduction

The morphology of hybrid inorganic-organic materials, ranging from nano- to micro-structures, impacts their transport properties. Tailoring morphology and transport properties of inorganic-organic membranes has been accomplished using the (i) chimie douce sol-gel methodology (*1,2*), (ii) mixed matrix approach with zeolites in rubbery or glassy polymers (*3*), and (iii) nano-particle approach (*4*).

Rule of mixture models (*5-7*) and molecular simulations (*8,9*) can be used to predict flux and separation properties for hybrid materials based on a perfect, defect-free interface (e.g. no influence of inorganic phase on organic phase or vice versa). Furthermore, simulation work has included the aspect ratio of the inorganic phase as a variable (*9,10*). However, models proposed to account for filler (*5*) and tortuosity (*11*) do not account for the presence of an interface/interphase due to interaction of organic and inorganic phases (polymer-particle interactions). Whilst these models tend to accurately describe macro-phase-separated materials, they have been unable to predict properties of nanocomposite materials (*4*). The surprising enhancement in flux and selectivity in nanocomposite membranes has been attributed to the effect of inorganic nanoparticles on the packing of nearby polymer chains (*4*).

Combinatorial chemistry and trained neural networks offer one route for expediently exploring the immense new area of multifunctional materials based on nano-sized inorganic and organic phases (*12,13*). In the quest for rapid materials design, one needs to verify predictions by limited experimentation and determine the key parameters of influence on the properties of interest. Examples of the parameters to be investigated for hybrid inorganic-organic materials include: (a) interfacial bonding, (b) particle morphology, (c) surface chemistry, (d) particle size, (e) particle loading, (f) particle dispersion, (g) local effects in the polymer (interphase), and (h) permeation characteristics.

In the present work we characterized a series of inorganic-organic materials based on 6FDA-6FpDA polyimide and silica. Previously, we reported the synthesis strategy used to obtain hybrid materials with mechanical integrity and varying morphology (*14*) and the effect that this synthesis approach has on gas permeability and selectivity (*2*). Results from these studies indicated that creation of accessible free volume via interaction of the inorganic and organic phases can lead to improvement in membrane performance. The present work explicitly investigates this result by measuring the ortho-positronium (oPs) accessible free volume using positron annihilation lifetime spectroscopy (PALS).

Previous PALS work in polyimide membranes has established the correlation between the oPs accessible volume and gas diffusion coefficient. Dolveck et al. (*15*) compared gas permeability and oPs annihilation parameters in polyimide membranes showing an increase in gas permeability with increasing relative free volume element size (τ_{oPs}) and concentration (I_{oPs}). Methyl grafting and heat treatment were used to vary the free volume and transport properties. Tanaka et al. (*16*) and Okamoto et al. (*17*) examined the correlation between oPs annihilation characteristics (τ_{oPs} and I_{oPs}) and gas diffusion coefficient (D) in 6FDA-based polyimides. Their results show a strong correlation between oPs accessible volume and log(D). The oPs accessible volume (f_{oPs}) is proportional to the average size of free volume element probed by oPs, v_{oPs}, multiplied by the relative number of elements I_{oPs}. This strong correlation in polyimides of similar chemical structure supports the notion that I_{oPs} is a relative measure of the concentration of free volume in the volume of material probed. Tanaka et al. (*16*) also examined the relationship between oPs accessible free volume and fractional free volume (FFV as defined by Bondi (*18*)). The result that D is correlated more closely to f_{oPs} than to FFV occurs presumably because FFV includes sites too small to be accessible to the penetrants used in the study (CO_2 and CH_4). Recently, Nagel et al. (*19*) correlated v_{oPs} and log(D) for H_2, He, N_2, O_2, CO_2, and CH_4 in poly(amide imides), poly(ester imides), and polyimides and also concluded that transport properties correlate much better with PALS free volume data than with free volume data estimated from group contribution methods.

Table I. Comparison of penetrant size and oPs probe size

Penetrant Size	oPs (20,22)	He	CO_2	O_2	N_2	oPS (21)	CH_4
van der Waals Radius (Å)	1.3	1.32	-	1.94	2.07	-	-
Kinetic Diameter (Å)	-	2.58	3.3	3.46	3.64	3.88	3.82

Table I gives a comparison of the van der Waals radii and kinetic diameters of the penetrants used in this study and includes the oPs probe (Bohr radius of 0.53Å (*20*), estimated kinetic diameter in polymers of 3.88Å (*21*)), which depending on method of calculation spans the typical penetrant size range. Greenfield and Theodorou (*22*) have used molecular simulations to suggest that penetrants of van der Waals radius from 1.2 to 2.25Å, and possibly greater, are probing the same characteristic free volume elements in glassy polymers. Molecular dynamics simulations and transition state theories of glassy polymers have suggested that the rate determining factor for penetrant diffusion is the time spent (on the order of 1 nanosecond) by the penetrant in a free volume element waiting for a neck to open to a neighboring element. The size (similar to typical penetrants) and dynamic nature of the oPs probe (typical lifetime of nanoseconds) accounts for the strong correlation between gas permeability and free volume element size as measured by PALS.

In the present work we used a thermally stable polyimide, 6FDA-6FpDA, with high gas selectivities as the matrix material and in-situ polymerization of silica nanodomains of various alkoxide chemistry in order to vary the nature of the inorganic-organic interface. A coupling agent, 3-aminopropyltriethoxysilane (APTEOS), was used to link the inorganic domains and the polymer matrix. The introduction of the inorganic domains into the final hybrid proceeds via sol-gel chemistry in the presence of a fully formed high molecular weight polyimide (14). The objective of the research is to improve polyimide membrane performance (flux and selectivity). Joly et al. (1) found that a hybrid tetramethoxysilane (TMOS) and PMDA-ODA polyimide showed improved gas separation properties over the pure polyimide. In that work the polyimide in the hybrid was 77% imidized and the TMOS domains were macro-phase-separated. The improvements in flux and selectivity were attributed to adsorption of polymer chains on the silica surface. Recent studies of hybrid or nanocomposite membranes (2,4,14,23,24) have linked the improvements in flux and selectivity to alterations in the free volume distribution of the polymer caused by the nano-sized inorganic domains.

Experimental

Materials. Details of the materials, synthesis and physical characterization of the 6FDA-6FpDA hybrid materials are outlined elsewhere (14). Briefly, in-situ cross-linking of a functionalized polyimide with a partially hydrolyzed alkoxide sol was used to produce the hybrid materials in this study. Polyimides were solution imidized using 4,4'-hexafluoroisopropylidene-diphthalic anhydride (6FDA), and controlled amounts of 4,4'-hexafluoroisopropylidene dianiline (6FpDA). Hybrid materials based on a 6FDA-6FpDA polyimide had pendant triethoxysilane groups present on the chain ends of the polyimide. The final molecular weight (M_w) of the polyimide was 63,000g/mol, as determined by GPC. Predetermined amounts of tetraethoxysilane (TEOS), phenyltrimethoxysilane (PTMOS), and methyltrimethoxysilane (MTMOS) were introduced into the polyimide matrix via sol-gel reactions. The alkoxysilane content in the hybrid material varied from 7.5 wt% to 22.5 wt% with respect to the total weight of the hybrid material. Samples designated as XL represent polyimides functionalized and cross-linked with 3-aminopropyltriethoxysilane. All membranes were cast from solution and heat-treated at 220°C. Figure 1 presents the hybrid synthesis scheme.

Gas Permeability. Gas permeation data for all materials studied were collected using the constant volume integral technique. The integral technique employs the time-lag method to measure the increase in pressure as a function of time and relate this gas accumulation to permeability, P. The gas transport properties for the series of hybrid inorganic-organic materials were evaluated from thin films having a nominal thickness of 80 μm. Gas permeation tests were initiated after the permeation equipment setup and film samples were degassed to a system pressure of 1-5 mtorr. Additionally, the entire permeation equipment was allowed to reach thermal equilibrium prior to testing. Gases employed in this

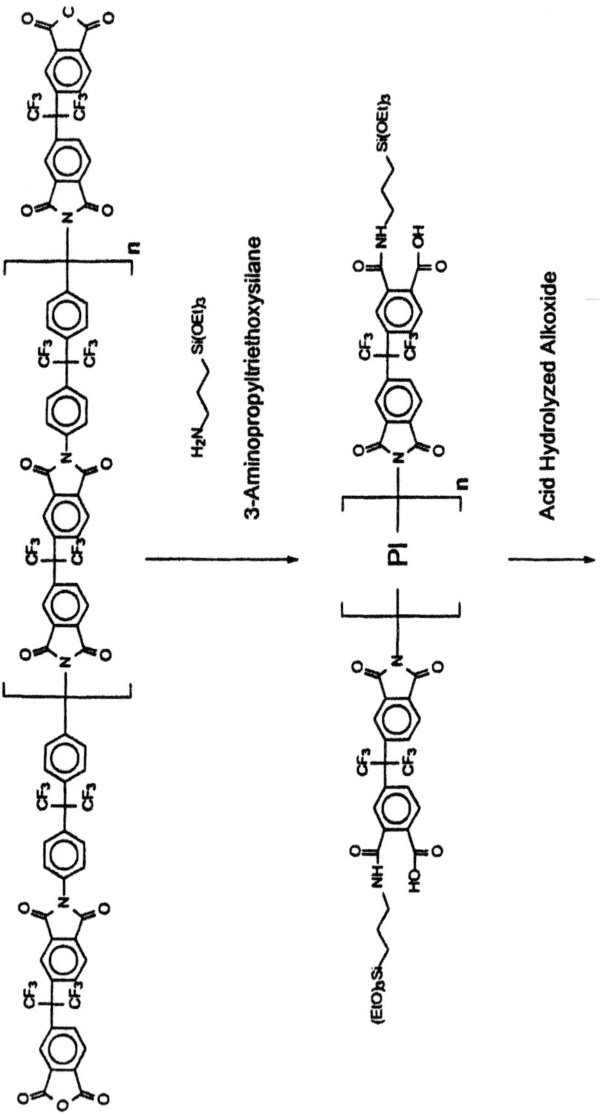

H2N
Si(OEt)3
3-Aminopropyltriethoxysilane
PI
(EtO)3Si
Acid Hydrolyzed Alkoxide

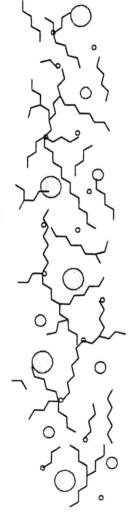

Figure 1. Hybrid synthesis scheme using a functionalized 6FDA-6FpDA polyimide and an acid hydrolyzed alkoxide.

240

study included He, O_2, N_2, CO_2, and CH_4 and had a purity of 99.999%. An absolute feed pressure of 4 atm was used in all permeation tests at 35°C.

Positron Annihilation. PALS measurements were made in air at room temperature using an automated EG&G Ortec fast-fast coincidence system. The timing resolution of the system was 240 ps determined using the prompt curve from a ^{60}Co source with the energy windows set to ^{22}Na events. Polymer films approximately 80 μm thick were stacked to a total thickness of 1 mm on either side of the 30 μCi ^{22}Na-Ti foil source. From three to eleven spectra for each sample were collected with each spectrum taking 1 hour to collect, and the results are the mean values for these spectra. The standard deviations reported are the population standard deviations for 3 to 11 spectra, and each spectrum consisted of approximately 1 million integrated counts. The PALS parameters for the sequential spectra did not vary as a function of contact with the radioactive source. The spectra were modeled as the sum of three decaying exponentials using the computer program PFPOSFIT. The shortest lifetime was fixed at 125 ps characteristic of para-positronium self-annihilation. No source correction was used in the analysis based on a fit for pure Al standards of 169 ps±2%, 99.3 ps ±0.3%, and 820 ps ± 0.7%. Only the oPs components (τ_3 or τ_{oPs} and I_3 or I_{oPs}) are reported as it is the oPs component that is related to annihilations in free volume elements of the polymer matrix. In general the PALS parameter τ_{oPs} (ortho-positronium pickoff lifetime) is associated with the average size of free volume elements, and the statistical weight of this component, I_{oPs}, is associated with the relative number of free volume elements. The free volume sites probed by PALS are of the order 2 to 10 Å in size, hence in hybrid films, inter- and intra-chain space as well as interfacial sites in this size range will be detected.

Transmission Electron Microscopy. Transmission electron microscope (TEM) images of 6FDA-6FpDA polyimides and hybrid materials were acquired from microtomed specimens on a Philips 420T electron microscope operated at 100 kV.

Density. Density measurements were made in iso-octane according to Archimedes' principle using a Mettler AJ100 analytical balance with a Mettler ME-33360 density determination kit. Sample mass ranged from 80 to 160 mg and all samples were dried under vacuum at 180°C prior to measurement.

Thermal Analysis. A Rheometric Scientific Mark IV dynamic mechanical thermal analyzer (DMTA) was used to measure the glass transition, T_g, in the polyimides and hybrid materials. Film samples having a length of 15 mm, a width of 4 mm, and a thickness of 80 μm were tested with a strain of 0.01%, a constant static force of 0.15N and a heating rate of 2°C min^{-1} in air. Data from the first heating scan are reported and the error in T_g is ± 0.5°C. A Perkin Elmer

Pyris 1 differential scanning calorimeter (DSC) was used to measure the T_g of pure TEOS and a TEOS-based 6FDA-6FpDA polyimide hybrid at a scan rate of 40°C min^{-1}.

Results and Discussion

Figure 2 shows the procedure for film formation. The post film formation heat-treatment was shown to reduce the amount of inorganic material in the hybrid. Table II compares the starting wt% alkoxide with the final wt% and the calculated vol% silica in the hybrids based on density. The amount of silica, the morphology of the silica domains, and the type of alkoxysilane are expected to affect the transport properties.

The morphology of the silica domains was examined by TEM. Table II summarizes TEM results reported previously (2,14). Nanodomains are typically defined as having dimensions ≤ 100nm. All of the 7.5 wt% alkoxide samples have uniformly distributed nanodomains as shown in Figure 3. Figure 4 illustrates the bimodal morphology for the 22.5 wt% MTMOS and PTMOS hybrids that consists of nanodomains (≤ 100 nm) and microdomains (≥ 1μm). Note that the 22.5 wt% TEOS hybrid contains nanodomains. Table III summarizes the measured permeability data for the range of penetrants. The critical volumes of these penetrants range from 57 to 99 cm^3/mol for He and CH$_4$, respectively. The permeability coefficients decrease on crosslinking. Figure 5 displays the measured decrease in permeability coefficient due to crosslinking. Crosslinking decreases the permeability coefficient for all penetrants by approximately 30 to 40%.

The APTEOS crosslinked polyimide, XL, will be used as the basis for comparison of the effect of alkoxide addition on penetrant permeability. Introduction of the various alkoxides increases the permeability coefficients for all penetrants at all levels of alkoxide addition. The exception is the PHe for the 7.5 wt% PTMOS hybrid. Figure 6 shows the trend of increasing PCH$_4$ due to alkoxide addition and illustrates that hybrid permeability depends on the content and nature of the alkoxide employed. It is evident from the permeability trends that the bimodal silica morphology observed via TEM does not hinder gas permeability, although hindrance may have been expected due to micron-sized filler and tortuosity effects (5,11).

Interestingly, the TEOS alkoxide, which forms nano-domains for all levels of addition, increases the permeability for additions up to 15 wt % (4.2 vol %), but then decreases permeability for 22.5 wt% (6.8 vol%). This initial increase and then decrease in permeability may be due to a competition between permeability enhancement due to inorganic-organic interaction and permeability suppression due to tortuosity effects at higher nanodomain content. This trend is consistent for all penetrants except He in the TEOS-based hybrids and indicates that there is an optimum concentration of nanodomains to enhance permeability in these hybrid systems.

Inorganic   ## Organic

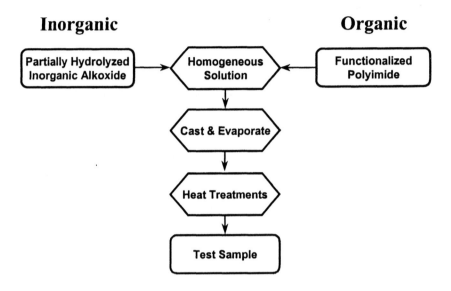

Figure 2. Schematic procedure for film formation.

TEOS   MTMOS   PTMOS

Figure 3. TEM images of the hybrid morphology of 6FDA-6FpDA polyimide-7.5 wt% alkoxide.

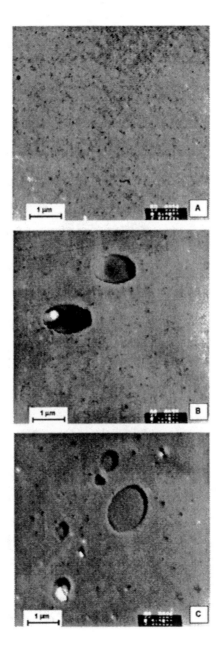

Figure 4. TEM images of the hybrid morphology of 6FDA-6FpDA polyimide-22.5 wt% alkoxide: (A) TEOS, (B) MTMOS, and (C) PTMOS.

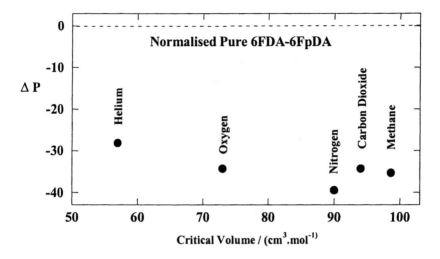

Figure 5. Decrease in permeability of APTEOS crosslinked polyimide (XL) relative to pure polyimide as a function of penetrant critical volume.

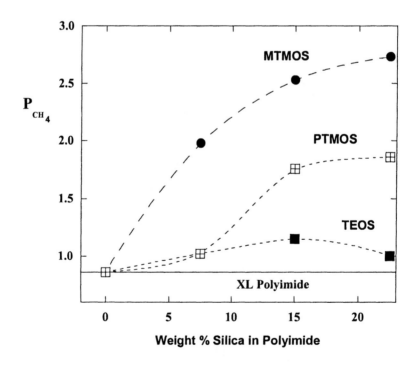

Figure 6. Increase in methane permeability of hybrid materials as compared to APTEOS crosslinked polyimide (XL).

Table II. Synthesis weight % silicate compared to final vol% silica in the hybrids, TEMs of silica domains, and measured and predicted densities

Synthesis wt % silicate	Silica content wt%	Silica content wt%	TEM observations	Exp. density (g/ml)	Cal. density (g/ml)
Pure polyimide	0	0	homogenous	1.477	-
APTEOS crosslinked polyimide XL	0	0	homogenous	1.454	-
7.5 wt% TEOS	2.74	1.99	≤ 25 nm silica domains	1.462	1.469
15 wt% TEOS	5.77	4.23	≤ 100 nm silica domains	1.477	1.463
22.5 wt% TEOS	9.15	6.77	≤ 100 nm silica domains	1.500	1.478
7.5 wt% MTMOS	4.34	6.18	≤ 50 nm silica domains	1.460	1.477
15 wt% MTMOS	8.99	12.5	Bimodal: 2mm + 25 nm silica domains	1.435	1.469
22.5 wt% MTMOS	14.0	19.1	Bimodal: 2mm + 25 nm silica domains	1.419	1.471
7.5 wt% PTMOS	5.66	6.4	100 nm silica domains	1.449	1.466
15 wt% PTMOS	11.6	13.0	<500 nm silica domains	1.440	1.476
22.5 wt% PTMOS	17.7	19.7	Bimodal: 1mm + <300 nm silica domains	1.426	1.481
TEOS		100		1.80	-
MTMOS	100	100		1.20	-
PTMOS	100	100		1.28	-

Density data are given in Table II. A 'rule of mixtures' can be used for a two component polymer/filler system in order to estimate the effective density of the polymer within the hybrid. If the silica domains alter the chain packing and free volume of the polymer, this should be reflected in the effective polymer density. The effective polymer density, $\rho_{polymer}$, may be estimated from the pure properties of the inorganic phase (it is assumed the inorganic phase is not altered by the presence of the organic phase) and the measured hybrid density as follows:

$$\rho_{polymer} = (\rho_{hybrid} - \phi_{inorganic}\,\rho_{inorganic})/\phi_{polymer} \qquad (1)$$

where $\rho_{inorganic}$ and ρ_{hybrid} are the densities for the pure silicate and the hybrid, respectively, and $\phi_{polymer}$ and $\phi_{inorganic}$ in the hybrid was estimated from Eq. 1 and compared to the pure polymer density measured by a gravimetric method. The results are compared in Table II. The uncertainty in the measured densities is 0.5%. The 'rule of mixtures' estimated densities of the polymer phase in these hybrid systems are approximately 1 to 2% denser than that measured for the APTEOS crosslinked polymer (XL). These results suggest that the presence of silica domains does not alter the packing of the polymer chains, especially in the bulk polymer far from an interface (2). However, as discussed by Merkel et al. (24), when the differences in calculated and measured densities are near to the uncertainty in the measurements, the result can imply that any change in polyimide free volume due to the presence of silica domains is subtle and not easily detected by the density method. In addition, because the calculation of FFV by group contribution theory is generally very sensitive to sample density measurements, a more sensitive method of probing the polymer and interphase chain packing is needed.

Table III: Permeability data for a range of penetrants in polyimides and hybrids

Synthesis wt % silicate	Vol %	PHe	PCO₂	PO₂	PN₂	PCH₄
Pure polyimide	0	120	49.2	12.9	2.71	1.33
APTEOS crosslinked polyimide XL	0	86.2	32.3	8.18	1.64	0.86
7.5 wt% TEOS	1.99	93.8	37.8	9.52	2.06	1.02
15 wt% TEOS	4.23	89.5	41.0	10.2	2.17	1.15
22.5 wt% TEOS	6.77	78.4	36.6	8.86	1.91	1.00
7.5 wt% MTMOS	6.18	118	61.3	14.8	3.37	1.98
15 wt% MTMOS	12.5	125	74.8	18.0	4.17	2.53
22.5 wt% MTMOS	19.1	124	77.9	18.7	4.39	2.73
7.5 wt% PTMOS	6.4	79.7	36.1	8.35	1.81	1.02
15 wt% PTMOS	13.0	91.2	52.9	12.2	2.66	1.76
22.5 wt% PTMOS	19.7	80.0	51.7	11.9	2.68	1.86

$P = 10^{-10}\,cm^3\,(STP)\cdot cm\,/\,cm^2\cdot s\cdot cmHg$

Alternative methods of probing the effect of silica domains on the packing and mobility of polymer chains, such as DMTA (14), indicate interfacial regions of higher T_g than the bulk polymer. Table IV shows the increase in T_g in the hybrids over that measured for the pure and crosslinked polyimide. Note that the greatest increases in T_g are seen for morphologies that consist of dispersed nanodomains of silica ≤ 100nm. Also note that although APTEOS crosslinking has been shown to reduce permeability as compared to the pure polyimide, there is no effect on the value of T_g as measured by DMTA; however, the breadth of the glass transition peak is greater in the crosslinked polyimide XL (*14*). Crosslinking occurs at the functionalized polyimide chain ends and would be expected to reduce the free volume associated with chain ends. This result suggests that the reduction in permeability is due to a change in penetrant accessible free volume.

The oPs probe was used to evaluate accessible free volume. PALS measurements are reported in Table IV. The PALS results indicate reduced free volume element size and number due to APTEOS crosslinking that may help explain the reduction in gas permeability over the range of penetrant sizes. Figure 7 shows the correlation between oPs lifetime, τ_{oPs}, and oxygen permeability, PO_2, for a range of polymers. The low permeability barrier polymer PVOH has an oPs lifetime of 1.4 ns that equates to an average free volume element radius of 2.23 Å, whilst the high permeability membrane polymer 6FDA-6FpDA polyimide has an oPs lifetime of 2.59 ns that equates to an average free volume element radius of 3.35 Å. It is worthy of note that alteration of the chain packing that causes a change in average free volume element radius of 1 Å results in three orders of magnitude variation in permeability for oxygen. Because the oPs lifetime in polymers is measurable to within 30 picoseconds (or 0.03 ns), PALS can detect even small variations in average free volume element size.

Table IV. PALS results and T_g results measured by DMTA

Synthesis wt % silicate	τ_{oPs} (ns) ±0.03	I_{oPs} (%) ±0.3	v_{oPs} ($\mathring{A}^3$) ±2	T_g (°C) ±1
Pure polyimide	2.59	15.4	158	322
APTEOS crosslinked polyimide XL	2.33	11.1	130	322
7.5 wt% TEOS	2.44	11.5	141	330
15 wt% TEOS	2.37	9.7	135	339
22.5 wt% TEOS	2.26	10.9	124	335
7.5 wt% MTMOS	2.54	14.9	152	329
15 wt% MTMOS	2.57	16.5	155	328
22.5 wt% MTMOS	2.68	17.0	166	328
7.5 wt% PTMOS	2.39	10.4	136	330
15 wt% PTMOS	2.35	10.5	132	327
22.5 wt% PTMOS	2.33	12.7	130	324

Construction of a plot similar to that presented by Tanaka et al. (*16*) is presented in Figure 8, which shows ln(P) for carbon dioxide, oxygen, and

248

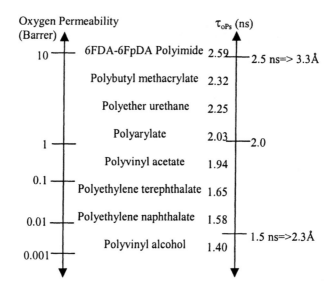

Figure 7. Schematic representing the correlation between oPs free volume element size and oxygen permeability for a range of polymers.

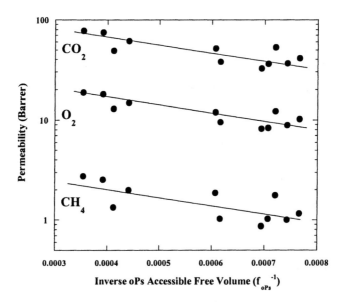

Figure 8. Plot of the logarithm of the permeability coefficients of various gases versus the reciprocal oPs accessible volume f_{oPs} for the polyimides and hydrids.

methane as functions of the oPs accessible free volume f_{oPs} where $f_{oPs} = v_{oPs}I_{oPs}$ for the polyimides and hybrids. The experimental relationship between permeability and oPs accessible volume can be used to predict the reduction in oxygen permeability caused by APTEOS crosslinking as well as to predict the permeability coefficients for these gases in the hybrid materials. Note that fractional free volume calculated from density measurements and group contribution theory would incorrectly predict an increase in both FFV and permeability due to APTEOS crosslinking.

The PALS results for the alkoxide polyimide hybrid samples support the notion that the introduction of silica domains alters the chain packing, resulting in higher free volume, and transport properties. The correlation between oPs accessible free volume f_{oPs} and permeability is surprising given the bimodal nature of the silica domains in some samples. Of particular note, however, is the TEOS-based series of samples containing nanodomains of silica for all levels of alkoxide addition but having an optimum concentration of 15 wt% (4.23 vol%) for highest permeability and highest free volume. Comparison with the T_g results from DMTA (Table IV) confirms that the 15 wt% TEOS hybrid also has the highest T_g for this series. Differential scanning calorimetery was used to determine whether the silica nanodomains had a T_g characteristic of TEOS and independent of the polyimide or whether the silica nanodomains were well integrated in the polymer. Figure 9 compares the T_g for pure TEOS (209 °C) to that for the 22.5 wt% TEOS hybrid (277 °C). There is a single T_g for the hybrid that indicates that the domains are well integrated with the polymer matrix, similar to those domains in the 7.5 wt% and 15 wt% TEOS hybrids. At this point in time it is unknown what determines the optimum TEOS content (~15 wt% or 4.2 vol%) that results in a level of crosslinking that limits segmental motion at the interface and props open free volume elements. Future work includes characterization of the silica formed from the various alkoxides as well as molecular modeling of the silica and the hybrids.

Table V compares the effect of silica nanodomains on permeability, T_g and oPs accessible free volume for the 7.5 wt% alkoxide hybrids. The MTMOS alkoxysilane is most effective at increasing free volume. It was previously postulated (2) that separation performance is improved in hybrid materials by simultaneously increasing free volume and decreasing polymer chain mobility (via judicious crosslinking). It is clear from Table V that the hybrid samples made with 7.5 wt % alkoxide have increased free volume. Ideal selectivities, α, of these samples for the gas pair O_2/N_2 are shown in Table V. Whilst nanodomains of MTMOS are most effective in increasing free volume and permeability, the ideal selectivities are not improved. The separation characteristics of these hybrid materials have been discussed elsewhere (2) and the loss in selectivity was attributed to decreases in solubility selectivity. This work has illustrated that silica nanodomains formed via sol-gel can be used to improve permeability in polyimide hybrids. Merkel et al. (4,24) have explored

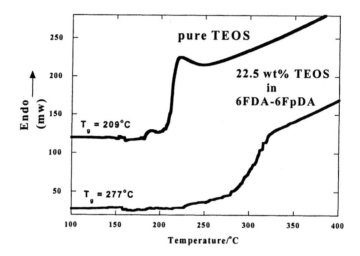

Figure 9. DSC results for pure TEOS and for 22.5 wt% TEOS-polyimide hybrid.

the effect of silica nanoparticles (mechanically added to polymer membranes) on free volume, flux, and selectivity in nanocomposite membranes. To the best of our knowledge, the present work is the first to examine these effects for nano-particle silica formed by the sol-gel process.

Table V. Permeability, PALS free volume, T_g, and ideal selectivity for hybrids made with 7.5 wt% alkoxide

Synthesis Silicate wt%	Silica vol %	Increase in Permeability (%)					Increase v_{oPs} %	Increase T_g	$\alpha O_2/N_2$
		He	CO_2	O_2	N_2	CH_4			
APTEOS polyimide XL	0	0	0	0	0	0	0	0	4.9
7.5 wt% TEOS	1.99	9	17	16	26	19	8	2	4.6
7.5 wt% MTMOS	6.2	37	90	81	106	130	17	2	4.4
7.5 wt% PTMOS	6.4	-8	12	2	10	19	5	3	4.6

Conclusions

Crosslinking of pure polyimide with APTEOS coupling agent causes a reduction in free volume element size and concentration as well as gas transport properties. The permeability for the gases studied can be accurately predicted from the measured oPs accessible free volume. In comparison to the crosslinked polyimide, incorporation of low levels of alkoxide (7.5 wt%) increases the average free volume element size of the hybrid materials. The type of alkoxide sol employed in the sol-gel process has an appreciable effect on the free volume distribution and gas transport properties in the hybrids. In general, permeability increases on crosslinking using low levels of alkoxide (7.5 wt%); the morphology of these hybrids consists of dispersed domains of size 25 nm (TEOS), 50 nm (MTMOS) and 100 nm (PTMOS). The mobility of the polymer chains is affected by the inorganic nanodomains as illustrated by the higher T_g in the hybrids compared to the pure and crosslinked polyimides. The data support the suggestion that the interaction between the inorganic nanodomains and the polymer chains contributes significantly to the free volume and gas transport properties of hybrid membranes.

References

1. Joly, C.; Goizet, S.; Schrotter, J.C.; Sanchez, J.; Escoubes, M. *J. Membr. Sci.* **1997**, *130*, 63.
2. Cornelius, C. J.; Marand, E. *J. Membr. Sci.* **2002**, *202*, 97-118.
3. Mahajan, R.; Zimmerman, C. M.; Koros, W. J. In *Polymer Membranes for Gas and Vapor Separation: Chemistry and Materials Science*; Freeman, B. D.; Pinnau, I. (Eds.), American Chemical Society: Washington, D.C., 1999; pp. 277-286.
4. Merkel, T. C.; Freeman, B. D.; Spontak, R. J.; He, Z.; Pinnau, I; Meakin, P.; Hill, A.J. *Science*, **2002**, *296*, 519-522.
5. Maxwell, C. *Treatise on Electricity and Magnetism*; Oxford University Press: London, 1873; Vol. 1. p. 440.
6. Robeson, L. M. *J. Membr. Sci.* **1991**, *62*, 165.
7. Zimmerman, C. M.; Singh, A; Koros, W.J. *J. Membr. Sci.*, **1997** *137*, 145-154.
8. Theodorou, D. private communication.
9. Cussler, E. L.; Hughes, S. E.; Ward, W. J.; Aris, R., *J. Membr. Sci.,* **1988**, *38*, 161-174.
10. Gusev, A. A.; Lusti, H. R. *Adv. Mater.* **2001**, *13 (21)*, 1641-1643.
11. Nielsen, L. J. *J. Macromol. Sci. Chem.* **1967**, A1(*5*), 929.
12. Meredith, J.C.; Karim, A.; Amis, E.J. *MRS Bulletin*, **2002**, *27 (4)*, 330-335.
13. Fraser, H. private communication.
14. Cornelius, C. J.; Marand, E. *Polymer* **2002**, *43*, 2385-2400.
15. Dolveck, J. Y.; Dai, G. H.; Moser, P.; Pineri, M.; Aldebert, P.; Escoubes, M.; Avrillon, R.; Mileo, J.C. *Materials Science Forum* **1992**, *105-110*, 1549-1552.
16. Tanaka, K.; Katsube, M.; Okamoto, K.-I.; Kita, H.; Sueoka, O.; Ito, Y. *Bull. Chem. Soc. Jpn.* **1992**, *65*, 1891-1897.
17. Okamoto, K.-I.; Tanaka, K.; Katsube, M.; Kita, H.; Sueoka, O.; Ito, Y. *Polymer Journal* **1993**, *25(3)*, 275-284.
18. Nagel, C.; Günther-Schade, K.; Fritsch, D.; Strunskus, T.; Faupel, F. *Macromolecules* **2002**, *35*, 2071-2077.
19. Bondi, A. *J. Phys. Chem.* **1964**, *68*, 441-451.
20. Schrader D. M. In *Positron and Positronium Chemistry*, Schrader D. M.; Jean, Y.C. (Eds.), Elsevier, Amsterdam, 1988, p.16.
21. Bartenev, G. M., Varisov, A. Z., Goldanskii, V. I., Mokrushin, A. D., Tsyganov, A. D. *Soviet Physics Solid State*, **1971**, *12*, 2806.
22. Greenfield, M. L.; Theodorou, D. N. *Macromolecules* **1993**, *26*, 5461-5472.
23. Marand, E., Cornelius, C. J., Meakin, P., Hill, A. J., *Polym. Sci and Eng.* **2001**, *85*, 297-298.
24. Merkel, T. C.; Freeman, B.D.; Spontak, R.J.; He, Z.; Pinnau, I.; Meakin, P.; Hill, A.J. *Chem. Mater.* **2003**, *15*, 109-123.

Chapter 17

Proton Conductivity and Vapor Permeation Properties of Polyimides Containing Sulfonic Acid Groups

Tatsuya Watari, Jianhua Fang, Xiaoxia Guo, Kazuhiro Tanaka, Hidetoshi Kita, and Ken-ichi Okamoto*

Department of Advanced Materials Science and Engineering, Faculty of Engineering, Yamaguchi University, 2–16–1 Tokiwadai, Ube, Yamaguchi 755–8611, Japan

Sulfonated polyimides (SPIs) were prepared from 1,4,5,8-naphthalenetetracarboxylic dianhydride (NTDA), three kinds of novel sulfonated diamine monomers such as 4,4'-bis(4-aminophenoxy)biphenyl-3,3'-disulfonic acid (BAPBDS), and common non-sulfonated diamines. Polyimide membranes with good water stability and high proton conductivity were developed. NTDA-BAPBDS polyimide membrane, for example, did not lose mechanical properties after being soaked in water at 80 °C for one month, while its proton conductivity was still at a high level (comparable to that of Nafion 117). Although the SPI membranes displayed very large diffusion and permeation coefficients of water vapor, they were still much smaller than those of Nafion 117.

Aromatic polyimides have found wide applications in many industrial fields (1, 2). Sulfonated six-membered ring polyimides have been identified to be promising proton conducting membrane materials for fuel cell application (3-8). Mercier and coworkers first synthesized a series of sulfonated copolyimides

© 2004 American Chemical Society

from naphthalene-1,4,5,8-tetracarboxylic dianhydride (NTDA), 2,2'-benzidine sulfonic acid (BDSA) and common non-sulfonated diamine monomers (3-6). These SPI membranes were tested in a fuel cell system and showed fairly good performance. However, the proton conductivity of these membranes was rather low (< 10^{-2} Scm^{-1}, in water) due to the low ion exchange capacity (IEC, defined as meq of sulfonic acid per g of dry polymer), which is essential for maintaining water stability of the membranes. Litt's group has also employed BDSA for preparation of various random and sequenced copolyimides (7, 8). However, the water stability of their membranes is still a problem.

Recently, we developed novel sulfonated aromatic diamines and related polyimides (9, 10). In this paper, their proton conductivity, water uptake and water stability were investigated and compared with those of BDSA-based polyimides. Water vapor sorption, diffusion and permeation for the SPI membranes are also important properties related to proton conductivity and fuel cell performance of the membranes. In this paper, these properties are also investigated.

Experimental

Preparation

4,4'-diaminodiphenyl ether-2,2'-disulfonic acid (ODADS), 9,9-bis(4-aminophenyl)fluorene-2,7-disulfonic acid (BAPFDS) and 4,4'-bis(4-aminophenoxy)biphenyl-3,3'-disulfonic acid (BAPBDS) were synthesized by direct sulfonation of the precursor diamines, namely, 4,4'-diaminodiphenyl ether (ODA), 9,9-bis(4-aminophenyl)fluorene (BAPF), and 4,4'-bis(4-aminophenoxy)biphenyl (BAPB) (9, 10).

As shown in Scheme 1, polymerization of NTDA and the sulfonated diamines was carried out by a "one-step" method in m-cresol in the presence of triethylamine (Et$_3$N) and benzoic acid. Et$_3$N was used to liberate the protonated amino groups for polymerization with NTDA, and benzoic acid functioned as a catalyst. This is a literature method which has been employed to prepare a series of BDSA-based polyimides (6). Random copolymerization of NTDA, the sulfonated diamines, and non-sulfonated diamines was also performed using this method. For comparison, BDSA-based polyimides were also prepared. Polyimide membranes (in triethylammonium salt form) were prepared by conventional solution casting method from their m-cresol or DMSO solutions. The as-cast membranes were soaked in methanol at 60 °C for 1 h to remove the residual solvent, and then the proton exchange was performed by immersing the membranes into 1.0 N hydrochloric acid at room temperature for 5-10 h. The membranes in proton form were thoroughly washed with de-ionized water and

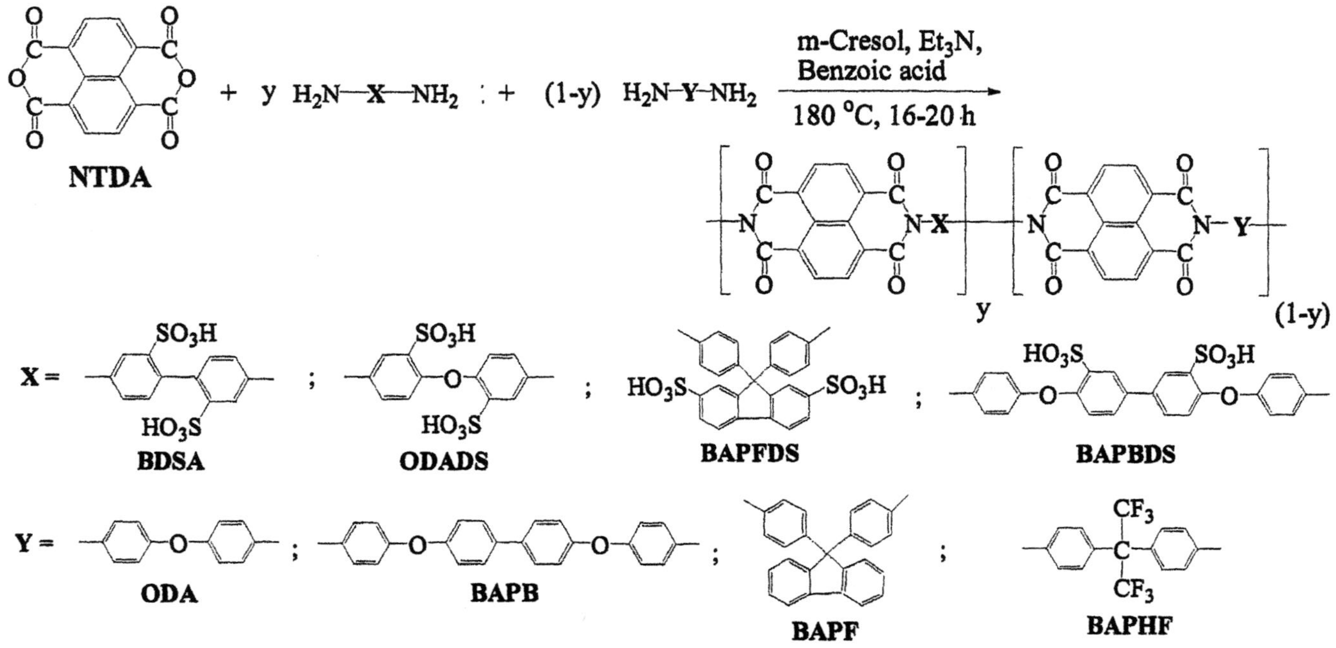

Scheme 1. Synthesis of SPIs (in triethylammonium salt form).

then dried in vacuum at 150 °C for 20 h. The thickness of the films was in the range of 20-40 μm.

Measurements

Proton conductivity was measured by an ac impedance method with two black platinum electrodes using a Hioki 3552 Hitester instrument over the frequency range from 100 Hz to 100 kHz. A sheet of membrane (1.0×0.5 cm^2) was placed between two pairs of blacken platinum plate electrodes and set in a Teflon cell. The distance between two electrodes was 0.5 cm. The cell was placed in either a temperature and humidity controlled chamber to measure the temperature and humidity dependence of proton conductivity (for measurement at relative humidity lower than 100%) or in distilled deionized water (for measurement in water). The resistance associated with the membrane conductance was determined from the high frequency intercept of the impedance with the real axis (*10*).

Water sorption experiments were performed by immersing three sheets of polyimide membrane (20-30 mg per sheet) into water at a given temperature for 5 h. Then the sheets were taken out, wiped with tissue paper, and quickly weighed on a microbalance. Water uptake was calculated from the following equation, water uptake = $(W_s - W_d)/W_d \times 100$ (%w/w), where W_d and W_s are the weights of the dry sheets and the corresponding water-swollen ones, respectively. The average value of three sheets was used as the water uptake of a polyimide membrane. Water vapor sorption was also measured using a volumetric method employing a vapor sorption apparatus (Nihon Bell, Bell-18SP).

Water vapor permeation experiments were performed using a vapor permeation (VP) cell. Nitrogen gas containing water vapor of a given activity was supplied to the upstream side of a flat membrane and the downstream side was evacuated through a vacuum line. The permeate was collected in a cold trap cooled with liquid nitrogen. The concentration averaged diffusion coefficient $\overline{D}$ was calculated from $\overline{D} = P/S$, where P and S are permeability and solubility coefficients, respectively. S was determined from water vapor sorption.

Results and Discussion

Preparation and Properties of Sulfonated Polyimides

The SPIs investigated are listed in Table I. The IEC values in Table I are the calculated ones. The completion of proton exchange was confirmed by the

Table I. IEC, Water Uptake, Proton Conductivity and Water Stability of Sulfonated Polyimide Membranes

Membrane	IEC [meq g^{-1}]	Water Uptake [%w/w] 50 °C RH 80%	Water Uptake [%w/w] 50 °C LW^a	Water Uptake [%w/w] 80 °C LW	σ [S cm^{-1}] 50 °C RH 80%	σ [S cm^{-1}] 50 °C LW	Membrane Stability Test T [°C]	Membrane Stability Test Time [h]	Membrane Stability Test Stability^b
NTDA-ODADS	3.37	38			0.039		50	10 min.	×
NTDA-ODADS/ODA(3/1)	2.70	36	113		0.024	0.21	80	5	×
							80	15	Dissolved
NTDA-ODADS/ODA(1/1)	1.95	25	101	87	0.017	0.12	80	25	○
NTDA-ODADS/BAPB(1/1)	1.68		-	57	0.011	0.10	80	200	○
NTDA-ODADS/BAPF(1/1)	1.71		-	69	0.011	0.09	80	13	×
NTDA-ODADS/BAPHF(1/1)	1.73		-	72		0.11	80	11	×
NTDA-BAPFDS	2.70	31	142	122	0.029	0.21	50	50	○
							80	5	Dissolved
NTDA-BAPFDS/ODA(4/1)	2.36		-	100	0.024	0.18	80	6	×
NTDA-BAPFDS/ODA(2/1)	2.09		-	76	0.020	0.12	80	20	○
NTDA-BAPFDS/ODA(1/1)	1.71		-	78	0.0094	0.12	80	26	○
NTDA-BAPFDS/BAPB(4/3)	1.68		-	56	-	0.10	80	27	○
NTDA-BAPBDS	2.63	30	85	130	0.013	0.24	80	720	○
NTDA-BDSA	3.47		-		0.038^d		50	A few sec.	Dissolved
NTDA-BDSA/ODA(1/1)	1.98	-	-	79	0.013	0.11	80	6	×
NTDA-BDSA/ODA(3/7)^c	1.26	22	41			0.018			
Nafion 117	1.03	11^e	24^e		0.03	0.11			

NOTE: ^{a}LW: in liquid water, ^{b}○: mechanical strength was maintained; ×: somewhat brittle,, ^{c}From Reference 11 at a fully hydrated state and room temperature, ^{d}From Reference 6 at 75% relative humidity and 25 °C., ^{e}From Reference 12 at 30 °C.

disappearance of peaks corresponding to the protons of triethylamine in the ^{1}H NMR spectra of the polyimides (for NTDA-BAPBDS, the IR spectrum was used instead of the ^{1}H NMR spectrum). Thermal stability of the SPIs (in proton form) was investigated by TG-MS measurements. For NTDA-ODADS, the weight loss starting from 275 °C is due to the decomposition of sulfonic acid groups judging from the evolution of sulfur monoxide and sulfur dioxide, indicating fairly good thermal stability of this polyimide. The sulfonic acid group of NTDA-BAPFDS started to decompose around 270 °C.

Water Vapor Sorption

Integral sorption and desorption of water vapor in SPI membranes displayed typical non-Fickian diffusion behavior, indicating that water-vapor-induced molecular relaxation of polymer took place during the experimental time scale. Water vapor sorption isotherms shown in Figure 1(a) were measured by successive differential (or interval) sorption experiments with enough time to establish sorption equilibrium. The sorption isotherms were sigmoidal and similar to type II BET adsorption. The water uptakes of SPIs largely depended on the IEC values. However, the number of sorbed water molecules per sulfonic acid group, λ, at a given water vapor activity a_w was very similar in magnitude among the SPIs investigated here, as shown in Figure 1(b). Furthermore, their sorption isotherms in form of λ vs. a_w were similar except for the range of $a_w >$ 0.9 and were not so different from that of Nafion 117. In the range of $a_w > 0.9$ and especially in liquid water, the water uptakes of the SPIs varied rather largely depending on the chemical structure. At the higher activities, the higher water uptakes tend to cause larger molecular relaxation which, in turn, results in still higher water uptakes. This is related to the water stability of the membranes and will be discussed later.

Proton Conductivity

Figure 2 shows the relative humidity dependence of the proton conductivity of the SPI membranes at 50 °C. The proton conductivities of Nafion 117 membrane measured in this study, which are close to those reported by Miyake et al. (13), are also shown in Figure 2. The proton conductivities and water uptakes of the membranes at 80 % relative humidity and 50 °C are listed in Table I. The proton conductivity of the polyimide membranes is strongly dependent on both the IEC and the relative humidity. For the same sulfonated diamine-based polyimide membranes, larger IEC values resulted in higher proton conductivity. With increasing relative humidity, the proton conductivity

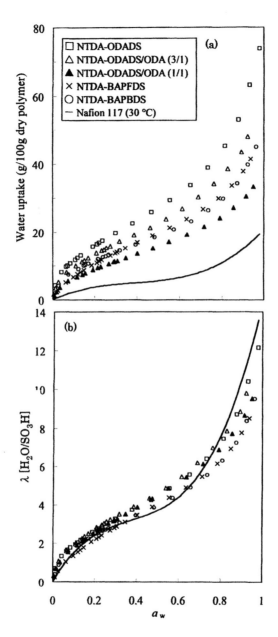

Figure 1. Sorption isotherms of water vapor at 50°C. (a) water uptake vs. water activity and (b) number of sorbed water molecules per sulfonic acid group.

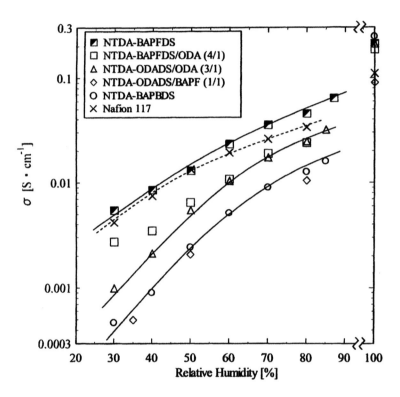

Figure 2. Relative humidity dependence of proton conductivity for SPI membranes at 50°C.

significantly increased for all of the polyimide membranes. Upon increasing the relative humidity from 30 to 80 %, the water uptake of the membranes increased by about 2.2 times, whereas the proton conductivities increased by more than 10 times, indicating strong dependence of proton conductivity on water vapor uptake. The dependence was stronger for the SPIs than for Nafion 117, especially at relative humidities below 40 % and above 90 %. Most of the polyimide membrane displayed poorer proton-conducting performance than Nafion 117 over the whole humidity range except for liquid water, in spite of the polyimides having much larger IEC values. In liquid water, most of the polyimide membranes displayed larger proton conductivities than Nafion 117. This is probably because Nafion has unique ion-rich channels (clusters), which are favorable for proton transport, whereas the polyimide membranes investigated here did not have a clear microphase separated structure. Nevertheless, the proton conductivities of the polyimide membranes are more than one order of magnitude higher than those reported in the literature (6).

Figure 3 shows the temperature dependence of proton conductivity of the polyimide membranes in water. For all the membranes the proton conductivity increased with increasing temperature. In addition, in water all the polyimide membranes displayed similar or higher proton conductivities than Nafion 117, indicating good proton conducting performance of these membranes. This is probably because the polyimide membranes in water are in a highly hydrated state and the transport of protons becomes quite easy, making the effect of channels (for facilitation of proton transport) less important.

Water-Stability of Membrane

Table I lists water uptake and water stability of the polyimide membranes at 80 or 50 °C. The water stability test was performed by immersing the membranes in distilled water and characterized by the time elapsed until the hydrated membranes lost mechanical properties. The criterion for the judgment of the loss of mechanical properties is that the membrane is broken when slightly bent. NTDA-BDSA homopolyimide membranes displayed the poorest stability among the membranes synthesized for this study. It completely dissolved in water at 50 °C within a few seconds. However, it took more than 2 h for NTDA-ODADS to completely dissolve in water. This indicates NTDA-ODADS is much more stable toward water than NTDA-BDSA. Another homopolyimide, NTDA-BAPFDS was insoluble in water at 50 °C and could maintain mechanical strength even after being soaked in water at 50 °C for 50 h, indicating significantly improved water stability relative to NTDA-ODADS. However, this polyimide could still dissolve in water at high temperature (80 °C) as a result of enhanced molecular relaxation of the polymer due to high

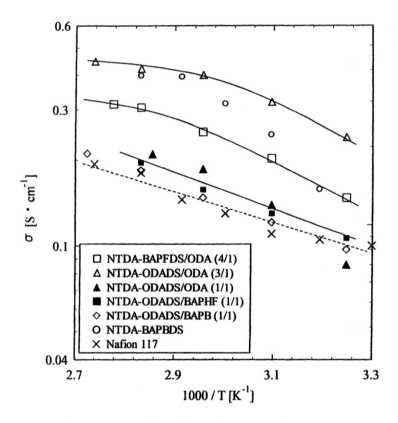

Figure 3. Temperature dependence of proton conductivity in water.

IEC. In contrast, the NTDA-BAPBDS polyimide membrane had very good water stability in spite of its high IEC (close to that of NTDA-BAPFDS). It could maintain mechanical properties even after being soaked in water at 80 °C for one month.

In order to obtain more information on the relationship between the chemical structure of the SPIs and their water stability, various copolyimides were prepared and their water stability was measured. As shown in Table I, decreasing the degree of sulfonation by incorporation of non-sulfonated diamines into the polyimide structure led to a significant improvement in water stability. All of the copolyimide membranes were insoluble at 80 °C. However, there is still a large difference in water stability between these copolyimide membranes. NTDA-BDSA/ODA(1/1), for example, became brittle (the membrane was broken when slightly bent) after being soaked in water at 80 °C for 5.5 h; whereas the NTDA-ODADS/ODA(1/1) membrane did not become brittle even after it was soaked in water at 80 °C for 25 h, indicating much better stability of the ODADS-based copolyimide membranes. Usually water stability of polymer membranes is strongly dependent on water uptake. However, in the present case, the difference in water stability between NTDA-ODADS/ODA(1/1) and NTDA-BDSA/ODA(1/1) is not because of the difference in water uptake. These two kinds of copolyimides had quite different chain flexibilities. The BDSA moiety is highly rigid, whereas ODADS is fairly flexible due to the existence of a flexible ether linkage, and therefore the backbone of NTDA-ODADS/ODA(1/1) should be much more flexible than that of NTDA-BDSA/ODA(1/1). Flexible chains undergo relaxation more easily than rigid ones, and this is likely the main reason that NTDA-ODADS/ODA(1/1) displayed much better water stability than NTDA-BDSA/ODA(1/1). This effect has also been observed for non-sulfonated five-membered ring polyimides. For example, rod-like polyimides showed much poorer hydrolysis stability than flexible ones (14). The chain flexibility of SPI is also affected by the non-sulfonated diamine moieties. NTDA-ODADS/BAPF(1/1) and NTDA-ODADS/BAPHF(1/1) copolyimides have less flexible structures than NTDA-ODADS/ODA(1/1), and their water stability is much poorer than that of the latter in spite of their lower IEC values and lower water uptakes. In contrast, for "flexible"-type copolyimides, the reduction of the IEC (and therefore the water uptake) by changing the non-sulfonated diamine moiety led to great improvement in water stability of the membranes. NTDA-ODADS/BAPB(1/1) displayed the best water stability among ODADS-based polyimide membranes because of its highly flexible structure and the lowest IEC. These results indicate that water stability is significantly affected by the flexibility of the whole polymer chain.

On the other hand, BAPFDS-based copolyimide membranes displayed rather good water stability in spite of the rigid structure of BAPFDS. NTDA-BAPFDS/ODA(2/1) had much better stability than NTDA-BDSA/ODA(1/1).

Furthermore, NTDA-BAPFDS/ODA(1/1) displayed much better stability than NTDA-ODADS/BAPF(1/1). These two kinds of copolyimides have the same IEC and chain flexibility, and the difference in their structures is only that the sulfonic acid groups are attached to different diamine moieties. This clearly indicates that the stability of polyimide membranes not only depends on the flexibility of the polymer chains and the IEC but also depends on other factor(s). A common structural feature between ODADS and BDSA is that the sulfonic acid groups are directly attached to the phenyl rings to which the amino groups are attached, and here they are noted as "type 1" sulfonated diamines. Unlike the case of ODADS and BDSA, the sulfonic acid groups of BAPFDS are attached to the bridged phenyl rings, and this is noted as a "type 2" sulfonated diamine (Scheme 2). Because the sulfonic acid group is a strong electron-withdrawing group, the electron density of the phenyl rings to which the amino groups are attached should be larger for BAPFDS than for ODADS or BDSA, i.e., BAPFDS is more basic than ODADS or BDSA. Aromatic diamines with higher basicity are generally more reactive with dianhydrides than those with lower basicity (15). Since hydrolysis is the reverse reaction of polymerization, polyimides derived from more basic diamines should have higher hydrolysis stability. As a result, the high basicity of BAPFDS is favorable for maintaining the stability of the imido rings, which offsets the unfavorable effect on water stability due to its rigid structure, and this might be the reason that BAPFDS-based polyimide membranes had much better water stability than BDSA-based ones with similar IEC. BAPBDS is a "type 2" and highly flexible sulfonated diamine, and therefore NTDA-BAPBDS polyimide membrane displayed very high water stability in comparison to others.

Water Vapor Permeation

Diffusion coefficients of water in the SPI membranes could not be determined by integral sorption and desorption experiments because of the non-Fickian diffusion behavior. Therefore, steady state water vapor permeation experiments were performed as a function of vapor activity at 50 °C, and the concentration averaged diffusion coefficient was calculated from $\overline{D} = P/S$ using water vapor sorption data (Figure 1(a)). The results are shown in Figure 4. With increasing vapor activity, both P and $\overline{D}$ increased significantly, because of the plasticization effect of sorbed water as well as the more relaxed polymer chain packing. Table II lists the P, $\overline{D}$ and S data of water for the non-sulfonated and SPIs and Nafion 117. Compared with the non-sulfonated polyimides, the SPIs had more than ten times larger S values and as a result had much larger $\overline{D}$ and P values (of the order of 10^{-7} cm^2/s and 10^5 Barrer, respectively). Compared with Nafion 117 at 50 °C, the SPIs had 2-3 times larger S values and smaller $\overline{D}$

265

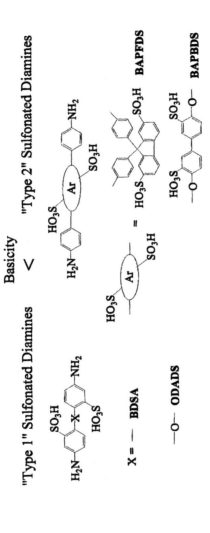

Scheme 2. Type 1 sulfonated diamines (BDSA and ODADS) and Type 2 sulfonated diamines (BAPFDS and BAPBDS).

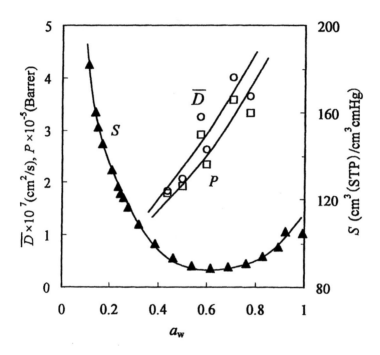

Figure 4. Activity dependence of P, D, and S of water vapor for a NTDA-ODADS/ODA (3/1) polyimide membrane at 50°C.

values by a factor of 6-10 and as a result smaller P values by a factor of 3-5 *(16)*. The much larger $\overline{D}$ values for Nafion 117 despite the smaller S values are due to its unique ion-rich channel structure *(17)*.

Table II. S, P and $\overline{D}$ of Water Vapor for SPIs at 50 °C

Polymer	a_w	$\overline{D}$	S	P	Ref.
6FDA-ODA[a]	0.5	*0.46*	*4.7*	*0.02*	*18*
	0.8	*0.37*	*5.2*	*0.02*	*18*
NTDA-ODADS	0.5	2.4	79	1.9	
/ODA(3/1)	0.8	4.3	77	3.3	
Nafion 117	0.5	10	74	7.4	19
(30 °C)	0.8	20	72	14	19

NOTE: [a] 6FDA: 2.2-bis(3,4-dicarboxyphenyl)hexafluoropropane dianhydride

NOTE: Units of $\overline{D}$, S and P are 10^{-7} cm^2/s, cm^3(STP)cm^{-3}cmHg^{-1} and 10^5 Barrer, respectively.

Conclusions

Water vapor sorption depended strongly on IEC, but the SPI membranes displayed similar sorption isotherms (in the form of λ - a_w), which were not so different from that of Nafion 117. The SPI membranes displayed a little larger relative-humidity-dependence of proton conductivity than Nafion 117. At relative humidities below 100 %, their proton conductivities were generally lower than that of Nafion 117. However, in liquid water, most of the SPIs showed similar or higher proton conductivities than Nafion 117. Water stability of the SPI membranes is greatly affected by the IEC, the flexibility of the polymer chains and the basicity of the sulfonated diamine moieties. As a result, the water stability of the homopolyimides was in the order: NTDA-BAPBDS ≫ NTDA-BAPFDS ~ NTDA-ODADS ≫ NTDA-BDSA. The SPI membranes displayed very large $\overline{D}$ and P values (of the order of 10^{-7} cm^2/s and 10^5 Barrer, respectively). However, these values were still much smaller than those of Nafion 117.

References

1. *Polyimides;* Ghosh, M. K.; Mittal, K. L., Eds.; Fundamentals and Applications; Marcel Dekker: New York, 1996.

2. *Polyimides and Other High Temperature Polymers;* Mittal, K. L., Ed.; Synthesis, Characterization and Applications; VSP: Utrecht, 2001; Vol. 1.
3. Faure, S.; Mercier, R.; Aldebert, P.; Pineri, M.; Sillion, B. *French Patent* 2,748,485, 1996.
4. Faure, S.; Cornet, N.; Gebel, G.; Mercier, R.; Pineri, M.; Sillion, B. In *New Mater. Fuel Cell Mod. Battery Syst. II;* Savadogo, O.; Roberge, P. R., Eds.; Proc. Int. Symp., 2nd; cole Polytechnique de Montr al: Montreal, Canada, July 6-10, 1997; pp 818-827.
5. Vallejo, E.; Pourcelly, G.; Gavach, C.; Mercier, R.; Pineri, M. *J. Membr. Sci.* **1999**, *160,* 127-137.
6. Genies, C.; Mercier, R.; Sillion, B.; Cornet, N.; Gebel, G.; Pineri, M. *Polymer* **2000**, *42,* 359-373.
7. Zhang, Y.; Litt, M.; Savinell, R. F.; Wainright, J. S. *Polym. Prepr.(Am. Chem. Soc., Div. Polym. Chem.)* **1999**, *40,* 480-481.
8. Kim, H.; Litt, M. *Polym. Prepr.(Am. Chem. Soc., Div. Polym. Chem.)* **2001**, *42,* 486-487.
9. Fang, J.; Guo, X.; Watari, T.; Harada, S.; Tanaka, K.; Kita, H.; Okamoto, K. *Macromolecules,* **2002**,*35,* 9022-9028.
10. Guo, X.; Fang, J.; Watari, T.; Tanaka, K.; Kita, H.; Okamoto, K. *Macromolecules* **2002**, *35,* 6707-6713.
11. Cornet, N.; Beaudoing G.; Gebel G. *Sep. Puri. Tech.* **2001**, *22-23,* 681-687
12. Zawodzinski, T. A. Jr.; Springer, T. E.; Davey, J.; Jestel, R.; Lopez, C.; Valerio, J.; Gottesfeld, S. *J. Electrochem. Soc.* **1993**, *140,* 1981-1985.
13. Miyake, N.; Wainright, J. S.; Savinell, R. F. *J. Electrochem. Soc.* **2001**, *148,* A898-A904.
14. Sroog, C. E. *Prog. Polym. Sci.* **1991**, *16,* 561-694.
15. Takekoshi, T. In *Polyimides: Fundamentals and Applications*; Ghosh, M. K.; Mittal, K. L., Eds.; Marcel Dekker: New York, 1996; p 7.
16. T. Watari, H. Wang, K. Kuwahara, K. Tanaka, H. Kita and K. Okamoto, submitted to *J. Membr. Sci.*
17. Gierke, T. D.; Munn, G. E.; Wilson, F. C. *J. Polym. Sci., Polym. Phys. Ed.* **1981**, *19,* 1687-1704.
18. Okamoto, K.; Tanihara, N.; Watanabe, H.; Tanaka, K.; Kita, H.; Nakamura, A.; Kusuki, Y.; Nakagawa, K. *J. Polym. Sci., Polym. Phys. Ed.* **1992**, *30,* 1223-1231.
19. D. Rivin, C.E. Kendrick, P. W. Gibson and N. S. Schneider, *Polymer,* **2001**, *42,* 623-635.

Chapter 18

Formation of Composite Membranes with Ultrathin Skins Using New Methods of Organic Film Formation: Gas-Selective Membranes

Merlin L. Bruening

**Department of Chemistry, Michigan State University,
East Lansing, MI 48824**

This chapter describes the formation of ultrathin (< 50 nm) gas-selective skins on porous alumina supports using either graft-on-graft deposition of hyperbranched poly(acrylic acid) (PAA) or alternating polyelectrolyte deposition (APD). The highly branched structure of the PAA films allows coverage of substrate pores with diameters as large as 0.2 μm, and derivatization of PAA with $H_2NCH_2(CF_2)_6CF_3$ results in an O_2/N_2 selectivity of 2.3, showing that films are free of defects. Formation of gas-selective membranes using APD involves alternating adsorption of poly(amic acids) and protonated poly(allylamine). Subsequent imidization by heating yields membranes with the same selectivities as bulk polyimides, even when skins are only 35-40 nm thick. Because PAA can be widely derivatized, and nearly any polyelectrolyte is suitable for APD, these techniques provide versatile methods for forming functional, ultrathin membranes.

© 2004 American Chemical Society

Introduction

Although membrane separations are attractive because of their simplicity and low energy costs, applications of membranes are often limited by insufficient flux, selectivity, or stability. This problem is especially challenging because the permeability of a material is frequently inversely related to its selectivity (*1,2*). The most common means for increasing flux while maintaining selectivity is to prepare a membrane containing a selective, ultrathin skin on a highly permeable support. Loeb and Sourirajan first demonstrated this concept by forming asymmetrically skinned cellulose acetate membranes (*3*). Subsequent development of composite membranes that consist of an ultrathin skin deposited on a separate porous support further increased the attractiveness of skinned membranes, because in these systems, only a small amount of selective material is needed. Thus, more expensive, high-performance materials can be used as membrane skins (*4*). Processes used to form skins for composite membranes include casting, interfacial polymerization, plasma grafting, and deposition of films from the air-water interface (*5-9*). However, in spite of successes in this area, depositing membrane skins that are selective, defect-free, and ultrathin (< 50 nm) remains difficult. Formation of defect-free membrane skins is critical to separations because even a small number of defects can negate selectivity (*10*).

This chapter reviews the use of two recently developed techniques for formation of ultrathin membrane skins. The first is the deposition of hyperbranched poly(acrylic acid) (PAA) (*11,12*). This technique is attractive because the hyperbranched structure of these films should allow coverage of relatively large pores as shown in Figure 1. Additionally, the –COOH groups of

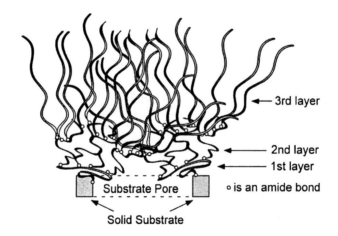

Figure 1. Idealized, schematic diagram of a hyperbranched PAA film covering a substrate pore. Reproduced with permission from reference 12. Copyright 2000 American Chemical Society.

PAA can be readily derivatized to provide specific functionality (*13*). Although this chapter focuses on the formation of gas separation membranes, PAA is also potentially attractive for its interactions with biomolecules such as proteins (*14,15*).

The second technique for membrane formation discussed herein is alternating polyelectrolyte deposition (APD) (*16*). Formation of membranes occurs simply through alternating immersions of a charged support into solutions containing polycations and polyanions as shown in Figure 2. We recently reviewed the formation of ion-selective membranes using this technique (*17*). This chapter focuses on the deposition of selective gas separation membranes through formation of films containing poly(amic acids) and subsequent imidization. By carefully controlling deposition conditions, APD allows formation of selective polyimide membrane skins with thicknesses as low as 35-40 nm (*18*). Future exploitation of this technique may allow further reductions in the thickness of the selective layer.

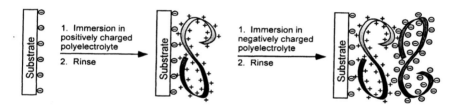

Figure 2. Schematic diagram of alternating polyelectrolyte deposition (16). Repetition of the procedure yields multilayer films. Intertwining of neighboring layers is not shown for figure clarity.

Hyperbranched Poly(acrylic acid) Skins

Synthesis of PAA films occurs as shown in Scheme 1 (*11*). To prepare a film on a porous alumina substrate, we first deposit 5 nm of gold on the surface and then immerse the substrate in a solution containing mercaptoundecanoic acid to form a monolayer with –COOH groups. Activation of the –COOH groups to anhydrides using ethyl chloroformate followed by reaction with amino-terminated poly(*tert*-butyl acrylate) (PTBA) yields a grafted layer of PTBA on the surface, and subsequent hydrolysis results in grafted PAA. Further grafting on each previous graft using the same activation, reaction, and hydrolysis procedure yields a hyperbranched PAA film.

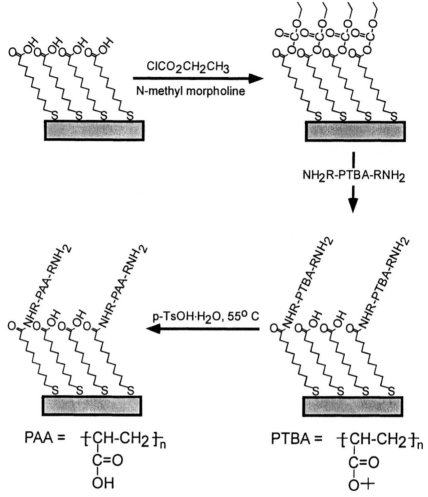

Scheme 1. Synthetic strategy for grafting one layer of PAA to a surface. Further grafting to previously grafted PAA chains yields a hyperbranched film. Reproduced with permission from reference 12. Copyright 2000 American Chemical Society.

Field-emission scanning electron microscopy (FESEM) images show that hyperbranched PAA films are capable of covering large pores in underlying supports. One of the reasons for performing initial studies with porous alumina was that this support has a high pore density and a regular pore structure that facilitates visualization of substrate coverage (*19*). Cross-sectional FESEM images such as that in Figure 3 clearly show that the support is fully covered, while pores, which in this case have diameters of about 0.2 μm, are not filled. Thinner films are also capable of covering underlying pores, but they are more difficult to see in SEM images.

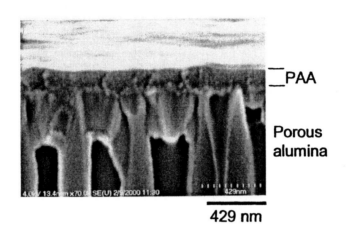

429 nm

Figure 3. Cross-sectional FESEM image of a 6-bilayer PAA film grafted onto a porous alumina (0.2 μm-diameter pores) substrate. Reproduced with permission from reference 12. Copyright 2000 American Chemical Society.

Gas-transport studies with derivatized PAA films also show that these materials form defect-free skins on porous substrates (*12*). As-deposited PAA films exhibit gas selectivities that are representative of Knudsen diffusion, i.e., flux is inversely proportional to the square root of the molar mass of the gas (*7*). However, selectivity increases significantly upon derivatization of films with $H_2NCH_2(CF_2)_6CF_3$. For example, O_2/N_2 selectivity increases to values above 2 and CO_2/CH_4 selectivity is about 8. Although these selectivities are not especially high, they are more than 2-fold higher than predicted values for Knudsen diffusion and show that films are free of defects. The selectivities of derivatized PAA are also consistent with those recently reported for bulk films of poly(1,1'-dihydroperfluoroacrylate), which is very similar in chemical composition to PAA derivatized with $H_2NCH_2(CF_2)_6CF_3$ (*20*). However, the permeability of derivatized PAA is somewhat lower than for poly(1,1'-dihydroperfluoroacrylate), perhaps because of incomplete derivatization (*21*).

One challenge in employing ultrathin PAA skins is that their synthesis is rather cumbersome because a gold layer must be sputtered on the surface and a number of grafting steps are required. In an effort to overcome these limitations, we began grafting PAA to poly(acrylic acid)/poly(allylamine hydrochloride)

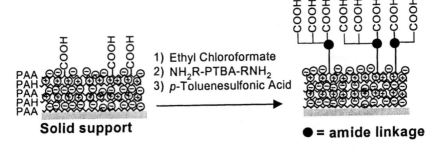

Scheme 2. Grafting of poly(acrylic acid) to a multilayer polyelectrolyte (PAA/PAH) film. Reproduced with permission from reference 20. Copyright 2000 American Chemical Society.

(PAH) films prepared by APD (Scheme 2) (*21*). Because the formation of multilayered polyelectrolyte films occurs simply through alternating immersions of a charged substrate in solutions containing PAA or PAH, and because only one PTBA grafting step is required to create a defect-free film, this procedure provides a convenient way to form ultrathin PAA skins. When derivatized with $H_2NCH_2(CF_2)_6CF_3$, films as thin as 30 nm show O_2/N_2 selectivities of 2.4. Thus, this deposition method yields ultrathin, defect-free skins. We should note, however, that these films were prepared on porous alumina containing 0.02 μm-diameter surface pores. This method is probably less capable of covering 0.2 μm-diameter pores than is deposition of hyperbranched PAA.

Although the gas selectivities of PAA and its derivatives are fairly modest, the derivatizability of PAA may allow its specialized application in areas such as affinity membranes. We recently demonstrated the use of PAA films in sensing and sample purification, and PAA membranes may also be attractive in this regard (*15,22*).

Skins Composed of Imidized Polyelectrolyte Multilayers

Typical multilayered polyelectrolyte membranes that we and others have tested, e.g., poly(styrene sulfonate) (PSS)/PAH and PAA/PAH, show only modest selectivity in gas-transport measurements (*23-26*). Because polyimide membranes are well-known for their gas-transport selectivities (*27-29*), we began preparing multilayer polyelectrolyte membrane skins using poly(amic acids) that could subsequently be imidized by heating to form selective membranes (Scheme 3). Several groups previously showed that imidization of poly(amic acid)-containing films is possible (*30-32*), and we set out to show that this procedure can yield highly selective membranes.

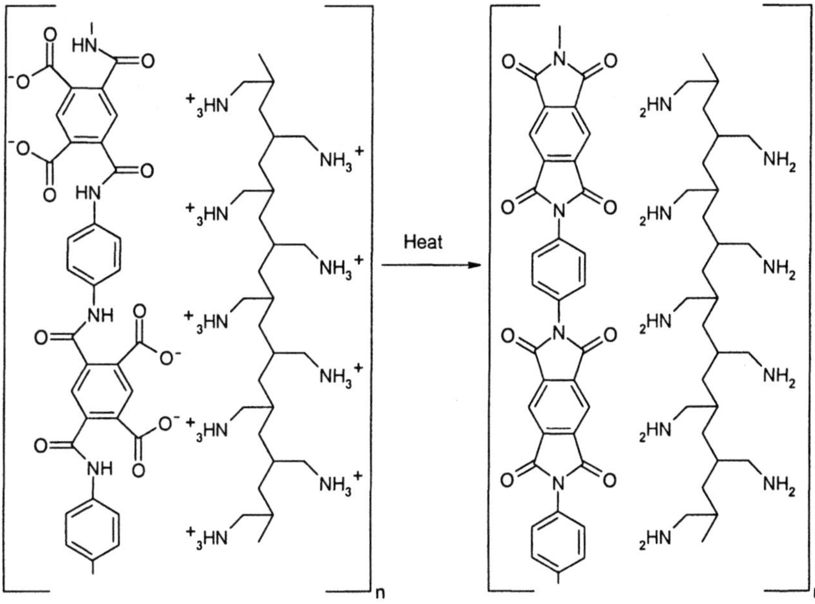

Scheme 3. Imidization of a PMDA-PDA/PAH film. Note that in an actual film there is more PMDA-PDA than PAH, and electrical neutrality is maintained.

Our first studies with poly(amic acid) membranes utilized the poly(pyromellitic dianhydride-phenylenediamine) (PMDA-PDA)/PAH system shown in Scheme 3. Reflectance FTIR spectroscopy of PMDA-PDA/PAH films deposited on Au-coated substrates showed that the extent of imidization depends on heating temperature, with about 85% imidization occurring when heating for 2 h at 180°C (*33*). Although imidization of these films occurs readily, PMDA-PDA has a rigid structure that does not allow large gas fluxes (*27-29*), so fabrication of gas separation membranes using these materials was not successful. Partially imidized PMDA-PDA/PAH films can provide, however, highly selective ion-transport membranes. Films heated at 165-180°C exhibited Cl^-/SO_4^{2-} selectivities as high as 1000 in diffusion dialysis.*(33)*

To synthesize gas separation membranes from polyelectrolyte multilayers, we utilized the fluorinated polyimides shown in Scheme 4 because these materials have O_2 permeabilities that are at least an order of magnitude higher than those of polyimides similar to PMDA-PDA (*27-29*). We began by preparing 6FDA-mPDA/PAH membrane skins on porous alumina. Because PAH is not likely to be a highly selective material for gas separation, we aimed to prepare films with a minimal PAH content. This can be accomplished by controlling the pH and solvent used during deposition of both the poly(amic acid) and PAH (*34,35*). To enhance the adsorption of the poly(amic acid), we utilized a deposition pH of 5 in a 30% water, 70% ethanol solvent containing

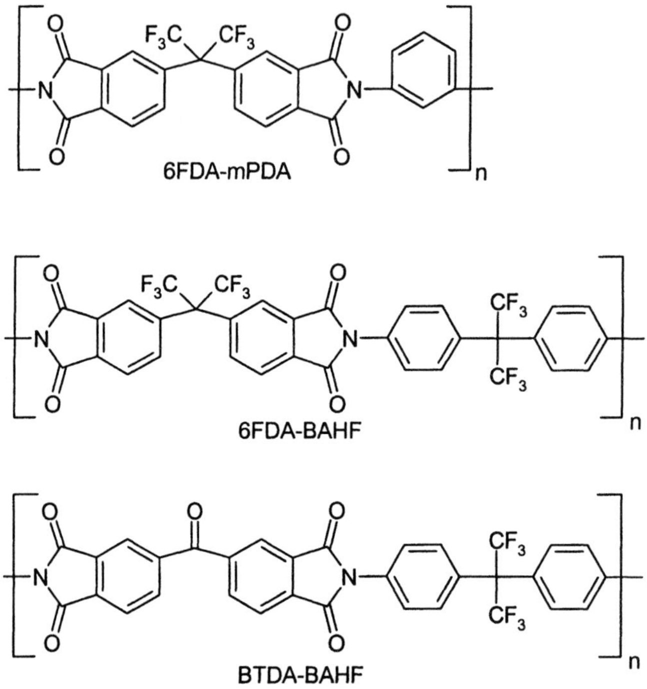

276

Scheme 4. Polyimide structures.

0.5 M NaCl. Under these conditions, the partially deprotonated poly(amic acid) is near its solubility limit, and thus is adsorbed in relatively large quantities on the surface (*34*). We deposited PAH from pH 5 water, because at this pH value, PAH is nearly fully protonated. This results in an extended chain conformation that yields very thin layers of adsorbed PAH (*35*). Overall, the use of these deposition solutions allows synthesis of films that contain 90% 6FDA-mPDA and 10% PAH as determined from ellipsometric thickness measurements performed after deposition of each polymer.

Remarkably, the gas permeabilities of several different imidized poly(amic acid)/PAH films are essentially identical to literature values for the corresponding bulk polyimides (*27-29,36*). Table I lists permeabilities and selectivities for three polyimide/PAH membrane skins. Selectivities reach values as high as 6.9 for O_2/N_2 and 68 for CO_2/CH_4, showing that these films are clearly free of defects. Using 6FDA-mPDA/PAH films, we found that about 7 polyelectrolyte bilayers must be deposited to achieve selectivities typical of bulk polyimides. This corresponds to a thickness of only 35-40 nm as shown by FESEM images and ellipsometry (*18*).

Studies using different imidization and film formation conditions show that membrane permeability and selectivity are a strong function of these conditions.

For example, while imidization of 6FDA-mPDA/PAH films by heating at 250°C for 2 h yields an O_2 permeability of 2.8 Barrers, imidization at 150°C results in a permeability of only 0.6 Barrers. The O_2/N_2 selectivity for these films also decreases from 6.9 to 1.9 on going to the lower imidization temperature. Reflectance FTIR spectra suggest that these films are ~85% imidized when heated at 150°C and essentially 100% imidized at 250°C. Thus, even a small amount of residual poly(amic acid) appears to decrease permeability and selectivity. Unheated films show O_2/N_2 selectivities of 0.9 and O_2 permeabilities of 0.6 Barrers.

Table I. Gas permeabilities[a] and selectivities of imidized[b] poly(amic acid)/PAH membranes.

Membrane Skin	O_2 Permeability	CO_2 Permeability	O_2/N_2 Selectivity	CO_2/CH_4 Selectivity
6FDA-mPDA/PAH (9.5 bilayers)	2.8±0.5	11±1	6.9	68
6FDA-BAHF/PAH (14.5 bilayers)	14±0.7	53±4	5.2	38
BTDA-BAHF/PAH (10.5 bilayers)	2.2±0.4	8.8±1.2	5.8	52

[a]Permeability is given in Barrers ($10^{-10}cm^3$ (STP)cm/(cm^2 s cmHg)) and was measured at room temperature (23-24°C). Film thickness was determined by FESEM and found to be consistent with ellipsometric thicknesses of films deposited on Si wafers.
[b]Films were imidized by heating at 250°C for 2 h under N_2.

Larger amounts of PAH in the film also decrease selectivity. To prepare 6FDA-mPDA membranes with a higher fraction of PAH, we deposited PAH at a pH of 9 from a solution that contained 0.5 M NaCl. The higher pH of the solution results in a lower fraction of amine groups that are protonated, while the salt is able to screen the charge on PAH. The net result of these deposition conditions is that there is less repulsion between the charged groups of PAH, and chains adopt a structure with more loops and tails. This structure yields larger amounts of adsorbed polymer (37). Using these deposition conditions, we prepared films containing 25-30% PAH. Compared to films containing only 10% PAH and having the same thickness, O_2/N_2 selectivity decreased from 6.9 to 2.2 and CO_2/CH_4 selectivity dropped from 68 to 5.8. Permeability coefficients for O_2 and CO_2 were about the same in the two cases. This example clearly shows that control over deposition conditions is essential for achieving membranes with high selectivities.

In both ion-transport and gas-transport membranes, one attractive feature of polyelectrolyte films is that several different polyelectrolytes can be deposited in the same film to provide specific membrane properties. In the case of polyimide membranes for ion separations, we first deposited a base PSS/PAH film and then deposited a few PMDA-PDA/PAH bilayers on this film to provide selectivity (*33*). This arrangement was very useful because the continuous base of PSS/PAH allowed full coverage of the porous support by an extremely thin (4 to 9 nm) polyimide layer. The PSS/PAH layer provided little resistance to ion transport, and thus flux through hybrid PSS/PAH + PMDA-PDA/PAH membranes was as high as 50% of that through a bare porous support. Similar hybrid membranes should be applicable to gas separation, although the minimal gas permeability of PSS/PAH will prevent its use as a base layer. Using hybrid systems, we are working toward developing gas-separation membranes with a selective layer that is 10-15 nm thick.

Conclusions and Future Directions

Deposition of hyperbranched poly(acrylic acid) and multilayered polyelectrolyte films on porous supports yields ultrathin, defect-free membrane skins. Hyperbranched PAA is especially capable of covering large pores and can span diameters as large as 0.2 μm. Additionally, both types of films can provide a very diverse set of membranes through either derivatization of PAA or variation of the constituents of multilayered polyelectrolyte films. Adsorption of multilayer polyelectrolytes is the more convenient technique for film formation, as it involves a simple series of alternating immersions and rinses. However, for both types of films, the time required for deposition needs to be reduced if large area membranes are to be formed. Because the membranes can be easily varied, they may prove most suitable for specialty applications, where a versatile, small-area membrane is useful.

Our future work with these membranes will include deposition of films on more practical polymeric supports, further reductions in the thickness of membrane skins, and development of nanofiltration membranes. Alumina is a brittle, expensive support that is considerably more useful for membrane characterization than implementation, so the introduction of polymeric supports will facilitate studies with these systems. Using hybrid multilayer polyelectrolyte films, we are aiming to develop composite gas-separation membranes with a selective skin that is only 10-15 nm thick. Such systems will allow considerably more flux than current polyimide membranes. Finally, although this chapter focused on gas-separation studies, multilayer polyelectrolyte membranes appear to be especially promising for nanofiltration (*33,38,39*). We are currently working in this area.

Acknowledgments

I am grateful to the National Science Foundation (CHE-9816108) and the Department of Energy Office of Basic Energy Sciences for financial support of this work. I also greatly appreciate the work of Bo Young Kim, Milind Nagale, and Daniel Sullivan on these projects.

References

1. Freeman, B. D. *Macromolecules* **1999**, *32*, 375-380.
2. Robeson, L. M.; Burgoyne, W. F.; Langsam, M.; Savoca, A. C.; Tien, C. F. *Polymer* **1994**, *35*, 4970-4978.
3. Loeb, S.; Sourirajan, S. U.S. Patent 3,133,132, 1964.
4. Pinnau, I.; Freeman, B. D. In *Membrane Formation and Modification*; Pinnau, I., Freeman, B. D., Eds.; American Chemical Society: Washington, D.C., 2000; pp 1-22.
5. Cadotte, J. In *Materials Science of Synthetic Membranes*; Lloyd, D. R., Ed.; American Chemical Society: Washington, DC, 1985; pp 273-294.
6. Le Roux, J. D.; Paul, D. R. *J. Membr. Sci.* **1992**, *74*, 233-252.
7. Kesting, R. E.; Fritzche, A. K. *Polymeric Gas Separation Membranes*; John Wiley & Sons: New York, 1993.
8. Zhang, L. H.; Hendel, R. A.; Cozzi, P. G.; Regen, S. L. *J. Am. Chem. Soc.* **1999**, *121*, 1621-1622.
9. Liu, C.; Martin, C. R. *Nature* **1991**, *352*, 50-52.
10. Henis, J. M. S.; Tripodi, M. K. *Science* **1983**, *220*, 11-17.
11. Zhou, Y.; Bruening, M. L.; Bergbreiter, D. E.; Crooks, R. M.; Wells, M. *J. Am. Chem. Soc.* **1996**, *118*, 3773-3774.
12. Nagale, M.; Kim, B. Y.; Bruening, M. L. *J. Am. Chem. Soc.* **2000**, *122*, 11670-11678.
13. Bruening, M. L.; Zhou, Y.; Aguilar, G.; Agee, R.; Bergbreiter, D. E.; Crooks, R. M. *Langmuir* **1997**, *13*, 770-779.
14. Franchina, J. G.; Lackowski, W. M.; Dermody, D. L.; Crooks, R. M.; Bergbreiter, D. E.; Sirkar, K.; Russell, R. J.; Pishko, M. V. *Anal. Chem.* **1999**, *71*, 3133-3139.
15. Xiao, K. P.; Kim, B. Y.; Bruening, M. L. *Electroanalysis* **2001**, *13*, 1447-1453.
16. Decher, G. *Science* **1997**, *277*, 1232-1236.
17. Bruening, M. L.; Sullivan, D. M. *Chem. Eur. J.* **2002**, *8*, 3832-3837.
18. Sullivan, D. M.; Bruening, M. L. *Chem. Mater.*, *in press*.
19. Martin, C. R. *Science* **1994**, *266*, 1961-1966.

20. Arnold, M. E.; Nagai, K.; Freeman, B. D.; Spontak, R. J.; Betts, D. E.; DeSimone, J. M.; Pinnau, I. *Macromolecules* **2001**, *34*, 5611-5619.
21. Xiao, K. P.; Harris, J. J.; Park, A.; Strautman, C.; Pradeep, V.; Bruening, M. L. *Langmuir* **2001**, *17*, 8236-8241.
22. Xu, Y.; Watson, J. T.; Bruening, M. L. *Anal. Chem., in press.*
23. Stroeve, P.; Vasquez, V.; Coelho, M. A. N.; Rabolt, J. F. *Thin Solid Films* **1996**, *284-285*, 708-712.
24. Leväsalmi, J.-M.; McCarthy, T. J. *Macromolecules* **1997**, *30*, 1752-1757.
25. Ackern, F. v.; Krasemann, L.; Tieke, B. *Thin Solid Films* **1998**, *327-329*, 762-766.
26. Kotov, N. A.; Magonov, S.; Tropsha, E. *Chem. Mater.* **1998**, *10*, 886-895.
27. Koros, W. J.; Walker, D. R. B. *Polymer J.* **1991**, *23*, 481-490.
28. Tanaka, K.; Kita, H.; Okano, M.; Okamoto, K. *Polymer* **1992**, *33*, 587-592.
29. Tanaka, K.; Okano, M.; Toshino, H.; Kita, H.; Okamoto, K. *J. Polym. Sci., Part B: Polym. Phys.* **1992**, *30*, 907-914.
30. Baur, J. W.; Besson, P.; O'Connor, S. A.; Rubner, M. F. *Mat. Res. Soc. Symp. Proc.* **1996**, *413*, 583-588.
31. Anderson, M. R.; Davis, R. M.; Taylor, C. D.; Parker, M.; Clark, S.; Marciu, D.; Miller, M. *Langmuir* **2001**, *17*, 8380-8385.
32. Liu, Y.; Wang, A.; Claus, R. O. *Appl. Phys. Lett.* **1997**, *71*, 2265-2267.
33. Sullivan, D. M.; Bruening, M. L. *J. Am. Chem. Soc.* **2001**, *123*, 11805-11806.
34. Dubas, S. T.; Schlenoff, J. B. *Macromolecules* **1999**, *32*, 8153-8160.
35. Shiratori, S. S.; Rubner, M. F. *Macromolecules* **2000**, *33*, 4213-4219.
36. Langsam, M. In *Polyimides Fundamentals and Applications*; Ghosh, M. K., Mittal, K. L., Eds.; Marcel Dekker: New York, 1996; pp 697-741.
37. Steeg, H. G. M. v. d.; Stuart, M. A. C.; Keizer, A. d.; Bijsterbosch, B. H. *Langmuir* **1992**, *8*, 2538-2546.
38. Stair, J. L.; Harris, J. J.; Bruening, M. L. *Chem. Mater.* **2001**, *13*, 2641-2648.
39. Balachandra, A. M.; Dai, J.; Bruening, M. L. *Macromolecules* **2002**, *35*, 3171-3178.

Chapter 19

Surface Modification of Polymeric Membranes by UV Grafting

Doo Hyun Lee, Hyun-Il Kim, and Sung Soo Kim*

School of Environmental and Chemical Engineering, Center
for Advanced Functional Polymers, Kyung Hee University, Yongin City,
Kyunggido 449–701, Korea

Polypropylene and polysulfone membranes were surface
modified by UV grafting with acrylic acid. ATR and XPS
analyses confirmed that acrylic acid was well grafted to the
membrane surface. Grafting of acrylic acid hydrophilized the
membrane surfaces, which enhanced the membrane water flux
up to 10 times. It also improved the rejection properties of the
membranes, which is an unusual trend compared to those of
conventional membranes. The hydrophilized membrane
surface exhibited strongly reduced membrane fouling. Various
parameters of the UV grafting process were investigated, such
as reaction time, monomer and photo-initiator concentration,
irradiating distance, reaction temperature, and solvent type.
The effect of each parameter was explored to optimize the
membrane performance of these materials.

© 2004 American Chemical Society

281

There have been many studies of membrane materials and membrane preparation processes aimed at developing membranes with better performance (*1*). Many candidate materials have been tested and practical methods of polymer and fiber processing have been adapted to membrane fabrication. Polymer surface modification techniques have also been developed and characterized, such as surface grafting, plasma modification, and corona discharge (*2*). In some membrane separation processes, surface properties often play more important roles than bulk properties. Therefore, membrane surface modification may be as important as the membrane preparation itself. Surface modification can enhance membrane performance and reduce fouling while retaining the original advantageous properties. Moreover, it can create functionality for specific purposes. Surface modification can be applied to almost every membrane application. The primitive technique of surface modification was physical application of surface-active agents to a membrane surface prior to use, which is a temporary solution. Membrane performance can be stabilized by permanent surface modification. Several methods have been developed so far for surface modification such as UV-induced grafting, plasma-induced grafting, plasma polymerization, corona discharge, γ-ray irradiation and ion beam irradiation.

Belfort *et al.* modified poly(arylsulfone) ultrafiltration membranes with various monomers by photochemical reaction (*3,4*). Nyström *et al.* grafted ethanol onto the surface of polysulfone membranes by UV irradiation in order to hydrophilize it (*5*). Ranby *et al.* applied glycidyl acrylate, glycidyl methacrylate, and acrylic acid to the surfaces of polyethylene (PE), polypropylene (PP), and polystyrene (PS) film by UV grafting technique (*6-8*). Plasma treatment was initiated by Yasuda, who plasma coated nitrile type monomers onto a silicone sheet in order to enhance the hydrogen/methane selectivity. He prepared composite membranes by applying various plasmas to metal, ceramic, and polymeric supports (*9,10*). Many studies have been performed to modify membrane surfaces by plasma processes. Belfort hydrophilized polyacrylonitrile (PAN) membranes by plasma treatment with helium and water (*11*). Uliana, Isayama and Miyasaka also plasma treated polyvinyl alcohol (PVA) membranes and PP membranes for various applications (*12,13*). Surface modification of membranes can also be used to reduce fouling of ultrafiltration and microfiltration membranes. Each surface modification technique has its own characteristics, and more investigations for surface modification are being performed (*14-16*).

In this study, surface modification of polypropylene and polysulfone membranes for microfiltration and ultrafiltration were performed. A UV grafting process was developed, and acrylic acid was used for hydrophilization of the membrane surface. The membrane performance was determined as a function of several parameters, such as irradiation time, distance of UV source from the polymer film, reaction temperature, kind of solvent, monomer and photoinitiator concentration. The modified membranes were characterized by surface and structure analyses. Performance enhancement and fouling reduction were also examined.

Experimental

Materials

Polypropylene (PP) microfiltration membranes (Celgard 2400®) were obtained from Hoechst-Celanese Co. and polysulfone (PSf) ultrafiltration membranes were prepared by conventional solvent casting method from a dope solution of Udel P-700, NMP, and PVP. Acrylic acid (AA) was purchased from Aldrich Co. The inhibitor in the monomer was removed just before UV irradiation by an inhibitor removal column supplied by Aldrich Co. Acetone, ethanol, and methanol were purchased from Aldrich Co., and they were used as solvents for the grafting reaction. Benzophenone was purchased from Aldrich Co., and was used as a photoinitiator. Dextran was purchased from Aldrich Co. and used as a test solute for membrane rejection tests.

UV Grafting Reaction

Monomer was used directly or dissolved in solvent. Benzophenone was added as an initiator for the photo-induced reaction. A membrane was soaked in the solution of monomer, solvent and initiator for 30 min then subjected to UV irradiation from a high-pressure mercury lamp to initiate the grafting reaction. The lamp power was set at 350 Watts. The UV beam was not filtered, and its wavelength range was 200 to 500 nm. The grafting time was varied from 30 sec to 4 min, and the distance between the light source and the sample was adjusted to control the power received by the sample. The reaction temperature was varied from 25 to 70 $^{\circ}$C.

Membrane Characterization

0.5 wt% dextran (MW = 20,000) solutions were prepared for membrane performance tests. Permeation was performed in the dead end filtration mode for 1.5 hr at 2 bar. The concentration of test solute in the permeate was determined by a UV spectrometer (HITACHI I-2000). For the fouling test, 0.3 wt% bovine serum albumin (BSA) solution was used and permeation was conducted for 1 hr at 2 bars. The dry weights of the membrane before and after the BSA test were measured to calculate the amount of BSA adsorbed to the membrane surface. The contact angle of each membrane surface with water was measured using a contact angle measurement system (FTA200) to characterize its hydrophilicity. 1 µL of water was dropped from a micropipet onto the leveled surface of the membrane to make a single drop. A CCD camera installed perpendicular to the surface captured the image of the drop at a magnification of 100, which helped us confirm that the membrane surface at the contact point is horizontal. The captured image was processed by an image analyzer system (ImagePro, Mediacybernetics Co.). Each contact angle was measured at least 5 times and averaged to avoid errors from surface roughness. Membrane structure was examined by scanning electron microscopy (SEM) using a Stereoscan 440 from Leica Co. Chemical analyses of the membrane surfaces were performed by FT-IR (ATR) (Perkin Elmer 2000) and XPS (ESCA 2000, VG Microtech).

Results and Discussion

Effect of UV Grafting Time

When acrylic acid was grafted to the PP and PSf membrane surfaces, UV grafting time was varied from 30 sec to 6 min. As shown in Figures 1 and 2, the carbonyl stretching peak at 1,720 cm^{-1} in the ATR spectra was observed for both PP and PSf membranes; this peak confirmed the presence of acrylic acid on the membrane surface. Even though the carbonyl stretching peak was confirmed from membrane surfaces that were washed with solvent to remove ungrafted acrylic acid, it was not clear that the carbonyl stretching peak directly represented grafted acrylic acid. XPS analysis was performed to confirm the grafting reaction by showing the formation of new chemical bonds at the surface. The C 1s spectra of untreated PP and UV grafted PP membrane are compared in Figure 3. An untreated PP sample exhibits only a C-C peak, whereas a UV grafted sample contained O=C-O and O=C peaks in addition to the C-C peak, which confirmed the grafting of acrylic acid to the membrane surface.

In ATR spectra, the carbonyl peaks in both samples increased with increasing grafting time. However, after optimum points, the carbonyl peaks decreased in both cases. The optimum wavelength range for benzophenone was 300 – 350 nm; UV irradiation below 300 nm might damage the grafted acrylic acid as well as the membrane matrix after long grafting times. As shown in Figure 4, the thickness of the sponge layer on PSf membranes decreased with grafting time, which reflected the damage of the sample by the UV irradiation. Unfortunately, preparation of SEM samples for cross sectional views was not possible for PP membranes.

In Figures 5 and 6, the effect of grafting time on flux and rejection of PP and PSf membranes is illustrated. Each membrane showed the best flux at the maximum absorbance seen in the FTIR in Figures 1 and 2. The pure water flux was compared with the solution flux for each membrane. The solution flux for

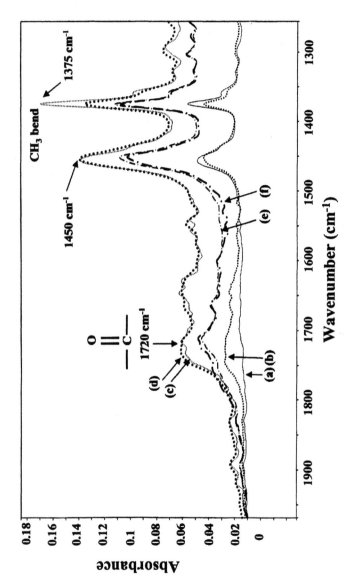

Figure 1. ATR spectra of PP membrane UV grafted with acrylic acid (irradiation distance: 16 cm, acrylic acid concentration: 0.3 mole/l, benzophenone concentration: 0.03 mole/l, temperature: 60 °C, solvent: acetone): (a) untreated, (b) 0.5 min, (c) 1 min, (d) 2 min, (e) 3 min and (f) 4 min.

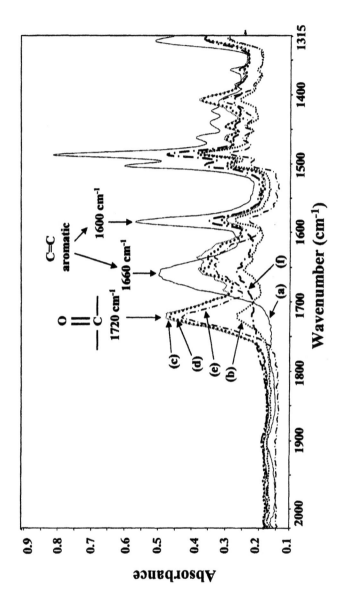

Figure 2. ATR spectra of PSf membrane UV grafted with acrylic acid (irradiation distance: 22 cm, acrylic acid concentration: 0.3 mole/l, benzophenone concentration: 0.02 mole/l, temperature: 60 °C, solvent: acetone): (a) untreated, (b) 0.5 min, (c) 1 min, (d) 2 min, (e) 3 min and (f) 4 min.

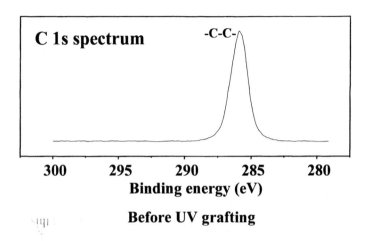

Before UV grafting

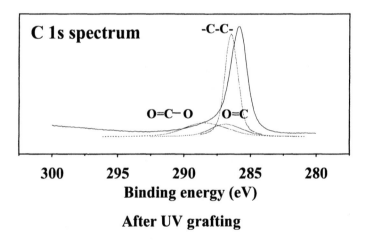

After UV grafting

Figure 3. XPS spectra of PP membrane UV grafted with acrylic acid (irradiation distance: 16 cm, acrylic acid concentration: 0.3 mole/l, benzophenone concentration: 0.03 mole/l, temperature: 60 °C, solvent: acetone).

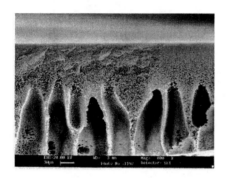
Original PSf

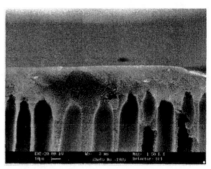
Grafted for 1 min

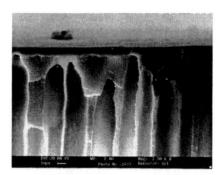
Grafted for 3 min

Figure 4. Cross sectional images of PSf membranes UV grafted with acrylic acid (irradiation distance: 22 cm, acrylic acid concentration: 0.3 mole/l, benzophenone concentration: 0.02 mole/l, temperature: 60 °C, solvent: acetone).

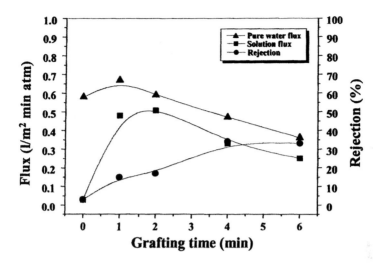

Figure 5. Effect of grafting time on performance of PP membrane UV grafted with acrylic acid (irradiation distance: 16 cm, acrylic acid concentration: 0.3 mole/l, benzophenone concentration: 0.03 mole/l, temperature: 60 °C, solvent: acetone).

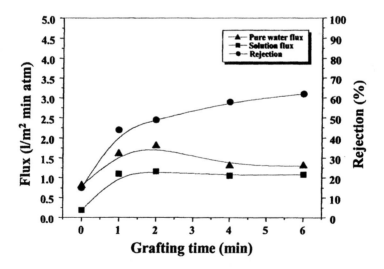

Figure 6. Effect of grafting time on performance of PSf membrane UV grafted with acrylic acid (irradiation distance: 22 cm, acrylic acid concentration: 0.3 mole/l, benzophenone concentration: 0.02 mole/l, temperature: 60 °C, solvent: acetone).

each membrane is smaller than the pure water flux, which can be attributed to membrane fouling by the dextran in the solution. The solution flux through untreated PP membranes was significantly reduced relative to the pure water flux value. Since PP is a hydrophobic material, membrane fouling was a serious issue for this sample. After UV grafting, the solution flux through the PP membranes greatly increased, and they reached values comparable to those of pure water. For both membranes, dextran rejection steadily increased with grafting time, which could be attributed to the damage of the microporous structure by the UV irradiation.

UV grafting with acrylic acid strongly reduced the contact angle with water as shown in Figure 7 for PP membranes. Surface roughness of microfiltration membranes could cause errors in measuring contact angle. In this work, the contacting image of the drop with the membrane surface was magnified by 1,000 times to assure correct measurements. Increasing grafting time reduced the contact angle of water drops on the membrane surface. However, there were also an optimum UV grafting time for minimizing the contact angle.

UV grafting with acrylic acid reduced the BSA adsorption for both PP and PSf membranes as shown in Figure 8. This result is believed to be associated with the increasingly hydrophilic character of the surface with increasing grafting time. PP membranes showed minimum adsorption at the optimum grafting time observed in other measurements.

Effect of UV Irradiation Distance

The distance between the sample and the UV source determined the beam intensity and influenced the membrane as shown in Figure 9. The hydrophobic PP membrane surface was hydrophilized at short distances with high beam intensity, and water flux increased. An increase in the source to sample distance reduced the beam intensity, the membrane surface was less hydrophilic after grafting, and the water flux was reduced. However, if the source to sample distance was too short, the sample was damaged, which decreased rejection, and it is shown in Figure 9 that there is an optimum distance for UV irradiation. For PP membranes, the optimum distance was 18 cm.

Effect of Monomer Concentration

Different amounts of acrylic acid were applied to the surface of PP membranes prior to the UV grafting reaction. As shown in Figure 10, an increase in the monomer concentration resulted in more grafting and a more hydrophilic surface, which increased water flux. However, if there was too much

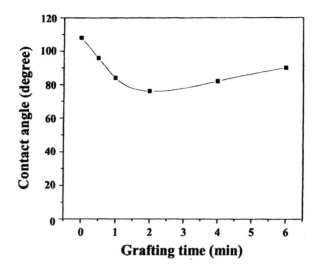

Figure 7. Effect of grafting time on contact angle of PP membranes UV grafted with acrylic acid (irradiation distance: 16 cm, acrylic acid concentration: 0.3 mole/l, benzophenone concentration: 0.03 mole/l, temperature: 60 °C, solvent: acetone).

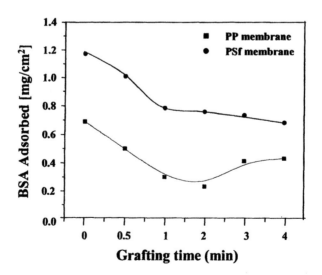

Figure 8. Effect of grafting time on BSA adsorption to PP and PSf membranes UV grafted with acrylic acid (irradiation distance: 22 cm, acrylic acid concentration: 0.3 mole/l, benzophenone concentration: 0.02 mole/l, temperature: 60 °C, solvent: acetone).

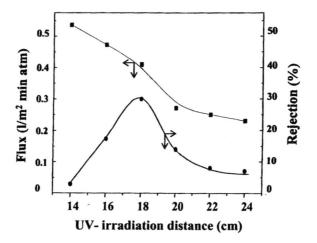

Figure 9. Effect of irradiation distance on performance of PP membranes UV grafted with acrylic acid (grafting time: 2min, acrylic acid concentration: 0.3 mole/l, benzophenone concentration: 0.02 mole/l, temperature: 60 °C, solvent: acetone).

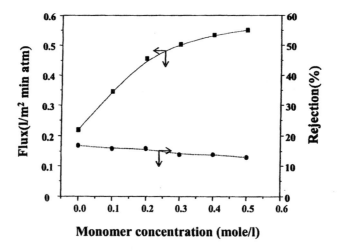

Figure 10. Effect of irradiation distance on performance of PSf membranes UV grafted with acrylic acid (grafting time: 2min, acrylic acid concentration: 0.3 mole/l, benzophenone concentration: 0.02 mole/l, temperature: 60 °C, solvent: acetone).

monomer, then monomer homopolymerization instead of grafting to the membrane surface occurred, and the rate at which flux increased with increasing monomer concentration was reduced. Rejection was low within the monomer concentration range considered. As shown in Figure 11, BSA adsorption was also reduced and reached a constant value at concentrations greater than 0.3 mole/l despite the increase of monomer concentration.

After the UV grafting reaction, the sample was solvent leached in a Soxhlet apparatus. Therefore, any unreacted monomer and ungrafted homopolymer were properly removed from the sample. The grafting ratio of each sample was determined by comparing the characteristic peak heights of acrylic acid and PP. Acrylic acid showed a characteristic peak at 1,710 cm^{-1} and PP has one at 710 cm^{-1}. The ratio of absorbance, A $_{1710 \text{ cm-1}}$/A $_{710 \text{ cm-1}}$ represented the relative grafting ratio. Relative grafting ratio increased with monomer concentration as shown in Figure 12. After the Soxhlet leaching, the relative grafting ratio was reduced by elimination of unreacted monomer and homopolymer. At high monomer concentration, Soxhlet leaching strongly reduced the relative grafting ratio. Therefore, there should be an optimum monomer concentration for effective grafting with little homopolymerization.

Effect of Photoinitiator Concentration

Different amounts of photoinitiator, benzophenone, were introduced into the UV grafting reaction of acrylic acid to the PP membrane. As shown in Figure 13, an increase in photoinitiator concentration enhanced the membrane performance to some extent by forming more radicals. However, too much radical formation inhibited the propagation reaction and terminated the chain reaction for grafting. Excessive photoinitiator resulted in a poor grafting reaction and deteriorated the membrane performance.

Effect of Reaction Temperature

UV grafting of acrylic acid to PP membranes was performed at various temperatures. An increase in reaction temperature helped the grafting reaction, and it enhanced the flux while slightly decreasing the rejection as shown in Figure 14. Figure 15 shows that BSA adsorption gradually decreased as reaction temperature increased. However, it must be remembered that a reaction temperature that is too high might damage the sample, so there is probably an optimum temperature for fast reaction, low BSA adsorption, and mimimal damage to the sample.

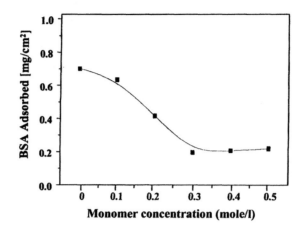

Figure 11. Effect of monomer concentration on performance of PP membranes UV grafted with acrylic acid (grafting time: 2min, irradiation distance: 16 cm, benzophenone concentration: 0.02 mole/l, temperature: 60 °C, solvent: acetone).

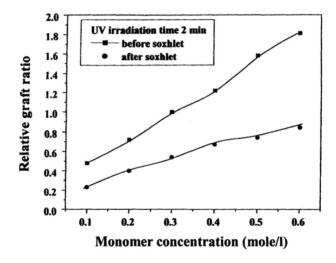

Figure 12. Effect of monomer concentration on relative grafting ratio of PP membranes UV grafted with acrylic acid (grafting time: 2min, irradiation distance: 22 cm, benzophenone concentration: 0.02 mole/l, temperature: 60 °C, solvent: acetone).

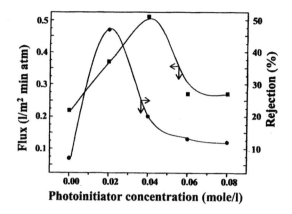

Figure 13. Effect of photoinitiator concentration on performance of PP membranes UV grafted with acrylic acid (grafting time: 2min, irradiation distance: 22 cm, acrylic acid concentration: 0.3 mole/l, temperature: 60 °C, solvent: acetone).

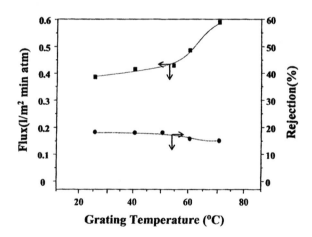

Figure 14. Effect of reaction temperature on performance of PP membranes UV grafted with acrylic acid (grafting time: 2min, irradiation distance: 22 cm, acrylic acid concentration: 0.3 mole/l, benzophenone concentration: 0.04 mole/l, solvent: acetone).

Effect of Solvent

UV grafting of acrylic acid to PP membranes was performed using various solvents, and performance variations were observed as shown in Figure 16. If acrylic acid itself was applied to the membrane without any solvent, the flux of the membrane was improved. Hsiue has reported that mobility of the monomer plays an important role in the grafting reaction (18). The mobility of monomer without any solvent should be low, and this is the likely reason that the grafting reaction was not successful. When acrylic acid was dissolved in several solvents for the grafting reaction, they showed better performance than the membrane treated with monomer without solvent. Three solvents were selected in this work; methanol, ethanol, and acetone, each at a concentration of 0.3 mole/L. Acetone is known to absorb UV light below 320 nm and can be excited even without any initiator (7). Therefore, acetone was the best solvent among those considered, and it enabled more grafting and better performance than the other solvents. The solubility parameter of the solvent was also an important factor. The solubility parameter difference between the solvent and acrylic acid made a difference in performance between methanol and ethanol. Reduction in BSA adsorption with different solvents was in the same order as separation performance enhancement, as shown in Figure 17.

Conclusion

UV grafting of acrylic acid to PP and PSf membranes was successfully developed. UV grafting enhanced membrane performance and reduced membrane fouling. The best performance and fouling resistance data for PP and PSf membranes after UV grafting are listed in Table I. The optimum grafting time resulted in the best performance without sample damage. Irradiating distance should also be controlled for the best grafting effects. Monomer and photoinitiator concentration should be optimized to prevent homopolymerization and termination. The grafting reaction was accelerated at high reaction temperature. The mobility of the monomer determined the reaction rate, and solubility differences and UV absorbing characteristics determined the performance and fouling resistance after UV grafting reaction. Acetone was the best solvent for the acrylic acid grafting reaction.

Acknowledgement

The authors wish to acknowledge financial support by Kyung Hee University and the Center for Advanced Functional Polymers.

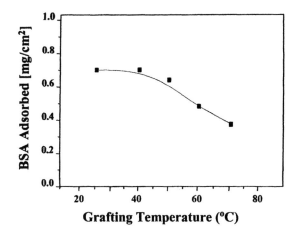

Figure 15. Effect of reaction temperature on BSA adsorption to PP membranes UV grafted with acrylic acid (grafting time: 2min, irradiation distance: 22 cm, acrylic acid concentration: 0.3 mole/l, benzophenone concentration: 0.02 mole/l, solvent: acetone).

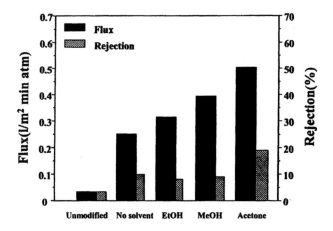

Figure 16. Effect of solvent on performance of PP membranes UV grafted with acrylic acid (grafting time: 2min, irradiation distance: 22 cm, acrylic acid concentration: 0.3 mole/l, benzophenone concentration: 0.02 mole/l solvent: acetone).

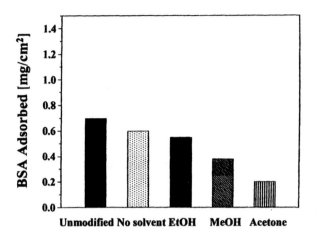

Figure 17. Effect of solvent on BSA adsorption to PP membranes UV grafted with acrylic acid (grafting time: 2min, irradiation distance: 22 cm, acrylic acid concentration: 0.3 mole/l, benzophenone concentration: 0.02 mole/l solvent: acetone).

Table I. Comparison of Flux, Dextran Rejection, and Protein Adsorption Before and After UV Grafting Acrylic Acid to PP and PSf Membranes.

Membrane	Flux (l/m² atm min)	Rejection (%)	BSA adsorbed (mg/cm²)
Untreated PP	0.03	3	0.7
UV grafted PP	0.53	18	0.2
Untreated PSf	0.19	15	1.2
UV grafted PSf	1.08	51	0.8

References

1. *Ultrafiltration and Microfiltration Handbook*; Cheryan, M., Technomic Pub. Co., Lancaster, PA, 1986; p 345.
2. *Polymer Surface Modification and Characterization*; Chan, C-M., Hanser Publishers, Cincinnati, OH, 1994; p 193.
3. Yamagishi, H.; Crivello, J. V.; Belfort, G. *J. Membrane Sci.* **1995**, *105*, 237-249.
4. Yamagishi, H.; Crivello, J. V.; Belfort, G. *J. Membrane Sci.* **1995**, *105*, 249-259.
5. Nyström, M.; Järvinen, P. *J. Membrane Sci.* **1991**, *60*, 275-296.
6. Allmér, K.; Hult, A.; Rånby, B. *J. Polym. Sci.* **1989**, *27*, 3419-3427.
7. Allmér, K.; Hult, A.; Rånby, B. *J. Membrane Sci.* **1988**, *26*, 2099-2111.
8. Allmér, K.; Hult, A.; Rånby, B. *J. Membrane Sci.* **1989**, *27*, 1641-1652.
9. Yasuda, H. *J. Membrane Sci.* **1984**, *18*, 273-284.
10. *Plasma Polymerization*; Yasuda, H., Academic Press, New York, NY, 1985; p277.
11. Ulbricht, M.; Belfort, G. *J. App. Polym. Sci.* **1995**, *56*, 325-343.
12. Vigo, F.; Nicchia, M.; Uliana, C. *J. Membrane Sci.* **1988**, *36*, 187-199.
13. Hirotsu, T.; Isayama, M. *J. Membrane Sci.* **1989**, *45*, 137-154.
14. Yokoyama, Y.; Tanioka, A.; Miyasaka, K. *J. Membrane Sci.* **1989**, *43*, 165-175.
15. Akhtar, S.; Hawes, C.; Dudley, L.; Reed, I.; Stratford, P. *J. Membrane Sci.* **1995**, *107*, 209-218.
16. Jonsson, G.; Pradanos, P.; Hernandez, A. *J. Membrane Sci.* **1996**, *112*, 171-183.
17. Jonsson, A.-S.; Jonsson, C. *J. Membrane Sci.* **1995**, *108*, 79-87.
18. Yang, J-M.; Hsiue, G-H. *J. App. Polym. Sci.* **1990**, *39*, 1475-1484.

Chapter 20

Effect of Microwave Irradiation on CO_2 Permeability in Cellulose Acetate Membranes

Y. Tsujita, Y. Nakai, H. Yoshimizu, and M. Yamauchi

Polymeric Materials Course, Department of Materials Science and Engineering, Nagoya Institute of Technology, Gokiso-cho, Showa-ku, Nagoya 466–8555, Japan

2.45 GHz microwaves are widely used for heating and drying materials. Water molecules in foods are heated up spontaneously by microwave radiation. This concept is applied to polymers containing OH groups to generate locally enhanced molecular motion of OH groups. Locally enhanced molecular motion might increase gas permeation and gas diffusion due to an increase of free volume accompanying the enhanced molecular motion. To investigate the ability to control gas permeation using microwave radiation, CO_2 permeation and diffusion coefficients in cellulose acetate were measured under constant microwave irradiation at 2.45GHz. Both permeability and diffusivity increased with increasing microwave power, up to 500W. In contrast, the solubility coefficient decreased as microwave power increased.

Separation of low molecular weight substances using chemical engineering process such as distillation can require large amounts of energy. Much attention has focused on gas separation membranes as a low energy separation technology (1, 2). Polymers are widely used in such applications because they can be fabricated in large surface areas as very thin, defect free membranes. The effect of polymer chemical structure on gas separation properties has been studied for a long time (1, 2). Recently, many polymeric membranes have been developed that contain high levels of free volume (3-14). Typical examples include poly(1-

© 2004 American Chemical Society

trimethylsilyl-1-propyne) and perfluorodioxole copolymers, which contain fractional free volumes of 0.343 and 0.30-0.33, respectively. These polymeric membranes are especially highly gas permeable, but they exhibit low permselectivity for the separation of permanent gases such as oxygen and nitrogen. However, not only high permeation rates but also high selectivity is required for gas separation membranes.

On the other hand, physical modifications of membranes have been explored as a route to enhance permeation properties. Physical modifications such as thermal quenching and pressure conditioning from the liquid state were tried for glassy polymers (15-17). Such processes effectively alter the physical rather than the chemical structure of the polymer. That is, they change the microvoid content, fractional free volume, and unrelaxed volume of glassy polymers, and variations in these properties can influence gas permeation performance.

We have studied the gas permeability of electro-conductive polypyrole membranes in the presence of external fields such as electric fields. Application of a direct electric field in the direction of the membrane thickness could change the O_2 permeability coefficient, and the amount of change in permeability depended on the amount of electric power applied to the membrane. Due to electroconductivity, the O_2 permeability coefficient was enhanced by applying a direct electric field to the membrane if the O_2 permeation and hole mobility were in the same direction. On the other hand, the O_2 permeability coefficient was reduced by reversing the direction of the direct electric field. It is probable that the interaction between the holes in the polypyrole membrane and O_2 gas was of a radical nature.

2.45 GHz microwave radiation has been widely used for heating and drying materials such as rubbers, foods, and prints (18). In particular, it has been applied to vulcanization of rubbers, foam processing, polymerization-solidification, adhesion, and so on. Water or water vapor in foods is heated spontaneously upon irradiation by 2.45 GHz microwaves. This concept is applied to polymers, such as cellulose and cellulose acetate, which contain OH groups. In principle, the microwave radiation can induce locally enhanced molecular motion of the OH groups attached to the polymer backbone. It is expected that the locally enhanced molecular motion might increase gas permeation and gas diffusion by increasing the free volume.

Experimental

Samples

A cellulose film 35.5 μm thick was kindly provided by Prof. E. Nakanishi. It was prepared by casting from a solution of LiCl and dimethylacetamide (DMAc).

Cellulose acetate was purchased from Aldrich Co. The molecular weight was 30,000, and the acetyl content was 39.8 wt% (degree of substitute 2.4). Tetrahydrofuran solutions of cellulose acetate were cast onto Petri dishes at room temperature and dried thoroughly at 30 °C for 24hrs or more and then annealed at 190 °C under vacuum for 6hrs. Transparent cellulose acetate films were obtained. Their thickness was about 45 μm, and their glass transition temperature was 180 °C.

Methods

Dielectric Dispersion

Dielectric dispersion curves were obtained at frequencies from 100Hz to 100kHz. Temperature was varied from -100 °C to 100 °C. The apparatus used was a dynamic dielectric spectrometer manufactured by Seiko Instrument (DES -100).

Permeability coefficient

A gas permeation apparatus, which could operate in the presence of microwave irradiation, was constructed from a conventional permeation apparatus and a microwave generator (see Figure 1). Gas permeation experiments were conducted under constant microwave irradiation at 2.45GHz. In a typical experiment, a flat membrane having an area of 1.54 cm^2 was set into the permeation cell. The total system, including the membrane, was evacuated under vacuum for 24 hours. The upstream pressure was fixed at 76 cmHg of CO$_2$. The downstream (i.e., low) pressure was recorded as a function of time while the membrane was exposed to constant microwave irradiation. The permeability coefficient was evaluated from the slope of the permeation curve after reaching steady state. We have taken care to homogeneously irradiate the polymeric membrane. To this end, a so-called "dummy road" (see Figure 1) was installed to prevent reflected radiation from reaching the membrane.

Diffusion coefficient

Diffusion coefficients were evaluated using the classic time lag method. After thorough drying, the polymeric membrane was subjected to constant irradiation. CO$_2$ at 76 cmHg was introduced to the upstream side of the membrane, and the time lag was determined after reaching steady state. The average diffusion coefficient (D) was calculated from the time lag (θ) obtained using eq (1):

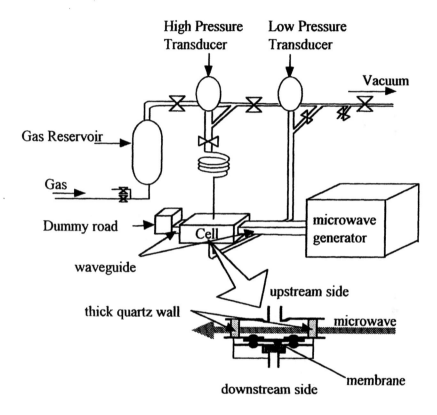

Figure 1. Apparatus for measuring gas permeability in the presence of microwave irradiation.

$$D=L^2/6\,\theta \qquad\qquad (1)$$

where L is the membrane thickness.

Results and Discussion

Dielectric Dispersion

Figure 2 presents the dielectric dispersion transition map of cellulose acetate in the form of an Arrhenius plot. According to this transition map, the dielectric loss peak at 2.45 GHz (i.e., the microwave frequency) is expected to appear at around 170 °C. However, the dielectric dispersion is quite broad, and it is likely that even at 25°C, some molecular motion is excited by exposing the sample to 2.45 GHz microwave radiation, and this molecular motion is probably related to the effect of microwave irradiation on the diffusion and permeability coefficients.

Permeation

CO_2 permeation curves of cellulose acetate membranes being irradiated with microwaves at power levels of 100, 300, and 500W are shown in Figure 3. One can observe enhancement of CO_2 permeation during microwave irradiation, indicative of a remarkable instantaneous response of permeation to microwave irradiation. The permeability enhancement increases with power, suggesting that one might control permeability by varying microwave power. The permeation curve reduced rapidly to the original permeation curve when the microwave radiation was stopped. These changes in permeation rates upon microwave irradiation were not observed in a polystyrene film, presumably because it does not contain any highly polar, strongly microwave responsive groups. Figure 4 presents the effect of microwave power level on average CO_2 gas permeability coefficient in a CA membrane at 76 cmHg and 25 °C. The permeability coefficient changes parabolically with microwave power, indicating a spontaneous response of local molecular motion of groups such as OH groups to the microwave radiation. The increased molecular motion resulting from microwave irradiation enhanced gas permeability coefficients.

As shown in Figure 5, the CO_2 permeability of cellulose increased by approximately one order of magnitude as microwave power increased from 0 to 500 W. Cellulose, which contains 3OH groups per repeat unit, exhibited a much stronger effect of microwave irradiation on permeability than cellulose acetate (see Figure 4). These results suggest that molecular motion of OH groups in the

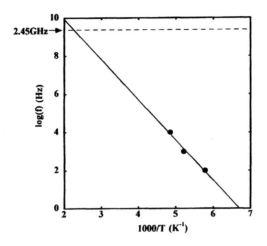

Figure 2. Dielectric dispersion transition map of cellulose acetate.

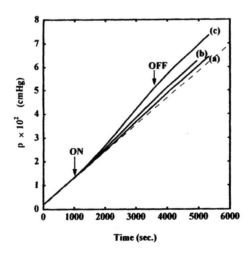

Figure 3. CO_2 gas permeation curve of cellulose acetate during microwave irradiation at 76cmHg and 25°C; (a) 100W, (b) 300W, and (c) 500W.

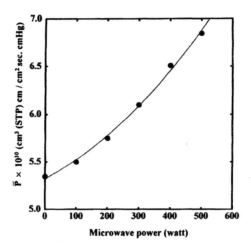

Figure 4. The effect of microwave power on average CO_2 permeability coefficient in a cellulose acetate membrane at 76cmHg and 25°C.

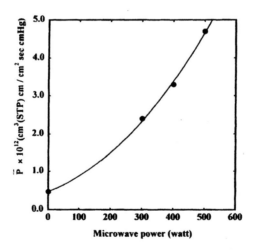

Figure 5. The effect of microwave power on average CO_2 permeability coefficient in a cellulose membrane at 76cmHg and 25°C.

polymer repeat unit can be activated by microwave irradiation, and the molecular motion of these groups can enhance gas permeation and diffusion.

Diffusion and Solubility

Diffusion coefficients during microwave irradiation were obtained by the time lag method. Figure 6 presents the effect of microwave power on average CO_2 gas diffusion coefficients in CA membranes at 1atm and 25 °C. Relative to the permeation data presented in Figure 4, there is a much stronger effect of microwave power on diffusivity. Consistent with the results presented in Figure 2, this increase in diffusivity is basically related to the enhanced local molecular motion of the OH groups in cellulose acetate during microwave irradiation.

One can calculate the effect of microwave irradiation on solubility coefficients (S) by using eq (2) and the permeability and diffusion coefficients obtained with and without microwave irradiation. The permeability coefficient is determined by the product of the diffusion coefficient and the solubility coefficient according to the solution-diffusion mechanism:

$$P=DS \qquad (2)$$

Solubility was reduced when the membrane was subjected to microwave irradiation. This might be explainable in terms of enhanced local molecular motion during microwave irradiation. This observation corresponds to lower gas solubility in the polymer which is in a state of more activated molecular motion, e.g., higher temperature. However, as shown below, the temperature increase in the polymer due to microwave irradiation is very small.

We are interested in the effect of microwave irradiation on permeability, diffusivity, and solubility. For a microwave radiation power of 500W, diffusivity increases by about 40% and solubility decreases by about 7%, which leads to an increase in permeability of about 30%. Assuming that the temperature dependence of solubility obeys an Arrhenius relation and using the reported heat of solution of CO_2 in a CA membrane (–6.5 kcal/mol (*19*)), a 7% increase in solubility corresponds to a temperature increase of only about 2 Kelvin. During microwave irradiation, resistance thermometer sensors generally indicate temperatures higher than the real temperature because the microwave electric field concentrates at the tip of the resistance thermometer sensor. Nevertheless, we measured the surface temperature of CA and Cellulose membranes using a metal-sheathed, mineral insulated platinum resistance thermometer sensor. For irradiation of 500W, the temperature increase of both membranes was about 5 Kelvin. Therefore, there is no remarkable increase in temperature due to exposing the polymer membranes to microwave radiation.

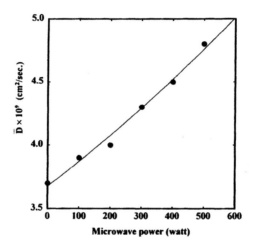

Figure 6. The effect of microwave power on average CO_2 diffusion coefficient in a cellulose acetate membrane at 76cmHg and 25°C.

Conclusion

2.45 GHz Microwave irradiation of polymer membranes containing strongly microwave responsive functional groups, such as OH groups, activates local molecular motion of these functional groups and enhances both gas permeability and diffusivity. The effect of microwave irradiation on transport properties is greater for cellulose than for cellulose acetate. There is little effect of microwave irradiation on polystyrene (which has no strongly microwave-responsive functional groups). Permeability and diffusion coefficients of cellulose acetate increased with microwave power up to 500W. In contrast, solubility decreased with increasing microwave power.

More research is needed to better understand the effect of microwave radiation on gas permeation and separation properties. Further studies of the frequency dependence of microwave radiation on transport properties are planned. Additionally, the influence of microwave radiation on polymers containing functional groups other than OH groups is not understood. Mixture gas separation properties of cellulose acetate membranes during microwave irradiation are expected to be different from those not subject to microwave irradiation, but this needs to be demonstrated.

Acknowledgement

We greatly acknowledge active discussions with Dr. M. Miyamoto, Central Research Laboratory, Nissan Chemical Co. Japan. Partial financial support was provided by a Grant-in-Aid for Scientific Research in Priority Area (B), "Novel Smart Membrane with Controlled Molecular Cavity", No.13133202 (2001) from the Ministry of Education, Culture, Sports, Science and Technology and for Scientific Research (B), No.11695047 (1999) from the Japan Society for the Promotion of Science.

References

1. Kesting, R. E. *Synthetic Polymeric Membranes;* A Structural Perspective Second Edition; Wiley-Interscience: New York, 1985.
2. Osada, Y.; Nakagawa, T., Eds.; *Membrane Science and Technology;* Marcel Dekker, Inc.: New York,1992.
3. Morisato, A.; Pinnau, I. *J. Membrane Sci.*, **1996**, *121*, 229-242.
4. Morisato, A.; Freeman, B. D.; Pinnau, I.; Casillas, C. G. *J. Polymer Sci.*, **1996**, *34*, 1925-1934.
5. Pinnau, I.; Casillas, C. G.; Morisato, A.; Freeman, B. D. *J. Polymer Sci.*, **1996**, *34*, 2613-2621.

6. Pinnau, I.; Casillas, C. G.; Morisato, A.; Freeman, B. D. *J. Polymer Sci.*, **1997,** *35,* 1483-1490.
7. Merkel, T. C.; Bonder, V.; Nagai, K.; Freeman, B. D. *Macromolecules*, **1999,** *32,* 370-374.
8. Toy, L. G.; Freeman, B. D.; Spontak, R. J.; Morisato, A.; Pinnau, I. *Macromolecules*, **1997,** *30,* 4766-4769.
9. Dixon-Garrett, S. V.; Nagai, K.; Freeman, B. D. *J. Polymer Sci.*, **2000,** *38,* 1078-1089.
10. Nagai, K.; Toy, L. G.; Freeman, B. D.; Teraguchi, M.; Masuda, T.; Pinnau, I. *J. Polymer Sci.*, **2000,** *38,* 1474-1484.
11. Bondar, V. I.; Freeman, B. D.; Yampolskii, Yu. P. *Macromolecules*, **1999,** *32,* 6163-6171.
12. Morisato, A.; Shen, H. C.; Sankar, S. S.; Freeman, B. D.; Pinnau, I.; Casillas, C. G. *J. Polymer Sci.*, **1996,** *34,* 2209-2222.
13. Nagai, K.; Mori, M.; Watanabe, T.; Nakagawa, T. *J. Polymer Sci.*, **1997,** *35,* 119-131.
14. Merkel, T. C.; Bondar, V.; Nagai, K.; Freeman, B. D. *J. Polymer Sci.*, **2000,** *38,* 273-296.
15. Tsujita, Y.; Hachisuka, H.; Imai, T.; Takizawa A.; Kinoshita, T. *J. Membrane Sci.*, **1991,** *60,* 103-111.
16. Hachisuka, H.; Takizawa, H.; Tsujita, Y.; Takizawa A.; Kinoshita, T. *Polymer*, **1991,** *32,* 2382-2386.
17. Hachisuka, H.; Tsujita, Y.; Takizawa, A.; Kinoshita, T. *J. Polymer Sci.*, **1991,** *29,* 11-16.
18. Shibata, C. *Industrial Microwave Power Engineering,* Dennkisyoinn, Japan, 1986.
19. Stern, S. A.; De Meringo, A. H. *J. Polymer Sci. Polym. Phys. Ed.*, **1978,** *16,* 735-751.

Chapter 21

Improving the Permselectivity of Commercial Cation-Exchange Membranes for Electrodialysis Applications

Sophie Tan, Alexis Laforgue, and Daniel Bélanger

Département de Chimie, Université du Québec à Montréal, C.P. 8888, Succursale Centre-Ville, Montréal, Québec H3C 3P8, Canada

Cation-exchange membranes bearing sulfonate groups (Neosepta CMX) were modified by chemical polymerization of aniline at the surface of the membranes. The doped polyaniline (PANI) adsorbed at the surface of the membrane consists of a positively charged layer which acts as an electrostatic barrier for multivalent cations. The resulting composite membranes (CMX-PANI) were characterized by electrodialysis, scanning electron microscopy (SEM), exchange capacity measurements (EC) and X-ray photoelectron spectroscopy (XPS). The presence the PANI layer was shown to improve the membrane permselectivity for protons vs. bivalent cations (Zn^{2+} and Cu^{2+}) by a factor of at least 20 after electrodialysis in acidic solutions. Optimization of the anilinium exchange time as well as the polymerization time were performed. It was also demonstrated that the blocking efficiency of the PANI layer depended on the thickness and uniformity of this layer.

© 2004 American Chemical Society

Introduction

Ion-exchange membranes are widely used in several electrochemical technologies such as fuel cells, electrolysis, and electrodialysis. Anion- and cation-exchange membranes allow selective transport of anions and cations, respectively, based on electrostatic repulsions. For instance, cation-exchange membranes (CEM) typically possess carboxylate (-COO⁻) or sulfonate (-SO₃⁻) groups which are negatively charged in sufficiently acidic solutions. Anion-exchange membranes (AEM) usually contain amine groups (-NR₃⁺ , R = H or alkyl chains) that are positively charged (*1,2*).

Electrodialysis is a process in which several pairs of CEMs and AEMs are placed alternatively between two electrodes. A current is applied between these two electrodes which forces the displacement of anions towards the anode and cations towards the cathode. The presence of CEMs and AEMs allows the separation of compounds in different compartments. Examples of applications for electrodialysis include the reduction of electrolytes in the food industry, the recovery of valuable electrolytes (pure NaCl from seawater, amino acids from protein hydrolysates, acids from metal pickling and rinsing baths, etc.), salt splitting and more (*1,3*).

The main limitations of electrodialysis are mostly related to the permselectivity of the membranes or to their stability after repetitive usage. In order to improve the permselectivity of cation-exchange membranes for specific ions, different approaches have been studied (*4*). One was based on a sieving effect by which smaller hydrated ions selectively cross the membranes (*4*). A second approach was to form a positively charged layer at the surface of a cation-exchange membrane to exclude transport of higher valence cations. Using the latter approach, Sata and co-workers have demonstrated that the presence of a polypyrrole (*5*) or polyaniline (PANI) (*6*) at the surface of a Neosepta CM-1 membrane improved its permselectivity for Na⁺ against Ca²⁺ after electrodialysis in neutral solutions.

In this study, commercial CEMs bearing sulfonate groups (Neosepta CMX from Tokuyama Soda) were modified by adsorbing a positively charged PANI layer on one side of the membrane. This modification is done in order to block the transport of bivalent metal cations in electrodialysis for the recovery of spent acids. We also report the characterization of the composite membrane.

Experimental

Materials and Chemicals

Neosepta CMX membranes (Tokuyama Corporation) were stored in 0.5 M NaCl prior to modification. $(NH_4)_2S_2O_8$ (EM Science), HCl (EM Science),

HNO$_3$ (EM Science), NaCl (BDH), H$_2$SO$_4$ (EM Science), ZnSO$_4$ 7H$_2$O (Anachemia) and CuSO$_4$ 5H$_2$O (Anachemia) were of A.C.S. reagent grade and used as received. Aniline (Aldrich) was distilled twice prior to use. Millipore water (18 MΩ) was used for the preparation of all solutions. The Neosepta CMX membrane properties are given in Table I.

Table I. Characteristics of the Neosepta CMX membrane given by the manufacturer (7)

Properties	Neosepta CMX
Composition	Poly(styrene-co-divinylbenzene)
Supporting Material	Poly(vinyl chloride)
Thickness	170-190 µm
Exchange Capacity	1.5-1.8 meq/g Na-form dry membrane
Burst Strength	5-6 kg/cm^2
Water Content	0.25-0.30 g H$_2$O/g Na-form dry membrane (equilibrated in 0.5 M NaCl)
Electric Resistance	2.5-3.5 Ωcm^2 (equilibrated in 0.5 M NaCl, at 25°C)

Modification of Cation-Exchange Membranes

The protocol used for the preparation of CMX-PANI composite membranes is based on a published procedure (6) and the detailed procedure used in this study is described elsewhere (8). Briefly, after exchanging the protons with anilinium species, an oxidant, 1 M aqueous (NH$_4$)$_2$S$_2$O$_8$, is added to induce polymerization. The modification was performed in a Teflon two-compartment cell in order to modify only a single face. Before use, the composite membranes were conditioned in a 1 M HCl aqueous solution to ensure complete protonation of PANI.

X-ray Photoelectron Spectroscopy (XPS)

After being conditioned in 1 M HCl, samples were rinsed with water and dried under vacuum at room temperature for 36 h. To compensate for charging effects, binding energies were corrected for covalent Cl2p$_{3/2}$ of the PVC support found in CMX membranes at 200.6 eV after deconvolution.

Electrodialysis (ED)

Electrodialyses were performed in a two-compartment cell containing, in the anodic compartment, a solution of 15 g Zn or Cu per liter of 0.5 M H_2SO_4 and in the cathodic compartment, 0.5 M H_2SO_4. The modified surface was placed facing the anolyte. A current density of 100 mA/cm^2 was applied between two platinum plates for 3 h. The metal concentration in both compartments was determined by atomic absorption spectroscopy. The total mass balance for cations was not calculated because the variation of proton concentration was too small to be detected at 0.5 M H_2SO_4. A smaller concentration needs to be used to allow detection of proton concentration variations. However, the percent metal leakage was corrected for the total mass of metal obtained after electrodialysis. It was calculated according to:

% leakage = (mg metal$_{cathodic}$) x 100 / (mg metal$_{cathodic + anodic}$). The total mass of metal in the cathodic and anodic compartments obtained in the latter expression corresponded to the initial amount of metallic ions added at the beginning of ED assays.

Exchange Capacity (EC)

After conditioning in 1 M HCl, H_2O and 1 M NaCl alternatively, the membranes were soaked for 24 h in a 1 M NaCl solution to ion-exchange H$^+$ with Na$^+$. The excess sodium chloride was then removed by immersion in water and the membranes were dried under vacuum for 48 h at room temperature in order to measure their weight in the Na$^+$ form. The Na$^+$ ions were released by immersion in a 1 M HCl solution during 24 h. The sodium concentration in the latter solution was determined by atomic emission spectroscopy.

Results and Discussion

Chemical oxidation of aniline using aqueous ammonium peroxodisulfate or $(NH_4)_2S_2O_8$ leads to the formation of the emeraldine form of polyaniline under optimal reaction conditions. The general structure of PANI is given below (9,10):

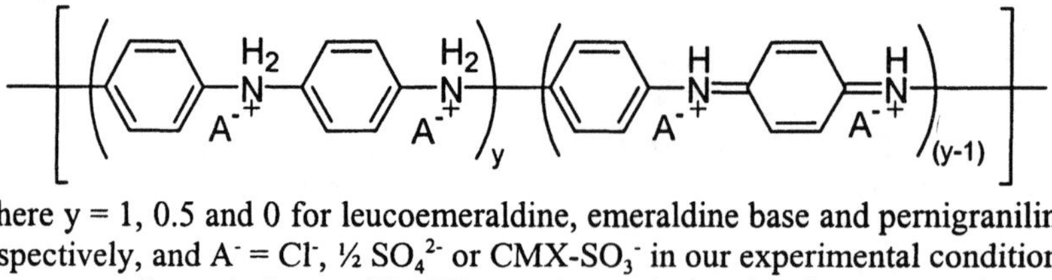

where y = 1, 0.5 and 0 for leucoemeraldine, emeraldine base and pernigraniline, respectively, and A$^-$ = Cl$^-$, ½ SO_4^{2-} or CMX-SO_3^- in our experimental conditions. The emeraldine salt form of PANI contains amine and imine groups that can both be protonated in acidic solutions, their respective pK_a values being 2.5 and 5.5 (10). Therefore, after immersing the membrane in 1 M HCl, a protonated polyaniline layer is expected to be formed at the surface of the membrane. Indeed, it was demonstrated by XPS that the PANI layer obtained under optimal polymerization conditions can be doped with chloride anions (8).

Characterization of the PANI Layer

Upon oxidation of aniline, the membrane presents a dark coloration associated with the presence of polyaniline. Due to the light brown color of the initial CMX membrane, it is difficult to determine whether the layer adsorbed at the surface is blue or green, characteristic colors of the emeraldine base and salt, respectively. The PANI seemed to adhere well to the membrane and cannot be removed even by scraping the surface with a spatula. To determine if PANI is present at the surface or within the bulk of the CMX membrane, SEM, EC and XPS characterization were performed.

The surface morphology was studied using SEM with micrographs taken at an angle of approximately 60°. Figure 1 shows that unmodified CMX membranes present a relatively smooth surface when compared to CMX-PANI composite membrane prepared under optimal conditions (anilinium exchange time: 1 h, polymerization time: 1 h). This first observation already suggests the presence of PANI at the surface of the membrane. In previous works (8,11), we have also deduced that there might be only a small amount of PANI within the membrane. In fact, the SEM micrographs of the cross-section of these membranes did not allow us to distinguish the PANI layer from the bulk of the membrane possibly because both layers are organic. It is actually believed that the PANI layer is partially interpenetrated within the CMX membrane considering the method used for modification. As for the thickness of the membrane, it was observed that after modification under optimal conditions, the thickness decreased by approximately 10 μm when compared to the CMX. This change in thickness was explained by the electrostatic interactions between the sulfonate groups and the positively charged PANI chains (11).

Figure 2 presents the XPS survey spectra of a bare CMX and a CMX-PANI membrane. The presence of PANI at the surface of the membrane is confirmed by the decrease in peak intensity corresponding to the fixed sulfonate groups, 978 eV ($O(KL_{2,3}L_{2,3})$ Auger peak), 531 eV (O1s), 228 eV (S2s) and 169 eV (S2p), and the enhancement of the N1s peak signal at 400 eV.

It should be noted that the opposite surface of the modified membranes showed neither any presence of PANI on the SEM micrographs nor on the XPS spectra (not shown). Both showed the same surface as the unmodified CMX membrane.

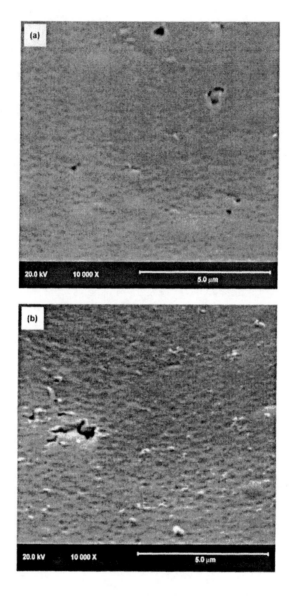

*Figure 1.   Scanning electron micrographs of the surface of a (a) CMX and (b)
CMX-PANI membrane (prepared under optimal conditions).*

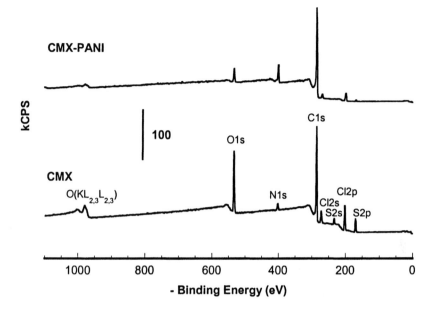

Figure 2. XPS survey spectra of a CMX and a CMX-PANI membrane (prepared under optimal conditions).

Blocking Efficiency of the CMX-PANI Membranes

A significant decrease in metal leakage is observed when comparing results obtained for the unmodified CMX and the modified membrane in Figure 3: the CMX membrane leads to metal leakage of over 10% whereas the modified membrane gives a value lower than 1%. Two modification parameters have been studied up to now: (1) the exchange time for incorporation of anilinium cations and (2) the polymerization time. Figure 3 shows the effect of immersion time (in a 10% aniline solution prepared in aqueous 1 M HCl) on the Zn(II) and Cu(II) leakage. Very little variation in blocking efficiency is observed with anilinium exchange time indicating that an optimal blocking efficiency is already attained after 1 h (with 1 h of polymerization). On the contrary, the polymerization time plays an important role in the PANI layer blocking efficiency. As observed in Figure 4, a polymerization time between 45 and 60 min leads to the lowest metal leakage.

Interestingly, Sata and co-workers also observed a loss in permselectivity for Na$^+$ vs. Ca^{2+} above 1 h of polymerization (*6*). They suggested that this could be caused by the overoxidation of PANI to form pernigraniline, the most oxidized state of PANI corresponding to the insulating form of PANI.

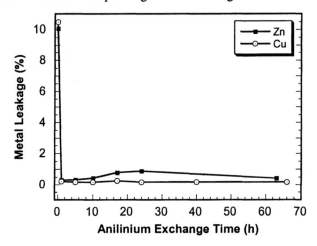

Figure 3. Zn(II) and Cu(II) leakage (%) for the unmodified and CMX-PANI composite membranes as a function of immersion time in aniline solution. t = 0 corresponds to the unmodified membrane. Polymerization time: 1 h.

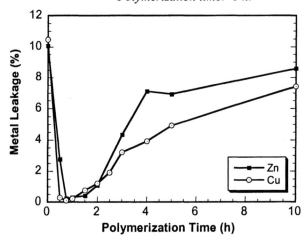

Figure 4. Zn(II) and Cu(II) leakage (%) for the unmodified and CMX-PANI composite membranes as a function of polymerization time. Anilinium exchange time: 1 h. t = 0 corresponds to the unmodified membrane. (Adapted from reference 8.)

Actually, the surface electronic conductivity of the membrane was found to decrease after 1 h of polymerization. Sata has demonstrated by doping PANI with bromine that no bromine was detected within the composite membrane after more than 10 h of polymerization.

Exchange Capacity

In order to understand the factors affecting the blocking efficiency of the PANI layer, a thorough characterization of the membrane was performed. First, exchange capacity values were measured as a function of anilinium exchange time and of polymerization time (Figure 5). EC data give an indication of positively charged amine groups present on and within the membrane. A portion of the positive charges on the polyaniline chains will be compensated by the fixed sulfonate groups on the membrane. Only the "free" sulfonate groups will be compensated by small ions such as H^+ or Na^+. Hence, the amount of PANI does not seem to change significantly when increasing the immersion time in the aniline solution over 1 h (Figure 5). This behavior could explain the electrodialysis data given in Figure 3. However, the EC results obtained as a function of polymerization time cannot directly explain the curve profile illustrated in Figure 4 for the metal leakage vs. polymerization time.

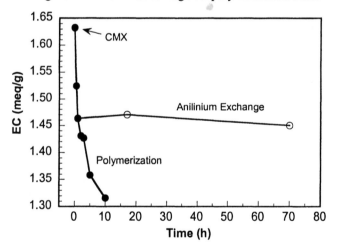

Figure 5. Exchange capacity of the CMX and CMX-PANI membranes as a function of immersion time in aniline solution (O) and of polymerization time (●).

The EC values decrease continuously with increasing polymerization time whereas the percent metal leakage for both Zn^{2+} and Cu^{2+} increases for polymerization times above 1 h. To further investigate the effect of polymerization time, an extensive characterization of the composite membrane has been completed and published elsewhere (8).

Evidence of PANI Degradation

Figure 6 illustrates the variation in surface content of different elements found for the CMX-PANI membranes as a function of polymerization time. The surface composition was determined using the appropriate XPS core level spectra (8). As mentioned previously, the PANI content at the surface of the membrane can be studied by monitoring the sulfur and nitrogen peaks. Figure 6a shows that the largest amount of PANI adsorbed at the surface of the membrane is found at 1 h of polymerization. Interestingly, this corresponds to the optimal polymerization time obtained from ED tests shown in Figure 4 which indicates that the blocking efficiency of the PANI layer might be related to the amount of PANI present at the surface. In fact, the EC and complete XPS study led us to believe that to obtain an efficient and permselective composite membrane, it is necessary to have a sufficiently thick and uniform PANI (8).

The evolution of the different chlorine components detected at the membrane surface was also studied and is depicted in Figure 6b. The unmodified CMX membrane is known to contain a large amount of poly(vinyl chloride) (PVC) since this polymer acts as a supporting material (1, 12). In addition, since the PANI chains are positively charged, chloride ions, introduced after conditioning the membrane in HCl, behave as dopants. Therefore, in the presence of a uniform PANI layer, the PVC content should be relatively low and if PANI is in fact doped, chloride ions should be in higher proportions. Again, Figure 6b indicates that an optimal PANI layer is obtained after 1 h of polymerization.

From the data given in Figure 6, we can conclude that as the polymerization time increases (above 1 h), a degradation of the PANI layer is observed. Cyclic voltammetry and UV-visible spectroscopy results are in agreement with this statement which demonstrated the presence of benzoquinone, a well-known degradation product of PANI (8, 9). The degradation of the PANI layer would explain the loss in permselectivity of the CMX-PANI composite membrane when these membranes are prepared at longer polymerization times. Therefore, this degradation, rather than the oxidation of aniline into its insulating form as suggested by Sata and co-workers (6), would explain why they observed the absence of the dopant, bromine, at long polymerization times and the loss of electronic conductivity at the surface of the composite membrane.

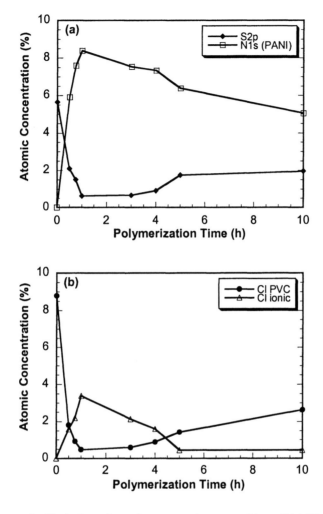

Figure 6. Variation of membrane atomic composition of (a) S2p and N1s, (b) Cl2p as a function of polymerization time. Anilinium exchange time: 1 h.

Moreover, SEM surface micrographs of a degraded membrane showed a smooth surface, very similar to the surface of an unmodified CMX membrane, and no cracks were observed (*8*). Partial delamination was not observed because (i) the modified membranes were pressed in a stainless steel grid for cyclic voltammograms analyses and (ii) no peaks associated with PANI were observed in the cyclic voltammograms indicating the absence of PANI in contact with the grid (8).

Conclusions

In this work, we have demonstrated that the permselectivity of CMX membranes for monovalent against bivalent cations can be improved by a factor of at least 20 when modified with a PANI layer and used in an acidic solution. These modified CMX membranes did not seem to have a significantly higher resistivity in comparison to an unmodified membrane since the cell voltage (during ED experiments) were found to be similar. To confirm this observation, further work will need to include ionic conductivity measurements. The PANI layer was characterized by SEM and XPS. The latter showed the degradation of PANI when prepared at polymerization times longer than 1 h.

Acknowledgments

The authors wish to thank Mr. Raymond Mineau (Département Sciences de la Terre, UQAM) for the SEM micrographs. This research was funded by the Natural Science and Engineering Research Council of Canada through a strategic grant (234959-00) and an equipment grant for an XPS spectrometer (to D.B. and nine others). S.T. acknowledges the "Fonds Québécois de Recherche sur la Nature et les Technologies" for a graduate student fellowship. The financial contribution of UQAM is also acknowledged.

References

1. Davis, T.A.; Genders, J.D.; Pletcher, D. *A First Course in Ion Permeable Membranes*; The Electrochemical Consultancy: England, 1997.
2. Mulder, M. *Basic Principles of Membrane Technology*; Kluwer Academic Publishers: Netherlands, 1996.
3. *About electrodialysis applications,* http://www.electrosynthesis.com
4. Sata, T. *J.Membr. Sci.* 1994, *93*, 117 and references therein.
5. Sata, T.; Funakoshi, T.; Akai, K. *Macromolecules* 1996, *29*, 4029.
6. Sata, T.; Ishii, Y.; Kawamura, K.; Matsusaki, K. *J. Electrochem. Soc.* 1999, *146*, 585.
7. Neosepta Ion Exchange Membranes, Product Brochure, Tokuyama Soda Inc., Japan.
8. Tan, S; Laforgue, A.; Bélanger, D. *Langmuir*, 2003, in press.

323

9. MacDiarmid, A.G.; Epstein, A.J. *Faraday Discuss. Chem. Soc.* **1989**, *88*, 317.
10. Hatchett, D.W.; Josowicz, M.; Janata, J. *J. Phys. Chem. B* **1999**, *103*, 10992.
11. Tan, S.; Viau, V.; Bélanger, D. *Electrochem. Solid-State Lett.* **2002**, *5*, E55.
12. Mizutani, Y.; Tesima, W.; Akiyama, S. ; Yamane, R. ; Ihara, H. U.S. Patent 3,451,951, 1969.

Chapter 22

Control of Transport Modes of Ions Using a Temperature-Responsive Charged Membrane

Mitsuru Higa and Tomoko Yamakawa

Applied Medical Engineering Science, Graduate School of Medicine, Yamaguchi University, Tokiwadai, 2–16–1, Ube City, Yamaguchi 755–8611, Japan

We describe the design and preparation of novel temperature-responsive charged membranes having a fast and reversible charge density response to temperature changes. The membrane consists of an interpenetrating network of two kinds of poly(vinyl alcohol)(PVA): a polymer with sulfonic acid groups and a polymer prepared by *in situ* polymerization of *N*-isopropylacrylamide in a PVA solution. Phase separation at temperatures above the lower critical solution temperature of poly(*N*-isopropylacrylamide) gives rapid and reversible changes in the membrane charge density. We demonstrate in a dialysis system, consisting of the membrane and mixed KCl and $CaCl_2$ solutions, that the membrane can control the transport modes of Ca^{2+} ions in two ways: *downhill* (transport along their own concentration gradient) and *uphill* (transport against their own concentration gradient), in response to temperature changes. The change of the transport modes allows the membrane to modulate the time-concentration profiles of bivalent ions in various forms to produce, for example, a saw wave.

There are many studies on stimuli-responsive hydrogel membranes that undergo abrupt changes in volume in response to external stimuli, such as pH (*1-4*), temperature (*5-13*), electric fields (*14-16*), light (*17-19*), antigens (*20*) and saccharides (*21-24*). Such membranes have potential applications in medicine, biotechnology, industry and in solving environmental problems because the permeability of solutes through the membranes can be controlled by these stimuli.

© 2004 American Chemical Society

For some applications, it would be very useful to have a membrane that could control not only the permeability, but also the transport modes (*downhill* versus *uphill*) of ions in response to external stimuli. In this study, *downhill* transport corresponds to transport of ions along their own concentration gradient in a system; *uphill* transport is ion transport against their own concentration gradient. There have been no studies on membranes having the ability to control the transport modes.

It has been reported that the transport modes in a dialysis system consisting of a charged membrane and mixed electrolyte solutions depend on the charge density of the membrane (*25, 26*). This indicates that a charged membrane whose charge density changes in response to external stimuli might control the transport modes.

The aim of this study is the design and preparation of a novel temperature-responsive membrane that has a fast and reversible charge density response to temperature changes. Our strategy to control the membrane charge density is to use reversible phase separation in a temperature-responsive (T-responsive) charged membrane consisting of an interpenetrating network (IPN) of polyanions and T-responsive polymers as shown in Figure 1. Aqueous solutions of poly(*N*-isopropylacrylamide) (polyNIPAAm) exhibit a lower critical solution temperature (LCST) at around 32 °C. Hence, polyNIPAAm chains in a poly(vinyl alcohol) (PVA) network are soluble in the water phase of the networks at temperatures below the LCST. At temperatures above the LCST, phase separation occurs; the polyNIPAAm chains form insoluble aggregates, with the charged groups on the ionic network concentrating in the water phases. Therefore, increases in the charge density of the membrane will occur at temperatures above the LCST.

We also demonstrate in a dialysis system, consisting of the membrane and mixed electrolyte solutions, that the membrane can control the transport modes of bivalent ions.

Experimental Section

Preparation of a Temperature-responsive Charged Membrane

Synthesis of a Temperature-responsive Polymer

A temperature-responsive polymer was prepared by *in situ* polymerization of N-isopropylacrylamide in a poly (vinyl alcohol) (PVA) solution as follows (*27*): PVA[Aldrich] (1.0g), NIPAAm[Wako Pure Chemical Industries, Ltd.] (11.0 g) and potassium persulphate (0.06 g) as an initiator were dissolved in 80ml of dimethyl sulfoxide (DMSO). The glass tube containing the solution was sealed by conventional methods and immersed in an oil bath held at 40 °C for 20 h. The reactant was poured into acetone to precipitate the T-responsive polymer.

Preparation of a Sample Membrane

The polymer obtained (1.54 g), a polyanion AP-2 [Kuraray Co. Ltd.] (0.32 g) and PVA (0.66 g) were dissolved in 34 ml of DMSO. AP-2 is PVA which contains 2 mol% 2-acrylamido-2-methylpropane sulfonic acid groups as a copolymer to provide negatively-charged sites. After casting the solution on a glass plate and drying it at 50 °C with a hot stage (NISSIN, NHP-45N), a free-standing membrane (0.1 mm thick) was obtained. The membrane was annealed for 20 min. at 160 °C with an electric drying oven (ADVANTEC, FS-320), and then it was crosslinked in an aqueous solution of 0.025 vol% glutaraldehyde, 0.1mol dm^{-3} HCl and 2 mol dm^{-3} NaCl at 25 °C for 24 h.

Determination of the Effective Charge Density

Our previous studies (*25, 26*) on ionic transport in a diffusion dialysis system consisting of a charged membrane and electolyte solutions show that the simulations performed by using the effective charge density obtained from the membrane potential measurements agree with the permeation experiments. In order to predict ionic transport through a T-responsive charged membrane at various temperatures, its effective charge density was estimated from membrane potential data with the same apparatus described elsewhere (*25*). The potential was measured at various temperatures using a dual chamber acrylic plastic cell, where the chambers are separated by the T-responsive charged membrane. One chamber was filled with KCl solutions of various concentrations (0.01, 0.03, 0.1 mol dm^{-3}), C_o. The other chamber was filled with KCl solutions whose concentrations were 5 times higher than those in the first chamber (*r*=5). From the measured membrane potentials, $\Delta\phi$, the charge density, X, was calculated using (*28, 29*)

$$\Delta\phi = -\frac{RT}{F}\ln\left(r\cdot\frac{\sqrt{X^2+(2C_o)^2}-X}{\sqrt{X^2+(2rC_o)^2}-X}\right) - \frac{RT}{F}W\ln\left(\frac{\sqrt{X^2+(2rC_o)^2}-XW}{\sqrt{X^2+(2C_o)^2}-XW}\right) \tag{1}$$

where $W = (\omega_K - \omega_{Cl})/(\omega_K + \omega_{Cl})$, ω_K and ω_{Cl} are the K$^+$ and Cl$^-$ mobilities in a membrane, respectively; and F, R and T are the Faraday constant, the gas constant and the absolute temperature, respectively. The parameters W and C_x were adjusted so that the left-hand side of eq 1 fits experimental data of $\Delta\phi$ at various KCl concentrations.

Permeation Experiments

Permeation experiments were performed in a dialysis system consisting of the T-responsive charged membrane and mixed KCl and CaCl$_2$ solutions. The membrane was fixed between the two chambers of the cell shown in Figure 2 at a high temperature (where the membrane was shrunken). When the membrane swells at low temperatures, its thickness increased by 200%, but its area increased by only 10%. The low level of the area change poses no problem for the measurements. One chamber of the cell was filled with a mixed salt solution of 3.0×10^{-3} mol dm^{-3} KCl and 5.0×10^{-4} mol dm^{-3} CaCl$_2$, and the other chamber a mixed salt solution of 3.0×10^{-2} mol dm^{-3} KCl and 5.0×10^{-3} mol dm^{-3} CaCl$_2$. The volumes of the chambers at the low- and the high-concentration sides were 100 cm^3 and 400 cm^3, respectively. The effective membrane area of the cell was 7.07 cm^2. The solution in the chamber at the low-concentration side was sampled to measure the concentration of K$^+$ and Ca^{2+} ions using an ion chromatograph (Hitachi Co. L-3710).

Results and Discussion

The Effective Charge Density Changes in Response to Temperature

Figure 3 shows the membrane potential change with temperature and the charge density obtained from the potential data in terms of eq 1. The value of the charge density changes from 0.03 mol dm^{-3} at 10 °C to 0.24 mol dm^{-3} at 50 °C. The effective charge density is proportional to the ion-exchange capacity divided by the water content. The temperature dependence of the water content and ion-exchange capacity, results which are not described here, indicates that the water content decreases from 65 vol.% at 10 °C to 30 vol.% at 50°C while the ion-exchange capacity does not change with temperature. Hence, the increase in the charge density is party due to the decrease of the water content. The charge density at 50 °C, however, was more than 8 times higher than that at 10 °C, while the water content just decreased by 50 %. This indicated that phase separation occurred in the membrane at temperatures around the LCST so that the charged groups concentrated in the water phase. The heterogeneous distribution of the charged groups due to the phase separation facilitated the increase of the charge density as shown in Figure 1. The membrane was clear at temperatures below the LCST, but cloudy at temperatures above. This phenomenon supports the notion that the phase separation occurred at temperatures above the LCST. The charge density did not change abruptly at temperatures around the LCST but increased gradually with increasing temperature. This property of the charge density

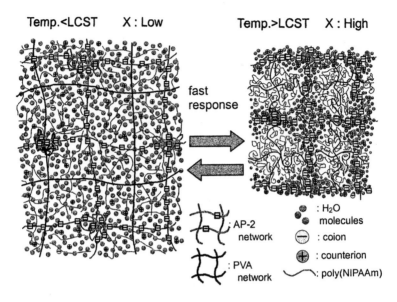

Temp.<LCST   X : Low   Temp.>LCST   X : High

fast
response

: AP-2
network

: PVA
network

: H₂O
molecules

: coion

: counterion

: poly(NIPAAm)

Figure 1. Suggested mechanism for temperature dependent charge densities of temperature-responsive charged membranes.

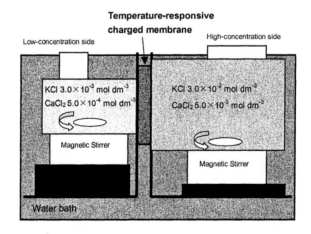

Temperature-responsive
charged membrane

Low-concentration side   High-concentration side

KCl 3.0×10^{-3} mol dm^{-3}
CaCl$_2$ 5.0×10^{-4} mol dm^{-3}

KCl 3.0×10^{-2} mol dm^{-3}
CaCl$_2$ 5.0×10^{-3} mol dm^{-3}

Magnetic Stirrer

Magnetic Stirrer

Water bath

Figure 2. Apparatus and system for permeation experiments.

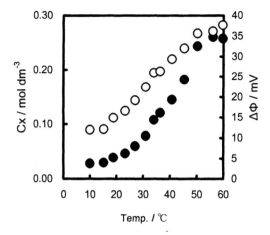

Figure 3. Membrane potential, $\Delta\phi$, and effective charge density, X, of the T-responsive charged membrane as a function of temperature. Open circles, membrane potential at $C_1=0.03$ mol dm^{-3} and $C_2=0.15$ mol dm^{-3}; solid circles, charge density calculated from potential data.

facilitated the application of the membrane to a polymer device for controlling the transport modes of specific ions in response to temperature changes.

In order to examine the temperature response of the membrane, we measured the charge density changes through stepwise changes in the temperature *(30)*. The value of the charge density changed from 0.03 mol dm^{-3} at 10 °C to 0.26 mol dm^{-3} at 50 °C within 1 minute. The temperature-charge density profiles were almost the same in each step, meaning that the membrane has a fast and reversible charge density response to temperature changes.

Control of Transport Modes in Response to Temperature Changes

Experimental Results

Our previous simulations (25, 26) indicate that such membranes can control the transport modes of specific ions in response to temperature changes. To examine the transport properties, we measured the time-dependent change in concentration of K^+ and Ca^{2+} ions in a dialysis system consisting of the membrane and mixed KCl and $CaCl_2$ electrolyte solutions. To perform saw wave-like concentration oscillations of only Ca^{2+} ions, in which the upper and the lower values would be, respectively, plus or minus 10% of the values of the initial concentration, we exposed the system to pulsatile temperature changes during the permeation experiment. The wavelengths of the temperature changes were controlled on the basis of the simulations (*31*). Figure 4 shows the data of the permeation experiments. The concentration of K^+ ions in the low-concentration chamber increased with time. This means that the ions were transported in the same direction as that of the concentration gradient between the two chambers (*downhill* transport), and the direction of K^+ ion transport was independent of the temperature. In contrast, the concentration of Ca^{2+} ions oscillated in the saw-wave form we expected. At high temperatures, the concentration of Ca^{2+} ions decreased with time, indicating that the ions were transported against their concentration gradient from the low- to the high-concentration chambers (*uphill* transport), driven by the diffusion of K^+ ions. At low temperatures, because of their downhill transport, the concentration of Ca^{2+} ions increased with time. Therefore, the Ca^{2+} ions had two transport modes in response to temperature changes: *downhill* and *uphill*. The sharpness of the edges of the saw wave in the time-concentration curve means that the transport modes changed quickly in response to stepwise changes in temperature.

For the data presented in Figure 4, we measured the concentration changes only in the low-concentration chamber. However, in other permeation experiments, whose results are not described here, we confirmed that the concentration of Ca^{2+} ions in the high-concentration chamber increased with time while that in the low-concentration chamber decreased.

Theoretical background of uphill transport

The concentration gradient of ions in the membrane

One reason why the *uphill* transport of Ca^{2+} ions occurs is because of the opposite direction of the concentration gradient in the membrane and that between the chambers. The ion concentrations in the membrane can be calculated by substituting the effective charge density of the T-responsive charged membrane at high temperatures into the equations governing Donnan equilibrium and electroneutrality (*25*). The concentration ratio of ith ion in a membrane $\overline{C}_i$ and in solution C_i is equal to the ionic valence power of the Donnan equilibrium constant K:

$$K^{z_i} = \frac{\overline{C_i}}{C_i} \qquad (2)$$

where z_i is the valence of the ions. The value of K in the low-concentration chamber is higher than that in the high-concentration chamber. For these reasons, the concentration of Ca^{2+} ions at the interface of the low-concentration chamber is larger than that of the high-concentration one under appropriate conditions (cf. Figure 5). To experimentally examine the opposite direction of the concentration gradient of Ca^{2+} ions, Higa et al. (32) measured ionic partitioning in a negatively-charged membrane immersed in a mixed KCl-CaCl$_2$ solution. When the ionic concentration in the outer solution is much smaller than the charge density, almost all of the counterions in the membrane are Ca^{2+} ions. However, the concentration of Ca^{2+} ions in the membrane decreases as the salt concentration in the solution increases because K^+ ions begin to enter the membrane, and ion-exchange of bivalent ions for univalent ions occurs. As a result, the concentration gradient of the bivalent ions in the membrane is in the opposite direction to that between the chambers.

The electroneutrality condition

The other reason why the *uphill* transport of Ca^{2+} ions occurs is because of the electroneutrality condition in the system. In our experimental systems, the electric current I is zero:

$$I = FS \sum z_i J_i = FS(J_K + 2J_{Ca} - J_{Cl}) = 0 \qquad (3)$$

where J_i is the flux of the ith ion, and F and S are the Faraday constant and the membrane area, respectively. Eq 2 can be rewritten as

$$J_K - J_{Cl} = -2J_{Ca} \qquad (4)$$

where we define $J_i > 0$ when the ion moves from the high-concentration chamber to the low-concentration chamber. When the charge density is much larger than the ionic concentration of the chambers (at high temperatures), the T-responsive charged membrane becomes a cation-exchange membrane. Hence, the flux of the anion is much smaller than that of the two cations. Thus, eq 4 becomes $J_K \approx -2J_{Ca}$. As mentioned above, the concentration gradient of K^+ ions in the

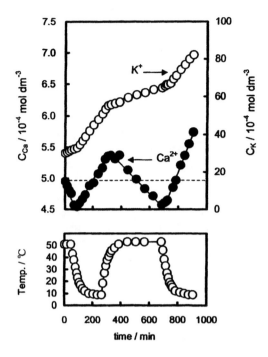

Figure 4. Ionic concentration changes at the low-concentration side in response to stepwise changes in temperature.

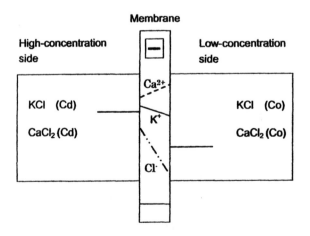

Figure 5. Schematic of concentration gradients of Ca^{2+}, K^+ and Cl^- ions in a negatively charged membrane when uphill transport of Ca^{2+} ion occurs.

membrane is in the same direction as that between the chambers while the concentration gradient of Ca^{2+} ions in the membrane is in the opposite direction to that between the chambers. Both ions move with the concentration gradient in the membrane. Therefore, $J_K>0$ and $J_{Ca}<0$. This means that Ca^{2+} ions and K^+ ions are transported in opposite directions to compensate for the transport of the other ion.

Conclusions

The T-response charged membrane prepared from an interpenetrating network of polyanions and temperature-responsive polymers exhibits a fast and reversible charge density change with temperature. The membrane can control ion transport quickly in response to temperature changes. Hence, the membrane modulates the time-concentration profile of bivalent ions in the system in various forms and at various values as long as the system has a sufficient concentration difference of the driving electrolytes. This means that the system has a negative feedback mechanism with regard to the concentration of multi-valent ions. Therefore, the T-sensitive charged membrane will be applied for self-regulating systems which can adjust the concentration of specific solutes in the systems in response to temperature: for example, 'intelligent' drug delivery and hemodialysis systems which have a homeostasis mechanism, and separation processes in a precise reaction system whereby the membrane can control reaction conditions such as pH and concentrations of reactants and products.

Many hydrogels undergo abrupt changes in water content in response to external stimuli, and the charge density of the gel membranes is a function of the water content. Therefore, the strategy detailed here for the control of the transport modes of specific ions can be applied to other external stimuli-responsive gels with charged groups.

Acknowledgments

This work was partially supported by the Ministry of Education, Science, Sports and Culture, Grant-in-Aid for Scientific Research on Priority Area (A), No. 13022245, Grant-in-Aid for Scientific Research (C), No. 13640581, the Salt Science Research Foundation, No. 0212 and Electric Technology Research Foundation of Chugoku.

References

1. Tanaka, T. *Phys. Rev. Lett.* **1978**, *40*, 820-823.
2. Tanaka, T. *et al. Phys. Rev. Lett.* **1980**, *45*, 1636-1639.
3. Siegel, R. A.; Firestone, B. A. *Macromolecules* **1988**, *21*, 3254-3259.
4. Hirokawa, Y.; Tanaka, T. *J. Chem. Phys.* **1984**, *81*, 6379-6380.
5. Hoffman, A. S. *J. Controlled Release* **1987**, *6*, 297-305.
6. Bae, Y. H.; Okano, T.; Hsu, R.; Kim, S. W. *Makromol. Chem., Rapid Commun.* **1987**, *8*, 481-485.
7. Okano, T.; Bae, Y. H.; Jacobs, Kim, H.; S. W. *J. Controlled Release* **1990**, *11*, 255-265.
8. Dong, L.-C.; Hoffman, A. S. *J. Controlled Release* **1991**, *15*, 141-152.
9. Chen, G.; Hoffman, A. S. *Nature* **1995**, *373*, 49-52.
10. Yoshida, R. *et al. Nature* **1995**, *374*, 240-242.
11. Aoki, T.; Kawashima, M.; Katono, H.; Sanui, K.; Ogata, N.; Okano, T.; Sakuraiet, Y. *Macromolecules* **1994**, *27*, 947-952.
12. Snowden, M. J.; Chowdhry, B. Z.; Vincent, B.; Morris, G. E. *J. Chem. Soc., Faraday Trans.* **1996**, *92*, 5013-5016.
13. Feil, H.; Bae, Y. H.; Feijen, J.; Kim, S. W. *J. Membrane Sci.* **1991**, *64*, 283-294.
14. Tanaka, T.; Nishio, I.; Sun, S.-T.; Ueno-Nishio, S. *Science* **1982**, *218*, 467-469.
15. Kwon, I. C.; Bae, Y. H.; Kim, S. W. *Nature* **1991**, 354, 291-293.
16. Osada, Y.; Okuzaki, H.; Hori , H. *Nature* **1992**, *355*, 242-244.
17. Ishihara, K.; Hamada, N.; Kato, S.; Shinohara, I. *J. Polym. Sci. Polym. Chem.* **1984**, *22*, 881-884.
18. Suzuki, A.; Tanaka , T. *Nature* **1990**, *346*, 345-347.
19. Suzuki, A.; Ishii, T.; Maruyama, Y. *J. Appl. Phys.* **1996**, *80*, 131-136.
20. Miyata, T.; Asami, N.; Uragami, T. *Nature* **1999**, *399*, 766-769.
21. Kokufuta, E.; Zhang, Y.-Q.; Tanaka, T. *Nature* **1991**, *351*, 302-304.
22. Miyata, T.; Jikihara, A.; Nakamae, K.; Hoffman, A. S. *Macromol. Chem. Phys.* **1996**, *197*, 1135-1146.
23. Kost, J.; Horbett, T. A.; Ratner, B. D.; Singh, M. *J. Biomed. Mat. Res.* **1985**, *19*, 1117-1133.
24. Kataoka, K.; Miyazaki, H.; Bunya, M.; Okano, T.; Sakurai. Y. *J. Am. Chem. Soc.* **1998**, *120*, 12694-12695.
25. Higa, M.; Tanioka, A.; Miyasaka, K. *J. Membrane Sci.*, **1990**, *49*, 145-169.
26. Higa, M.; Kira, A. *J. Phys. Chem.*, **1992**, *96*, 9518-9523.
27. Kurihara, S.; Sakamaki, S.; Mogi, S.; Ogata, T.; Nonaka, T. *Polymer*, **1996**, *37*, 1123-1128.
28. Teorell, T. *Proc. Soc. Exp. Biol. Med.*, **1935**, *33*, 282-285.
29. Meyer , K. H.; Sievers, J.-F. *Helv. Chim. Acta*, **1936**, *19*, 649-664.
30. Higa, M.; Yamakawa, T. *to be submitted*.
31. Higa, M.; Yamakawa, T. *to be submitted*.
32. Higa, M.; Tanioka, A.; Kira, A. *J. Chem. Soc. Faraday Trans.* **1998**, *94*, 2429-2433.

Chapter 23

Ion-Exchange Membranes from Blends of Sulfonated Polyphosphazene and Kynar FLEX PVDF

Ryszard Wycisk[1,3], Roy Carter[1,3], Peter N. Pintauro[1,3], and Catherine Byrne[2]

[1]Department of Chemical Engineering, Tulane University at New Orleans, New Orleans, LA 70181
[2]Science Research Laboratory, Inc., 15 Ward Street, Somerville, MA 02143–4241
[3]Current address: Department of Chemical Engineering, Case Western Reserve University, 10900 Euclid Avenue, Cleveland, OH 44106

Ion-exchange membranes were prepared by blending sulfonated poly[bis(3-methylphenoxy)]phosphazene (SPOP) with Kynar® FLEX, a copolymer of vinylidene fluoride and hexafluoropropylene, which was followed by crosslinking with use of ultraviolet or e-beam radiation. It was found that, in general, Kynar FLEX was immiscible on a molecular level with the SPOP. The polymers had, however, acceptable degree of compatibility and no sign of macroscopic phase separation was observed for a relatively wide composition range. The degree of compatibility increased with increasing IEC of the SPOP. Also, when sulfonic groups of the SPOP were converted to the tetrabutylammonium (TBA) form, highly homogeneous, transparent blends were obtained. These, however, could not be crosslinked effectively. The Na-form of SPOP was then used in further studies. A linear correlation was observed between the conductivity and swelling of the blended and crosslinked membranes. A conductivity increase from 0.01 to 0.075 S/cm was accompanied by an increase in swelling from 20 to 80%. Additionally, membranes with the

© 2004 American Chemical Society

same swelling but prepared from SPOP with higher IEC showed higher conductivity. No significant difference in the swelling versus conductivity dependence was observed between membranes crosslinked with benzophenone and UV light and those crosslinked with electron-beam radiation. Preliminary DMFC tests showed methanol crossover rates to be 3 times lower (6 times lower if thickness corrected) for the blended membranes as compared to Nafion 117.

Polyorganophosphazenes offer new possibilities as potential membrane materials. These polymers belong to a class of hybrid polymers; their backbone is inorganic while the side groups are of organic nature. Numerous papers have been published on various aspects of polyphospazene synthesis, properties, and applications but there are only a few reports on their use as ion-exchange or proton-exchange membranes (*1-5*).

Our previous studies (*2-6*) have shown that ion-exchange membranes composed of sulfonated and crosslinked poly[bis(3-methylphenoxy) phosphazene] possess high proton conductivity (0.01-0.1 S/cm) and low methanol diffusivity (10^{-8}-10^{-7}cm^2/s). These properties make this polymers a candidate materials for use in direct methanol fuel cells. Additionally, at low and moderate sulfonation degrees the polymer had acceptable mechanical properties and it could be crosslinked using benzophenone (BP) and UV irradiation. However, at ion-exchange capacities greater than 1.2 mmol/g the polymer was brittle which rendered the membranes difficult to handle and electrodes for fuel cell testing could not be hot-pressed to them. The search for a solution to this problem was directed in two different approaches. The first one was the synthesis of a new, more sophisticated polyphosphazene, and the second path was focused on blending of SPOP with some other, non-polyphosphazene polymer. The present paper deals with the second path.

Several different polymers were tested for their blending capability with sulfonated poly[bis(3-methylphenoxy)phospahzene (SPOP): polyimides, polysulfone, polyphenyleneoxide, polyacrylonitrile, and various fluoropolymers.

Recently, membranes have been prepared by blending SPOP with Kynar FLEX, a copolymer of vinylidene fluoride and hexafluoropropylene, manufactured by Elf Atochem North America, Inc. The formulation used in the present work contained ca. 10 % of hexafluoropropylene units.

Proton conducting membranes containing Kynar FLEX can be attractive for direct methanol fuel cells because: (i) the mechanical properties of the blended film are better compared to pure SPOP films, (ii) the thermoplastic properties of the Kynar FLEX polymer should make hot-pressing electrodes easier during MEA fabrication, (iii) the blend is expected to have even lower methanol crossover because the fluoropolymer restricts SPOP swelling and is virtually impermeable to water and methanol, and (iv) the amount of expensive polyphosphazene in the film is lowered, thus reducing the overall cost of the membrane.

In this paper we present preliminary data on the properties and morphology of blended SPOP/Kynar FLEX ion-exchange membranes.

Experimental

Sulfonation of the polyphosphazene. A known weight (2g) of poly[bis(3-methylphenoxy) phosphazene] (MW=700,000 g/mol) was dissolved in 80 ml of 1,2-dichloroethane (DCE). An appropriate amount (0.3-2.0 ml) of SO_3 in DCE was then added to the polymer solution at 0°C in a dry nitrogen atmosphere. The resulting precipitate was stirred for about 3 hours, followed by the addition of NaOH dissolved in a water/methanol mixture to terminate the reaction. After solvent evaporation, the sulfonated polymer was soaked in water and then treated sequentially with 0.01 M NaOH, water, 0.1 M HCl, and water. Next, the sulfonic groups were converted into the appropriate salt form using a 1M aqueous solutions of either NaCl or TBACl (tetrabutylammonium chloride) followed by through washing with water. In the final step the SPOP membrane was dried at 60°C.

Membrane preparation with UV crosslinking. SPOP, Kynar FLEX 2821 and benzophenone (BP) were dissolved in dimethylacetamide (DMAc). The solution was cast into a PTFE dish and kept in an oven at 60°C overnight. The dry membrane was removed from the dish and placed in a vacuum oven at 60°C for 24 hours to remove traces of DMAc. The SPOP was then crosslinked by exposing the membrane to UV radiation of 365 nm wavelength (15 mW/cm^2) for a period of 12 h on each side.

Membrane preparation with e-beam crosslinking. SPOP and Kynar FLEX 2821 were dissolved in dimethylacetamide (DMAc). The solution was cast into a PTFE dish and kept in an oven at 60°C overnight. The dry membrane was removed from the dish and placed in a vacuum oven at 60°C for 24 hours to remove traces of DMAc. The SPOP was then crosslinked by exposing the membrane to a specified dose of electron beam (e-beam) radiation using Science Research Laboratory's EB-10 RF accelerator running at a power of 650 watts and an energy of approximately 4 MeV. The sample chamber consisted of an aluminum base and a thin (1.5 mm) aluminum cover with inside dimensions of 200x300 mm sealed with Viton o-rings. Zero grade nitrogen was passed through the chamber constantly at a rate 280 cm^3/min. The total applied dosage ranged from 40 to 160 Mrad.

Ion-exchange capacity measurements. A known weight of dry polymer sample (0.2-0.4 g, with sulfonic groups in SO_3H form) was placed into 50 ml of

2.0 M NaCl solution at 25°C and shaken occasionally for 48 h. Three 10 ml samples were then removed and the amount of H^+ released by the polymer was determined by titration with 0.01 M NaOH. The ion-exchange capacity (IEC) of the sample was calculated according to the following equation:

$$IEC \text{ [mmol/g]} = 0.05 * v / m_d \qquad (1)$$

where v [ml] was the endpoint volume of 0.01 M NaOH (average of the three titrations), and m_d [g] was the dry weight of the sample.

Conductivity Measurements. Proton conductivity of the water swollen membranes, with sulfonic groups in SO_3H form, was measured by an AC impedance method (real axis intercept of the impedance spectrum) in the frequency range 1 Hz to 100 KHz using a lock-in amplifier (EG&G Model 5210) and potentiostat (EG&G Model 273). The cell employed was a two-probe type, similar to that described in (7). Measurements were taken at 25°C.

Swelling Measurements. Membrane sample (sulfonic groups in SO_3H form) was equilibrated in water at 25°C for 24 h. After removal from the water bath, the membrane surface was blotted dry with filter paper and the sample was weighted on an electronic balance (m_w). After thorough drying under vacuum and over P_2O_5 for 24 h, the membrane was re-weighed (m_d). The equilibrium water swelling (S) was calculated according to the following formula:

$$S \text{ [g/g]} = (m_w - m_d) / m_d \qquad (2)$$

Tensile Tests. Tensile measurements were performed on an Instron 5567. The specimens were tested in water-swollen state at 25°C and the extension rate was 50 mm/s. Based on the stress-strain curve, the ultimate strength and elongation to break were determined.

SEM micrographs. Dry membranes were manually fractured after reaching thermal equilibrium in liquid nitrogen. Specimens were sputter-coated with gold to a final thickness of approximately 2 nm and imaged on a JEOL JSM-820 scanning electron microscope at 15 kV.

X-ray Measurements. Wide-angle x-ray scattering (WAXS) spectra were obtained using Scintag XDS 2000 diffractometer operated at 45 KV and 40 mA, with Cu anode. The scan range was 5-40 deg. and the scan rate was 2.00 deg./min. Membrane samples in the dry state and swollen with water were placed horizontally on zero background quartz plates. The measurements were taken at 25°C.

Results and Discussion

The membrane preparation procedure consisted of the following four steps: (i) dissolving the components (SPOP in a salt form and Kynar FLEX) in DMAc, (ii) casting a film and evaporating the solvent, (iii) crosslinking the SPOP with benzophenone and UV light or e-beam radiation, and (iv) converting the sulfonic groups back to the acid form.

Several batches of sulfonated poly[bis(3-methylphenoxy)phosphazene] (SPOP) in appropriate salt forms were prepared. Their IEC ranged from 1.6 to 3.5 mmol/g. Blends with Kynar FLEX were then prepared. Half of the blends was crosslinked with use of benzophenone and UV light and the other half was crosslinked with e-beam radiation.

Blending can be treated as a dilution of the polymers. In order for the blended membrane to show appropriate electrochemical performance it is necessary to use an SPOP of significantly higher ion-exchange capacity than that in the final blend. For example, if the desired IEC of the blended membrane is 1.2 mmol/g and the Kynar FLEX content is to be 50% than the required IEC of the SPOP has to be 2.4 mmol/g. It may occur that the sulfonated polymer of that high IEC is water-soluble and the blended membrane may degrade in aqueous solutions. Therefore, additional crosslinking or grafting may be required to ensure stability of the blended membrane.

Figure 1 shows a relationship between the ion-exchange capacity of the SPOP and its water affinity. The starting, unsulfonated material has an IEC of 0 mmol/g. When half of the number of aromatic rings is monosubstituted with SO_3H, the IEC is 3 mmol/g. Finally, when all the aromatic rings are mono-substituted, an IEC of 4.8 mmol/g is reached. The most important finding was that at an IEC greater than 2.1 mmol/g, the SPOP became water soluble. Below that limit the polyphosphazene only swelled in water but did not dissolve. It may be concluded, that using SPOP of IEC greater than 2.1 will require crosslinking or grafting not only to reduce swelling but, primarily, to prevent the sulfonated component from being leached out from the blend in aqueous solutions.

Two crosslinking procedures were used in the present research. The first one was benzophenone-assisted UV-crosslinking (8). When benzophenone is added to the blend, followed by irradiation with UV light of wavelength 340 nm, a reactive triplet state forms. This state can be considered as a di-radical, which can abstract hydrogen from a benzylic carbon of the polyphosphazene and create macroradicals and ketyl radicals. Recombination of two macroradicals results in crosslink formation. The second procedure used was e-beam crosslinking. Exposure of the blend to an electron beam led to hydrogen abstraction and bond cleavages with subsequent macroradical formation. It was assumed that crosslinking would be a dominant effect. In Figure 2 the advantages and disadvantages of both crosslinking methods are summarized. The fact that the use of e-beam did not require, at least theoretically, addition of benzophenone or any other activator made this technique the preferred one. From initial measurements of swelling and conductivity it was found that the useful effective

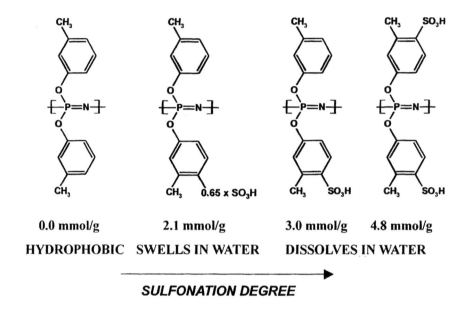

CH3
CH3
CH3
CH3
SO3H
O
P=N
O
CH3
CH3
0.65 x SO3H
CH3
SO3H
CH3
SO3H
0.0 mmol/g
2.1 mmol/g
3.0 mmol/g
4.8 mmol/g
HYDROPHOBIC
SWELLS IN WATER
DISSOLVES IN WATER
SULFONATION DEGREE

Figure 1. The dependence between the sulfonation degree of SPOP and its water affinity.

UV-CROSSLINKING CROSSLINKING	E-BEAM
PROS	PROS
relatively inexpensive hardware	crosslinking and grafting
more specific	high penetration depth
	no gradient
CONS	no photoinitiator required
limited penetration depth	
possible gradient	CONS
presence of photoinitiator	expensive hardware
miscibility problems	some degradation possible

Figure 2. Comparison of the two crosslinking methods used in this work.

IEC of the blended membranes should range from ca. 1.0 to 1.2 mmol/g. These capacities could be obtained in two ways. First, by blending SPOP of moderate IEC (1.4-2.1 mmol/g) with Kynar FLEX as minor component, or second by blending SPOP of high IEC (2.5-4.0 mmol/g) with Kynar FLEX as the major component. This second option seemed very interesting because having Kynar FLEX as major component should result in a significant reduction of the SPOP water affinity and, therefore, reduced swelling.

In this work, the following three groups of membranes were synthesized: group 1 – containing SPOP of IEC=1.6 mmol/g; group 2 – with SPOP of IEC=2.1 mmol/g; and group 3 - with SPOP of IEC=3.5 mmol/g. All membranes were crosslinked with either UV-light and benzophenone or e-beam irradiation. Membranes belonging to group 1 contained ca. 70% of SPOP; those belonging to group 2 contained ca. 50% of SPOP, and those belonging to group 3 contained ca. 30% of SPOP. It was found that, in general, Kynar FLEX was immiscible on a molecular level with the SPOP. For certain SPOP/Kynar compositions, SPOP IEC, and SPOP salt form, an acceptable degree of compatibility was obtained and those blends showed no signs of macroscopic phase separation. The degree of compatibility increased with increasing IEC of SPOP. Also, when sulfonic groups of the SPOP were converted to tetrabutylammonium (TBA)-form, highly homogeneous, transparent blends were obtained.

Photographs of three different SPOP/Kynar FLEX membrane samples are shown in Figure 3. Each membrane is about 150 μm in thickness. The left one is a blend of SPOP in an acid form; macroscopic phase separation is visible. The second blend was prepared from the Na-form of SPOP. It seems to be quite homogeneous. The third sample, the most transparent one, was prepared by blending SPOP in a TBA-form.

Two effects can account for the improved miscibility of the TBA-form of SPOP with Kynar Flex. First, the bulkiness of the tetrabutylammonium cation contributed to the reduction of the strong polar interactions between sulfonic groups. Second, because both SPOP and Kynar FLEX are acidic in nature and TBA is basic, some sort of compatibilization can be expected.

Blending of polymers is always associated with the problem of compatibility of the components. In most cases it is not possible to achieve a true, molecular level miscibility, because the entropic contribution to the free energy of mixing is negligible. When blending ionic and uncharged polymers, the situation is even more complex. The ionic groups tend to cluster, making blending difficult. Thus, in order to obtain a blend with very fine domains, either the ionic charges should be effectively masked or the polymers should possess some level of mutual interactions (*9*). In the case discussed above, TBA probably serves both purposes.

In Figure 4, SEM micrographs are shown of freeze-fractured blends prepared from SPOP in three salt forms: H^+, Na^+, and TBA. Blending of the acid form (left picture) resulted in significant phase separation and domains with sizes greater than 10 μm. When SPOP was in Na-form (middle picture), the

342

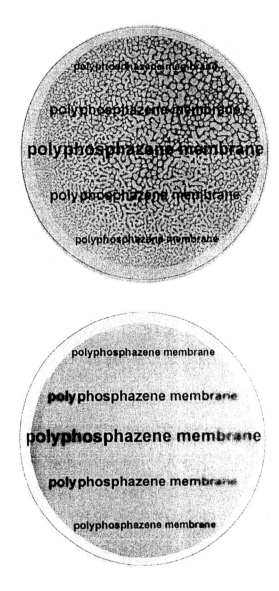

Figure 3. Photographs of three membrane samples prepared from SPOP in different salt forms (from the left: SO_3H, SO_3Na, SO_3TBA). Clarity of the underlying text is indicative of level of miscibility.

Figure 3. *Continued.*

344

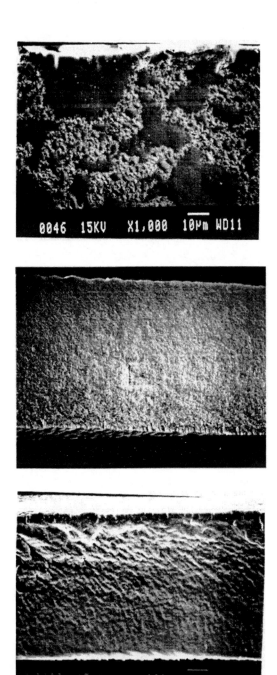

Figure 4. SEM micrographs of three membrane samples prepared from SPOP in different salt forms (from the top: SO₃H, SO₃Na, SO₃TBA).

domain size ranged from a fraction to several µm. The use of SPOP in the TBA-form (right picture) resulted in a featureless morphology. This was indicative of a very high level of compatibility; the size of the SPOP domains was smaller than 0.1 µm.

Despite the morphological advantage of the TBA-form of SPOP its use was postponed because of two reasons. First, it was not possible to fully convert SPOP back into the acid form, and, secondly, it was not possible to effectively crosslink SPOP in the TBA form. The reasons for this behavior are not fully understood, but it is believed that, in the second case, the bulky TBA groups cause some sort of steric blockage to such an extent that macroradical recombination processes are hindered. For this reason, membranes prepared from blends of SPOP with sulfonic groups in Na-form were investigated only. Prior to further testing, the sulfonic groups of the blended membranes were converted to the acid form.

Another important property investigated was crystallinity. A wide-angle X-ray scattering technique was used to semi-quantitatively access the degree of crystallinity in the starting components as well as in the blends. Examples of typical diffractograms obtained for the raw polyphosphazene, Na-form SPOP (IEC=2.1 mmol/g), Kynar FLEX, and a hydrated (wet) 50:50 % blend are presented in Figure 5. It can be seen that the unsulfonated, raw polyphosphazene is a semicrystalline polymer with a highly complicated diffraction pattern. After sulfonation, however, all the three-dimensional order was lost. Kynar FLEX is a semicrystalline polymer with an amorphous halo maximum at ca. 18 deg., and a crystalline peak localized at ca. 20 deg. Crystallinity of the Kynar FLEX was preserved after blending - the sharp peak on the diffractogram was still present. The fact that the diffractogram of the water swollen blended membrane had similar appearance as the one recorded for a dry membrane was an indication of possible separation between the SPOP and the Kynar FLEX phases. At the same time, the preserved crystallinity contributed to the improvement in mechanical properties of the blend as compared to the pure SPOP. In general, the blends were less brittle and tougher.

Swelling and proton conductivity are among the most important characteristics of proton-exchange membranes. These parameters were measured in water at 25°C and the results are presented in Figure 6. It can be seen that there exists, approximately, a linear correlation between the swelling of SPOP/FLEX membranes and proton conductivity. A swelling increase from 20 to 80 % is accompanied by an increase in conductivity from 0.01 to 0.075 S/cm. No correlation was observed between the effective IEC of the blended membranes and their proton conductivity, although membranes of the same swelling, but prepared from SPOP with higher IEC, showed higher conductivity. It can also be seen from Figure 6, that there was no significant difference in the swelling versus conductivity behavior of membranes crosslinked with benzophenone, UV light, and those crosslinked with electron-beam radiation. It should be stated, however, that membranes prepared from the SPOP of the highest IEC (3.5 mmol/g) and to lesser extent those prepared from SPOP of

346

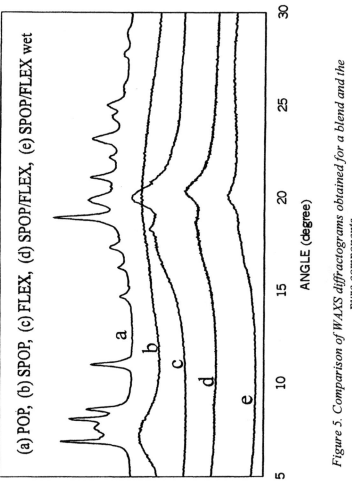

Figure 5. Comparison of WAXS diffractograms obtained for a blend and the pure components.

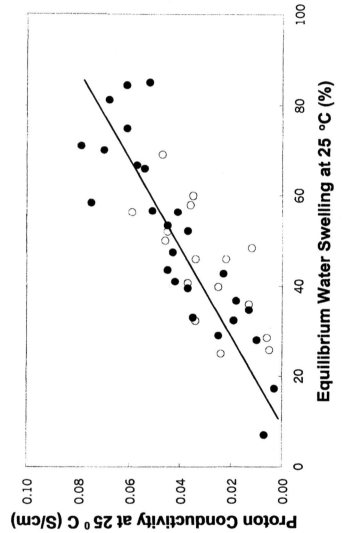

Figure 6. Correlation between eqilibrium water swelling and conductivity. Closed circles represent UV-crosslinked membranes and the open circles represent e-beam crosslinked ones.

IEC=2.1 mmol/g were unstable. When placed in an aqueous solution the SPOP leached out which resulted in a gradual loss of the membrane conductivity. This was an indication of insufficient crosslinking of the highly sulfonated polyphosphazene.

Blending SPOP with a flexible and tough polymer was thought to improve mechanical properties of the final membrane. Tensile tests were performed to verify this hypothesis. An example of tensile tests is presented in Table I. The blends were prepared from Kynar Flex and the sodium form of SPOP with an IEC=2.1 mmol/g. The extension rate was 50 mm/s. Specimens were tested in the water swollen state at 25°C. Based on the stress-strain curve recorded, the ultimate strength and elongation at break were determined. It can be seen from Table I that the incorporation of Kynar FLEX into sulfonated polyphosphazene leads to a material with mechanical properties significantly better than those of pure SPOP. The tensile strength increased from 3.4 to 6.4 MPa when the Kynar FLEX content was increased from 40 to 60 wt.%. At the same time the extension at break increased from 150% to 186%.

Table I. Results of tensile tests performed on water-swollen blends

Kynar FLEX Content (%)	Effective IEC (mmol/g)	Ultimate Strength (MPa)	Elongation at Break (%)
40	1.25	3.39	150
45	1.15	4.68	152
50	1.05	5.42	166
55	0.95	6.03	178
60	0.85	6.38	186

As the final point of discussion, an example of fuel cell test results is presented. The experiments were performed at Giner Inc. in Waltham, MA. The liquid methanol-air fuel cell was operated at 60°C with methanol concentrations of 0.5 and 1.0 M. The membrane-electrode assembly (MEA) was prepared by hot pressing. The cathode layer consisted of unsupported platinum and the anode was composed of unsupported platinum-ruthenium with a Teflon binder. Three I-V curves are presented in Figure 7. The upper curve was obtained with Nafion 117 and was treated as a reference. The two lower curves are the results obtained for a SPOP/Kynar FLEX e-beamed membrane (sulfonic groups in the acid form). As can be seen, the power output of the SPOP/Kynar FLEX membrane was somewhat lower than that for a Nafion membrane. The membrane conductivity was about four times smaller than that of Nafion leading to an increase in IR drop during fuel cell operation. Methanol crossover, measured with the use of a CO_2 sensor in the exit air stream, was ca. 3 times lower for the blended membrane than for Nafion. The thickness-corrected methanol crossover was even lower – ca. 6 times. It is expected that with

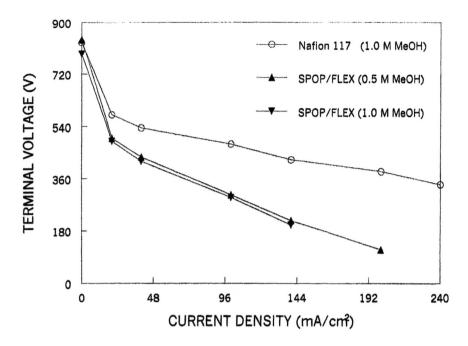

Figure 7. Current-voltage curves obtained in a liquid methanol-air fuel cell test station at 60°C. The membrane tested was an e-beam crosslinked SPOP/Kynar FLEX blend. Data obtained with NAFION 117 membrane is plotted for comparison.

SULFONATION
PROBLEM - heterogeneous reaction (non-uniform SPOP at lower sulfonation degrees)
RECOMMENDATION - IEC > 2 mmol/g

BLENDING
PROBLEM - phase separation (domain-type blends, limited UV light penetration, low mechanical strength)
RECOMMENDATION - highest IEC possible, SPOP in SO_3TBA form

CROSSLINKING
PROBLEM - SPOP leaching or membrane swelling too high
RECOMMENDATION - low IEC, SPOP in SO_3Na form, high Kynar FLEX loading, high irradiation dosage

Figure 8. The most important conclusions on each of the three steps of membrane preparation.

350

appropriate optimization of the morphology and crosslinking degree of the blended membranes and with improvement of the MEA preparation technique, much better fuel cell performance can be achieved.

Conclusions

Blended membranes composed of sulfonated polyphosphazene and Kynar FLEX were synthesized. The membranes had good proton conductivities and mechanical properties. Successful blending and crosslinking was dependent on the ion-exchange capacity and the counter-ion form of the sulfonated polyphosphazene. A summary of the most important challenges in the membrane preparation procedure along with some possible recommendations are presented in Fig. 8.

A linear correlation was observed between the swelling and proton conductivity of the membranes. A conductivity increase from 0.01 to 0.075 S/cm was accompanied by an increase in swelling from 20 to 80%. Additionally, membranes of the same swelling but prepared from SPOP with higher IEC showed a higher conductivity. No significant difference in the swelling versus conductivity behavior was observed between membranes crosslinked with benzophenone, UV light, and those crosslinked with electron-beam radiation.

Future research will focus on improving the blended membrane morphology by optimizing membrane casting conditions and on increasing the efficiency of crosslinking.

Acknowledgements

This work was funded by the Army Research Office, Grant No. DAAD19-00-1-0517, and by an Army Research Office STTR program, through a grant to Science Research Laboratory, Inc. We would like to thank Giner, Inc., Waltham, MA for preparing membrane electrode assemblies and performing the fuel cell tests. We also thank Elf Atochem North America, Inc. for supplying the Kynar FLEX®.

References

1. Wycisk, R.; Pintauro, P.N. *J. Membr. Sci.* **1996**, *119*, 155.
2. Tang, H.; Pintauro, P.N.; Guo, Q.; O'Connor, S. *J. Appl.Polym.Sci.* **1998**, *71*, 387.
3. Jones, L.; Pintauro, P.N.; Tang, H. *J. Membr. Sci.* **1999**, *162*, 135.

4. Guo, Q.; Pintauro, P.N.; Tang, H.; O'Connor, S. *J. Membr. Sci.* **1999**, *154,* 175.
5. Tang, H.; Pintauro, P.N. *J. Appl. Polym. Sci.* **2000**, *79*, 49.
6. Carter, R.; Evilia, R.; Pintauro, P.N. *J. Phys. Chem. B* **2001**, *105*, 2351.
7. Zawodzinski, T.A., Neeman, M., Sillerud, L.O., Gottesfeld, S. *J. Phys. Chem.* **1991**, *95*, 6040.
8. Graves, R.; Pintauro, P.N. *J. Appl. Polym. Sci.* **1997**, *68*, 827.
9. Smith, P.; Hara, M.; Eisenberg, A. In *Current Topics in Polymer Science, Volume II;* Ottenbrite, X.; Utracki, X.; Inoue, X., Eds.; Hanser Publishers: New York, NY, 1992, p 255.

Chapter 24

Behavior of Absorbed Water in and Oxygen Permeability of Hydrophilic Membranes with Bulky Hydrophobic Side Chain Groups

Shuichi Takahashi, Akiko Saito, and Tsutomu Nakagawa

Department of Industrial Chemistry, Meiji University, 1–1–1 Higashimita, Tama-ku, Kawasaki 214–8571, Japan

Polymeric membranes exhibiting high water permeability and high water-dissolved oxygen permeability are required for materials such as contact lenses (CL) or artificial lungs, which must be swollen with water to fit into the human body. In this study, the effect of composition of hydrophilic groups in copolymer membranes on water content or water-dissolved oxygen permeability was investigated in detail. In copolymers containing 4-vinylpyridine (4VP), we attempted to modify the copolymer membranes by quaternization to improve the wettability and permeability. We also investigated the effect of the type of quaternization ammonium salt on copolymer membrane properties. The water content and water-dissolved oxygen permeability of each copolymer membrane increased with increasing quaternization ratio.

© 2004 American Chemical Society

In the polymer medical field, polymeric membranes exhibiting high permeability for water and water-dissolved oxygen are required for materials such as contact lenses (CL) or artificial lungs, since membranes in these applications are swollen with water inside the human body. It is commonly known that the wettability of a CL has an effect on its comfort. Therefore, research to improve CLs should focus on the development of polymers having both high gas permeability and high water content. In general, there are two approaches for preparing hydrophilic membranes with high oxygen permeability. The first involves the development of higher water content materials. High water content CL materials increase the supply of oxygen to the cornea. The second approach involves copolymerization of methacrylate monomers containing a bulky pendant group with hydrophilic monomers (1-4).

In this study, the effect of composition of the hydrophilic group in the copolymer membranes on water content and water-dissolved oxygen permeability was investigated in detail. The copolymers were synthesized by bulk radical copolymerization of a monomer having a bulky hydrophobic group with the hydrophilic monomers 4-vinylpyridine (4VP) and N-vinyl-2-pyrolidone (NVP). Furthermore, the effect of the type of quaternary ammonium salt on the copolymer membranes was also investigated.

Experimental

Material and Synthesis of Copolymer

The monomer structures used for membrane preparation and the polymerization steps for the copolymers used are shown in Figure 1. 3-methacryloxypropyl tris(trimethylsiloxy)silane (SiMA), 4VP and NVP were purified by distillation under vacuum prior to use. All copolymers were synthesized by a radical polymerization technique with AIBN as the initiator. These copolymers, copoly(SiMA-NVP) and copoly(SiMA-4VP), were finally cast on a horizontal Teflon plate. The mole fraction of 4VP and NVP in the copolymer was determined by elemental analysis of nitrogen using a combustion method.

Modification of Copolymer Membranes

The copolymers were modified with quaternary ammonium salts using methyl iodide in 5 wt% EtOH solution for 24 hours. The chemistry of this modification of the copolymer membranes is also shown in Figure 1. In this modification reaction, some copoly(SiMA-4VP) membranes were utilized. The modified copolymers were cast onto a hot plate at 333 K. The quaternization ratio was calculated by determining the amount of iodide in the final membrane using a combustion method.

354

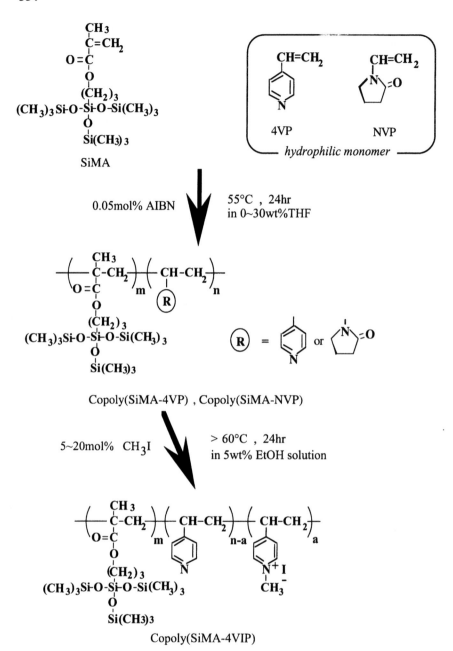

Figure 1. Synthesis and structure of copolymer and modified copolymer.

Characterization of Used Polymer Membranes

The water content was calculated as follows:

$$\text{Water content [\%]} = [(W_1 - W_0)/ W_0] \times 100 \qquad (1)$$

where W_1 is the weight of a water-swollen membrane, and W_0 is that of a completely dry membrane. Water-swollen membranes were blotted to remove excess water, and the measurement of weight (W_1) was repeated until a constant value was obtained. The glass transition temperature (T_g) was determined using a Perkin-Elmer DSC-7 at a heating rate of 20 K/min. Additionally, the phase transition peaks of water absorbed in the membrane were evaluated using the Perkin-Elmer DSC-7. These peaks were estimated to be in the temperature range from 153 to 293 K at heating and cooling rates of 5 K/min. DSC measurements were repeated at least 3 times. The d-spacing was measured in the dried and hydrated states by wide angle X-ray diffraction (WAXD: RIGAKU RINT-1200) using Bragg's equation; $\lambda = 2d \sin\theta$ (Cu – Kα = 1.54 Å). In the analysis of WAXD, the two broad peaks obtained were classified as the intermolecular distance and the intramolecular distance according to the paper reported by Nakamae et al. (5). Intermolecular and intramolecular distances show the relative distance between two main chains and between two side chains, respectively. In the copolymer membranes modified by quaternization, the contact angle, θ, of water on the polymer membrane surface was also measured to track changes in the membrane surface due to modification.

Oxygen Permeability Measurement in the Dry and Wet States

The oxygen permeability in the dry state was determined using a vacuum-pressure method at 308 K (6). The upstream pressure was up to 1 atmosphere. The coefficient of variation in the gas permeability coefficient was less than ±3%. The water-dissolved oxygen permeability in the wet state was measured using an oxygen electrode according to the experimental method developed by Minoura et al. (7). This apparent water-dissolved oxygen permeability coefficient includes the effect of the boundary layer. Therefore, the true permeability coefficient, which eliminated the effect of the boundary layer, is calculated using five data points measured at different membrane thicknesses. The water-dissolved oxygen permeability coefficients were calculated using the following equation and measurements from five membranes ranging in thickness from 50 ~ 250 μm (7-8).

$$1/P_g = 1/P_{true} + (1/R)(1/d) \qquad (2)$$

The above equation yields the true water-dissolved oxygen permeability coefficient (P_{true}) at infinite thickness from the apparent one (P_g) of a membrane of thickness d. The term (1/R) is the sum of the resistances of the boundary layers.

Results and Discussions

Characterization of Copolymer Membranes

Basic characterization data for copoly(SiMA-NVP) and copoly(SiMA-4VP) membranes are recorded in Table I and Table II, respectively. In Table I, the NVP content in the copoly(SiMA-NVP) membranes is different from that measured in the preparation before polymerization because the reactivity of SiMA was much faster than that of NVP. The T_g and water content increased with increases in the NVP content of the copolymer. In contrast, the intramolecular distance decreased as NVP content increased. In Table II, the water content and the intramolecular distance of copoly(SiMA-4VP) membranes showed the same tendency with hydrophilic comonomer concentration as the copoly(SiMA-NVP) membranes. The T_g also increased as 4VP content increased. In copoly(4VP 70, 80 and 90), however, two transition peaks were detected by DSC because these copolymers formed a microphase separated structure. All of the synthesized copolymer membranes were colorless and transparent in the dry state. Also, the water content of all copolymer membranes was low due to the presence of bulky hydrophobic groups on the side chains.

Oxygen Permeation Properties in the Dry and Wet States

The relationship between sorbed water into the copolymer membranes and the hydrophilic monomer mole fraction in the copolymer membranes is shown in Figure 2-a, b. The vertical axis and horizontal axis represent the amount of water per gram of polymer and the mole fraction of hydrophilic monomer, respectively. In this study, water in the copolymers was classified into 3 states (9-12). The state of the water in the water-swollen membrane was determined by DSC at a scanning rate of 5 K/min. The weight of freezing water ($W_{bulk\ and\ bound}$) was calculated using the following equation:

Table I Characterization of Copoly(SiMA-NVP) membranes.

Sample	NVP content [mol%]	T_g [°C]	Water content [wt%]	Intermolecular distance d-spacing [Å]	Intramolecular distance d-spacing [Å]
PSiMA	0.00	4.70	0.70	7.45	5.05
Copoly(NVP10)	7.10	6.20	0.12	7.29	4.95
Copoly(NVP20)	17.1	14.8	1.3	7.27	4.88
Copoly(NVP30)	29.1	28.2	3.2	7.08	4.88
Copoly(NVP40)	41.4	31.3	13	7.28	4.80
Copoly(NVP50)	49.8	44.6	18	7.32	4.74
PNVP	100	180	soluble	-	-

Table II Characterization of Copoly(SiMA-4VP) membranes.

Sample	NVP content [mol%]	T_g [°C]	Water content [wt%]	Intermolecular distance d-spacing [Å]	Intramolecular distance d-spacing [Å]
PSiMA	0.00	4.70	0.70	7.45	5.05
Copoly(4VP40)	37.6	17.8	2.1	7.32	4.89
Copoly(4VP50)	48.0	32.3	2.8	7.32	4.87
Copoly(4VP70)	69.5	26.2 , 140	7.9	7.32	4.62
Copoly(4VP80)	80.1	32.6 , 151	11	7.37	4.50
Copoly(4VP90)	88.9	31.6 , 153	22	7.49	4.48
P4VP	100	153	35	-	4.39

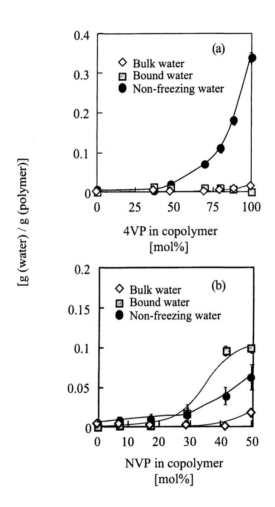

Figure 2. Relationship between absorbed water and the hydrophilic monomer mole fraction in (a) copoly(SiMA-4VP) and (b) copoly(SiMA-NVP) membranes.

$$W_{bulk\ or\ bound} = \frac{Q^h}{\Delta H} \qquad (3)$$

where Q^h is the heat absorbed in the cooling process, which is calculated from the peak area of the DSC scan. ΔH can be estimated using the difference in the heat capacities of ice and super cooled water, ΔC_p.

$$\Delta H(T) = \Delta H(T : 273) - \int_T^{273} \Delta C_p\, dt \qquad (4)$$

where $\Delta C_p = C_p$ (super cooling water) - C_p (ice). We assumed that the water in the membrane and the bulk water have the same relationship between transition temperature and melting enthalpy. In the cooling process, two exothermic peaks were observed at about 230 K and 260 K. The high temperature peak was ascribed to freezing bound water, as Nakamura et al. previously reported (13). In the heating process, the single melting peak, which has a shoulder on the low temperature side, was observed at about 273 K. Assuming that the area of the single melting peak in the heating process is nearly equal to the sum of the areas of the two exothermic peaks in the cooling process, the weight of bound and bulk water in the membrane was determined by eq 3 and eq 4. Non-freezing water is strongly oriented on the polar surface and is not detected by DSC measurements. It is presumed that the bound water is merely oriented by the surrounding non-freezing water. The non-freezing water was estimated by subtracting the weight of the bound and bulk water, which are detected by DSC measurement, from the weight of all of the water in the swollen copolymer membrane. The calculated error limits are shown in Figures 2 and 4.

Oxygen permeation properties of copoly(SiMA-4VP) and copoly(SiMA-NVP) membranes in the dry and wet states are compared in Figures 3-a and 3-b. The oxygen gas permeability coefficients of both series of materials decreased as the polar hydrophilic group content increased. That is, permeability decreases as the concentration of bulky, hydrophobic SiMA units decreases. One contributing factor to the decrease in permeability is the decrease in polymer chain mobility with decreasing concentration of SiMA, as indicated by the glass transition data presented earlier.

Fundamentally, the transport in these membranes is through a solution-diffusion mechanism. The transport of water through the membrane is attributed to both diffusion and solubility. The oxygen gas permeability coefficients of copoly(SiMA-4VP) membranes in the dry state decreased with increasing 4VP content and dramatically decreased at 4VP concentrations greater than 50 mol%. However, water-dissolved oxygen permeability coefficients decreased gradually with increasing 4VP content. The water-dissolved oxygen permeability coefficient of a 4VP-homopolymer membrane in the wet state was one order of magnitude higher than that of the same material in the dry state. In this membrane, only the amount of non-freezing water increased sharply with an increase in 4VP content (Figure 2-a). Therefore, the presence of non-freezing water in the polymer probably causes a relaxation of the polymer chains, which leads to higher permeability coefficients. Additionally, as shown in Figure 3-b, the oxygen gas permeability coefficients of copoly(SiMA-NVP) membranes in the dry state decreased gradually with an increase in NVP content. The water-dissolved oxygen permeability coefficients increased slightly up to 20 mol% NVP content but gradually decreased thereafter. In the copoly(SiMA-NVP) membranes, the non-freezing water also increased slightly up to 20 mol% NVP content and the bound water increased rapidly at concentrations higher than 20 mol% NVP content (Figure 2-b). Therefore, the water-dissolved oxygen permeability of these membranes in the wet state was probably influenced by relaxation of polymer chains by non-freezing water at compositions up to 20 mol% NVP content. The decrease in permeability at compositions above 20 mol% NVP content was attributed to bound water, which interfered with diffusion. The interference with oxygen diffusion by bound water counteracts the contribution that oxygen solubility has with increasing water content.

Properties of Modified Copolymer Membranes

Poly SiMA containing bulky hydrophobic groups on the side chain has high gas permeability, but it is very hydrophobic. Therefore, we attempted to modify copolymer membranes by quaternization to improve the wettability and permeability.

Characterization of the modified copolymer membrane by quaternization is listed in Table III. The T_g and water content increased with increasing quaternization ratio. The water content of each copolymer membrane increased with an increase in the quaternization ratio.

The relationship between the quaternization ratio and the contact angle of water in each copolymer is also shown in Table III. In copoly(4VIP40) and copoly(4VIP50) membranes, the contact angle scarcely changed with an increase in quaternization ratio. However, in the 6.44 mol% sample of quaternized copoly(4VIP70), wettability increased significantly (*i.e.* contact angle decreased strongly). We believe that the other membranes with low quaternization ratios were modified only in the bulk of the membrane. However, a membrane with a high level of quaternization was modified not only in the bulk but also on the surface. Figures 4-a, 4-b, and 4-c show the behavior of water in each modified membrane. In all of the modified membranes, the bulk water rarely changed with an increase in the quaternization ratio. However, the non-freezing water gradually increased with an increase in the quaternization ratio. Moreover, in

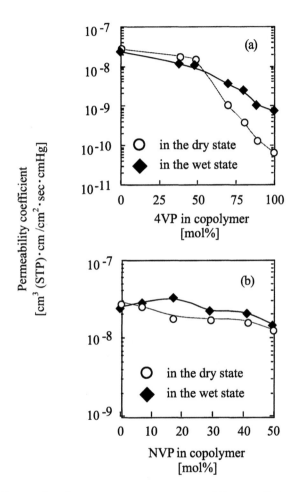

Figure 3. Comparison between oxygen gas permeability coefficients in the dry state and water-dissolved oxygen permeability coefficients in wet state of (a) copoly(SiMA-4VP) and (b) copoly(SiMA-NVP) membranes at 35°C.

362

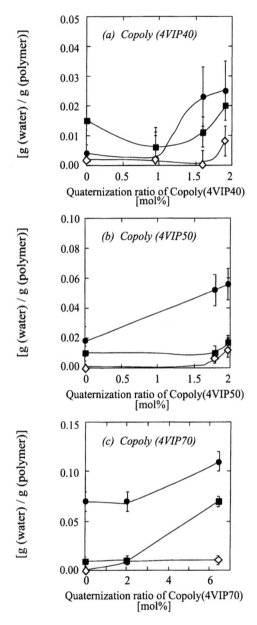

Figure 4. Relationship between classified water in the copolymer and quaternization ratio for (a) copoly(4VIP40), (b) copoly(4VIP50) and (c) copoly(4VIP70) : ◊ *is bulk water,* ■ *is bound water,* ● *is non-freezing water, respectively.*

Table III Characterization of Copoly(SiMA-4VP) membranes.

Sample	Quaternization ratio [mol%]	T_g [°C]	Water content [mol%]	Contact angle [θ]
Copoly(4VP70)	0.00	26.5 , 141	7.9	96±1
Copoly(4VIP70-A)	2.02	21.5 , 152	9.1	95±1
Copoly(4VIP70-B)	6.44	39.7 , 150	19	73±1
Copoly(4VP50)	0.00	22.1	2.8	93±1
Copoly(4VIP50-A)	1.79	16.7	7.2	96±1
Copoly(4VIP50-B)	1.97	24.7	8.6	94±1
Copoly(4VP40)	0.00	11.5	2.1	95±1
Copoly(4VIP40-A)	0.95	13.7	2.8	95±1
Copoly(4VIP40-B)	1.60	15.7	3.4	94±1
Copoly(4VIP40-C)	1.91	29.1	5.2	97±1

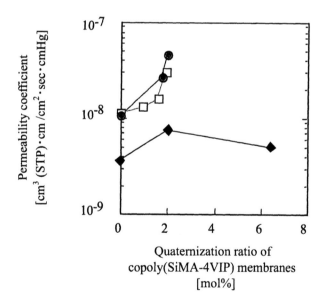

Figure 5. The effect of the quaternization ratio on the permeability coefficients for water-dissolved oxygen in copoly(SiMA-4VIP) membranes at 35°C :
◆ *is copoly(4VIP70),* ◉ *is copoly(4VIP50),* □ *is copoly(4VIP40), respectively.*

the 6.44 mol% sample of copoly(4VIP70), the amount of bound water sharply increased. Figure 5 shows the effect of the quaternization ratio on the water-dissolved oxygen permeability coefficients at 308 K. The water-dissolved oxygen permeability increased with modification by quaternization. We believe that this increase in the water-dissolved oxygen permeability depends on the increase in the amount of non-freezing water that accompanies quarternization. In the 6.44 mol% sample of quaternized copoly(4VIP70), however, the water-dissolved oxygen permeability decreased. We consider that an increase in the amount of bound water with increasing quarternization, as shown in Figure 4-c, interfered with the diffusion of oxygen.

Conclusions

Water entering the membrane breaks the hydrogen bonds and facilitates chain mobility in the copolymer membrane. For water-dissolved oxygen permeability in the wet state, the non-freezing water and bound water increase diffusion by increasing polymer chain mobility. In the polymers which have a bulky hydrophobic side-chain, however, the interference with oxygen diffusion due to bound water counteracts the increase in oxygen solubility with increasing water content.

The water content of each copolymer membrane increased with an increase in the quaternization ratio. Moreover, the water-dissolved oxygen permeability also increased as the extent of quaternization increased. This increase in water-dissolved oxygen permeability appears to be strongly related to the increase in non-freezing water that accompanies quaternization. In the 6.44 mol% sample of quaternized copoly(4VIP70), however, the water-dissolved oxygen permeability decreased because the amount of bound water was increased by quaternization.

References

1. Lai, Y. C.; Valint, P. L. Jr. *J. Appl. Polym. Sci.* **1996**, *61*, 2051-2058.
2. Lai, Y. C. *J. Appl. Polym. Sci.* **1995**, *56*, 317-324.
3. Lai, Y. C. *J. Appl. Polym. Sci.* **1996**, *60*, 1193-1199.
4. Kunzler, J.; Ozark, R. *J. Appl. Polym. Sci.* **1995**, *55*, 611-619.
5. Minoura, N.; Fujiwara, Y.; Nakagawa, T. *J. Appl. Polym. Sci.* **1979**, *24*, 965-973.
6. Nakagawa, T.; Hopfenberg, H. B.; Stannett, V. *J. Appl. Polym. Sci.* **1971**, *15*, 231-245.
7. Refojo, M. F.; Leong, F. L. *J. Membr. Sci.* **1979**, *4*, 415-426.

8. Nakagawa, T. *J. Jpn. C.L. Soc.* **1988**, *30*, 1-12.
9. Hatakeyama, T.; Hatakeyama, H.; Nakamura, K. *Thermochimica Acta.* **1995**, *253*, 137-148.
10. Hatakeyama, T.; Yoshida, H.; Hatakeyama, H. *Polymer* **1987**, *28*, 1282-1286.
11. Hatakeyama, T.; Yoshida, H.; Hatakeyama, H. *Thermochimica Acta.* **1995**, *266*, 343-354.
12. Takahashi, S.; Nakagawa, T.; Yoshida, M.; Asano, M. *J. Membr. Sci.* **2002**, *206*, 165-177.
13. Nakamura, K.; Hatakeyama, T.; Hatakeyama, H. *Tex. Res. J.* **1981**, *51*, 607-613.

Chapter 25

Chemically Modified Polysulfone Hollow Fibers with Zwitterionic Sulfoalkylbetaine Group Having Improved Blood Compatibility

Akon Higuchi[1], Hirokazu Hashiba[1], Rika Hayashi[1], Boo Ok Yoon[1], Mitsuo Hattori[2], and Mariko Hara[1]

[1]Department of Applied Chemistry, Seikei University, Tokyo 180–8633, Japan
[2]Health Care Center, Seikei University, Tokyo 180–8633, Japan

Hydrophilic polysulfone membranes (SPE-PSf) were prepared by covalently bonding poly(sulfoalkylbetaine) (SPE) on the surface. The immobilized amount of SPE on SPE-PSf membranes was controlled by the SPE monomer concentration in the reaction solution. The SPE-PSf membranes were significantly more hydrophilic than unmodified polysulfone or other surface-modified polysulfone membranes due to the long hydrophilic side chain of SPE, which contributes to the hydrophilic nature of the modified PSf membranes. SPE-PSf membranes showed lower protein adsorption from a plasma solution than polysulfone and other surface-modified membranes due to the highly hydrophilic surface of the SPE-PSf membranes. The SPE-PSf membranes showed significantly lower number of adhering platelets on its surface than polysulfone and other surface-modified membranes. The hydrophilic surface of SPE-PSf membranes may suppress of platelet adhesion on the SPE-PSf membranes.

© 2004 American Chemical Society

Polysulfone hollow fibers blended with poly(vinylpyrrolidone) [PVP] have been widely used as hemodialysis membranes, because the pore size and pore distribution of the polysulfone membranes can be easily controlled by changing the composition of the polysulfone casting solution. These polysulfone membranes can effectively remove low-molecular weight proteins, such as β_2-microglobulin (MW = 11,500 g/mol) and endotoxin (MW = 5,000 - 20,000 g/mol). Because polysulfone itself has poor blood compatibility, modification of the polysulfone membranes is required for use as hemodialysis membranes. Polysulfone membranes blended with PVP are currently used as hemodialysis membranes in the medical industry, because PVP is a hydrophilic polymer having neither a hydroxyl group nor ionically charged groups and, therefore, shows good blood compatibility (1).

Other types of modification have also been reported for improving the blood compatibility of polysulfone membranes (2-4). Ishihara *et al.* prepared a phospholipid polymer having a 2-methacryloyloxyethyl phosphorylcholine (MPC) unit (2). The MPC polymer containing a zwitterionic group could be blended with polysulfone by a solvent evaporation method during membrane processing. The number of platelets adhering to the polysulfone membranes blended with the MPC polymer was reduced, and the change in the morphology of adherent platelets was suppressed (2).

Another typical zwitterionic group is N,N-dimethyl-N-methacryloxyethyl-N-(3-sulfopropyl)ammonium betaine (SPE). Viklund and Irgum prepared a porous sulfoalkylbetaine polymer with zwitterionic SPE, which showed reversible protein adsorption relative and was used as a chromatographic monolith for protein separation (5).

Surface-modified polysulfone hollow fibers having several hydrophilic groups were previously prepared in our laboratory (6-9) and by other researchers (3,4,10). A variety of hydrophilic groups such as $-CH_2CH_2CH_2SO_3^-$, $-CH(CH_3)CH_2OH$, $-CH_2N(CH_2CH_3)_3^+$, $-CH_2NHCH_2CH_2NH_2$ and $-CH_2OH$ were introduced on the surface of the hydrophobic polysulfone membranes by chemical reaction (6-9). We recently succeeded in introducing an aliphatic double bond on the surface of polysulfone hollow fibers using N-succinimidylacrylate (11). In this study, the polymerization of zwitterionic monomer of SPE on the surface of polysulfone hollow fibers is reported. Platelet adhesion on polysulfone membranes conjugated with SPE polymer on the surface in human plasma was investigated. The blood compatibility of the modified membranes was evaluated by comparison of unmodified and other surface-modified polysulfone membranes.

Experimental

Materials

The membranes used for chemical modification were commercially available polysulfone (PSf) ultrafiltration hollow fibers (SI-1, Asahi Chemical

Co., Ltd.). The inside and outside diameters of the fibers were approximately 0.75 mm and 1.3 mm, respectively. The zwitterionic monomer N,N-dimethyl-N-methacryloxyethyl-N-(3-sulfopropyl) ammonium betaine (SPE) was kindly supplied from Dr. P. Koeberle (Rasching Chemie GmbH).

Mouse monoclonal anti-human albumin (300-06551, Nippon Bio-Test Laboratories, Inc., Tokyo), mouse monoclonal anti-human fibrinogen (F4639, Sigma-Aldrich, Inc., MO), and goat anti-human fibronectin (F1509, Sigma-Aldrich, Inc., MO) were used as primary antibodies. Goat F(ab)$_2$ anti-human immunoglobulin peroxidase conjugate antibody (AHI1304, Biosource International, CA), rabbit H+L anti-mouse immunoglobulin peroxidase conjugate antibody (014-17611, Wako Pure Chemical Industries, Ltd., Tokyo), and rabbit anti-goat immunoglobulin peroxidase conjugate antibody were used as secondary antibodies. Block Ace^{TM} (UK-B80) was purchased from Funakoshi Co., Ltd. TMB (3,3',5,5'-tetramethylbenzidine) peroxidase substrate (TMB Microwell Peroxidase Substrate System, 50-76-00, Kirkegaard & Perry Laboratories, Guildford UK) was used as received.

Other chemicals were of reagent grade and were used without further purification. Ultrapure water was used throughout the experiments.

Synthesis of NSA

The active ester of N-succinimidylacrylate (NSA) was synthesized from N-hydroxysuccinimide and acryloyl chloride following the method of Adalsteinsson *et al.* (*12-14*). The recovered yield was 65±8%.

Surface Chemical Modification of PSf Membranes

Chloromethylation of the PSf membranes (Cl-PSf) was performed by dipping in a solution of chlorodimethyl ether, hexane, and SnCl$_4$ as the Friedel-Crafts catalyst at 25 °C for 15 min as described previously (*9,11*) (Scheme 1). The apparatus for the chemical modification of the membranes was described recently (*6,9*), and the same procedures for the chemical modification were employed in this study.

Ethylenediamination of the chloromethylated PSf membranes (i.e., Scheme 1) was performed by dipping the chloromethylated membranes into ethylenediamine for 15 min at 25 °C.

Ethylenediaminated PSf membranes (EDA-PSf, fiber length = 100 cm) were immersed in 250 ml of phosphate buffer solution (0.02 M, PBS, pH 7.4) containing 0.327 g of NSA for the preparation of NSA-PSf-1 membrane, 0.1635 g of NSA for NSA-PSf-2 membrane, and 0.109 g of NSA for NSA-PSf-3 membrane at 37 °C. The reaction was then incubated at 37 °C for one hour to

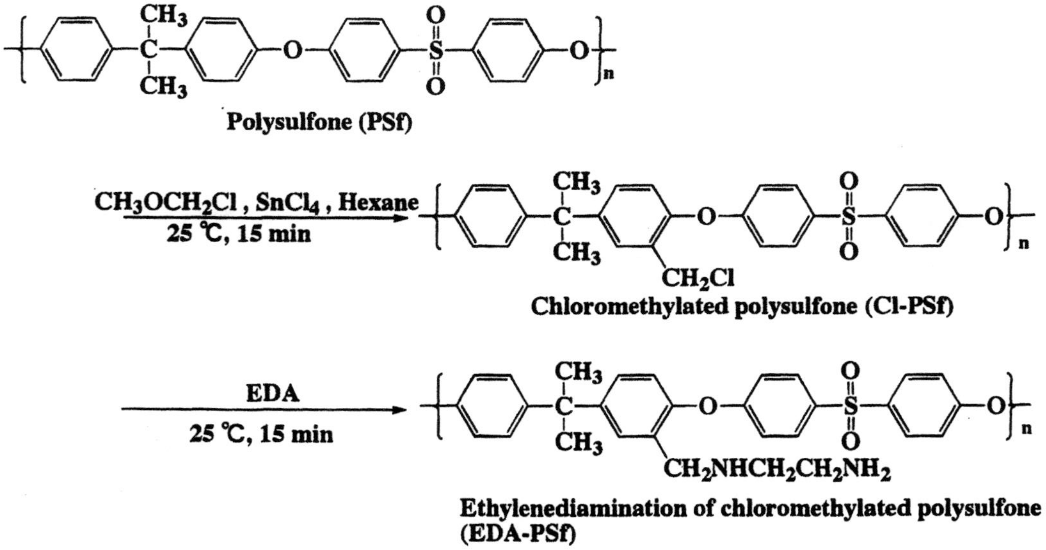

Polysulfone (PSf)

CH_3OCH_2Cl, $SnCl_4$, Hexane
25 ℃, 15 min

Chloromethylated polysulfone (Cl-PSf)

EDA
25 ℃, 15 min

Ethylenediamination of chloromethylated polysulfone (EDA-PSf)

Scheme 1

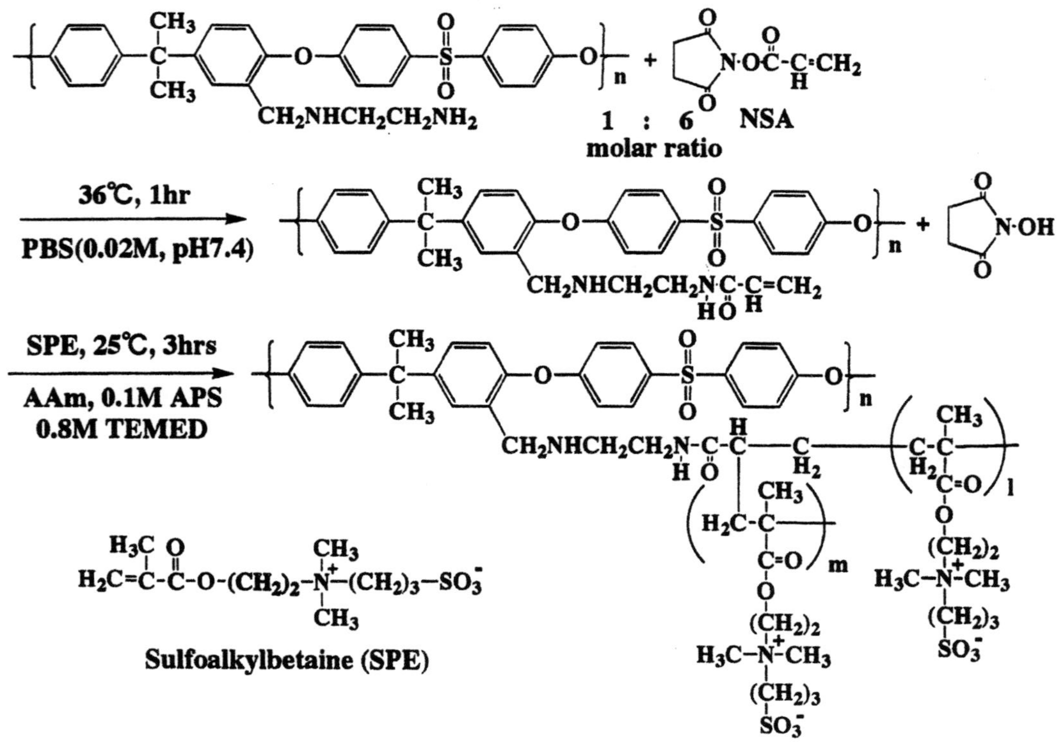

370

Scheme 2

introduce vinyl groups onto the surface of the membranes (Scheme 2). After the reaction, the membranes conjugated with NSA were washed in a phosphate buffer solution for one hour and then in water for one hour. The NSA-conjugated membranes (NSA-PSf) were immersed in 150 ml of phosphate buffer solution containing 10, 200, 500, 1,000, 2,000 or 3,000 moles of SPE to NSA moles of the NSA-PSf membranes together with 0.01 ml of 0.1 M aqueous ammonium persulphate (APS) and 0.01 ml of 0.8 M aqueous N,N,N',N'-tetramethylethylenediamine (TEMED) as redox initiators. Polymerization of SPE was performed at 25 °C on the surface of the membranes for 3 hours (SPE-PSf, see Scheme 2). SPE-PSf-X-Y (X=1, 2 or 3 and Y=10, 200, 500, 1000, 2000 or 3000) indicates SPE-PSf membranes prepared from NSA-PSf-X membranes using Y moles of SPE to NSA moles of the membranes.

Characterization of Surface-modified PSf Membranes

The degree of chloromethylation of the Cl-PSf membranes was measured by ^{1}H-NMR (400 MHz, JNM GX-400, JEOL, Ltd.). The degree of ethylenediamination of the EDA-PSf membranes was estimated using a standard titration method (11). The advancing and receding water contact angles (11,15) were measured in air at 25±2 °C using a Langmuir-Blodgett trough (NL-LB200S-NWC, Nippon Laser & Electronics Lab.) as described previously (11).

Protein Adsorption Assay on the Membranes

Small samples of PSf membranes and surface-modified PSf (Cl-PSf, EDA-PSf, NSA-PSf and SPE-PSf) membranes were cut into separate membranes and were immersed in a phosphate buffer solution (PBS, 0.02M, pH 7.4) of (1) 5,000 ppm bovine serum albumin (BSA), (2) 5,000 ppm bovine γ-globulin, (3) 300 ppm fibrinogen or (4) 50% platelet-poor plasma (PPP) containing 0.15 mol/L NaCl for 120 min at 37 °C. The membranes were then rinsed five times with PBS. The membranes were then inserted into a glass tube containing 1 wt% aqueous solution of sodium dodecyl sulfate (SDS) and the glass tubes containing the membranes and SDS solution were shaken for 60 min at room temperature to remove the proteins adsorbed on the membranes. The amount of proteins adsorbed on the membrane surface was calculated from the concentration of proteins in the SDS solution using a protein analysis kit (Micro BCA protein assay reagent kit) (2).

Specific protein adsorption from human plasma on the membranes was evaluated using the antigen-antibody reaction using enzyme-immunoglobulin conjugate as described previously (11).

Platelet Adsorption on the Membranes

A 35 ml sample of fresh human blood (woman, 32 years old) was collected using five vacuum tubes (7 ml, Venoject II, Terumo, Co.) containing 10.5 mg of EDTA·2Na and centrifuged at 1000 rpm for 10 min to obtain platelet-rich plasma (PRP) or at 2800 rpm for 15 min to obtain platelet-poor plasma (PPP) (2,16,17). Fresh PRP or PPP samples were used in all the studies. PSf and the modified membranes were cut, placed into 24-well tissue culture plates and equilibrated with 0.15 mol/L NaCl solution at 37 °C for one hour. The 0.15 mol/L NaCl solution was removed and then 1 ml of fresh PRP was introduced. The membranes were incubated with PRP at 37 °C for two hours. PRP was decanted, and the membranes were rinsed again with 0.15 mol/L NaCl solution. Finally, the membranes were treated with 3 wt% glutaraldehyde in 0.15 mol/L NaCl solution for two days at 4 °C. The samples were washed again with 0.15 mol/L NaCl solution, subjected to a drying process by passing them through a series of graded alcohol-NaCl solutions (0, 25, 50, 75, and 100 %) and dried in vacuum for 10 hours at room temperature. The dried membranes were sputter-coated with gold and were examined using a JSM-5200 scanning electron microscope (SEM, JEOL, Ltd.).

The number of adhering platelets on the membranes was calculated from 4 SEM pictures at a magnification of 500 from different places on the same membranes. These procedures were performed on each membrane using four independent membranes (totally n = 16), and the results were averaged to obtain reliable data.

Results and Discussion

Chemical Modification of PSf

Chloromethylation and ethylenediamination of the PSf membranes were performed as previously reported (see Scheme 1) (9). The degree of chloromethylation was estimated to be 2.6% of PSf repeat units from the peak of ^{1}H-NMR spectra at δ=4.5 ppm. The degree of ethylenediamination was estimated from the ion-exchange capacity of the EDA-PSf membranes and was 0.8% of PSf-repeat units. The immobilized amount of NSA on PSf (NSA-PSf, see Scheme 2) membranes was estimated from the amount of NSA consumed in the NSA reaction solution detected by UV absorption at 275 nm and was 0.88 ± 0.26 μmol/cm^2 (NSA-PSf-1), 0.18 ± 0.05 μmol/cm^2 (NSA-PSf-2) and 0.10 ± 0.03 μmol/cm^2 (NSA-PSf-3) depending on the composition of NSA in the reaction solution (i.e., 0.13, 0.065, and 0.044 g/100 ml of PBS, respectively). The immobilized amount of SPE on PSf (SPE-PSf, see Scheme 2) membranes was also estimated from the amount of SPE consumed in the SPE reaction solution detected by UV absorption at 260 nm. The immobilized amount of SPE on PSf (i.e., SPE-PSf) membranes was controlled by the amount of its monomer (i.e., vinylpyrrolidone) in the reaction solution.

Figure 1 shows the dependence of the polymerized number of SPE (N) on the molar ratio of SPE to NSA [C(SPE)/C(NSA)] of NSA-PSf-1 in the reaction solution when the reaction time was 3 h. The polymerized number of SPE was calculated from the immobilized SPE amount divided by the immobilized NSA amount on the SPE-PSf membranes and was an average number for the polymerization of SPE with the SPE-PSf membranes. The polymerized number of SPE increased with the concentration of SPE in the reaction solution. However, the polymerized number of SPE tended to reach a saturated polymerization number at a high molar ratio of SPE to NSA.

Water Contact Angles of Surface-modified PSf Membranes

The advancing and receding water contact angles on PSf and surface-modified PSf membranes were measured. The advancing water contact angle of SPE-PSf was the same as that of PSf or other surface-modified PSf membranes (i.e., 89-90±3 degrees), while the receding water contact angle of SPE-PSf (i.e., 26±2 degrees) was smaller than that of PSf or the other surface-modified PSf membranes (i.e., 32-34±2 degrees). This result indicates that the SPE-PSf membranes were more hydrophilic than PSf and other surface-modified membranes prepared in this study. This can be explained by the long hydrophilic side chain of SPE, which contributes to formation of the hydrophilic surface of the PSf membranes.

Protein Adsorption on Surface-modified PSf Membranes

Plasma protein adsorption on materials is a key phenomenon during thrombogenic formation. The amount of plasma proteins adsorbed on the materials has been reported to be one of the most important factors in evaluating the blood compatibility of materials (2).

Figure 2 shows the adsorption of single proteins on PSf and surface-modified PSf membranes from albumin, γ-globulin or fibrinogen solution containing 0.15 mol/L NaCl at pH 7.4. It was found that the SPE-PSf membranes had a lower protein adsorption of albumin, γ-globulin, and fibrinogen than the PSf, Cl-PSf, EDA-PSf, and NSA-PSf membranes. This result can be attributed to the hydrophilic surface of the SPE-PSf membranes, because a hydrophilic surface is known to reduce the protein adsorption on the membranes (5-8). SPE-PSf-1-500 showed the best characteristics for anti-protein adsorption. This is because SPE-PSF-1-500 (27 $\mu mol/cm^2$) had significantly more immobilized SPE than SPE-PSF-2-500 (1.3 $\mu mol/cm^2$) or SPE-PSF-3-500 (0.62 $\mu mol/cm^2$).

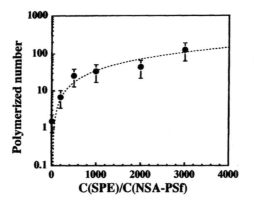

Figure 1. Dependence of polymerized number of SPE on the concentration molar ratio of SPE to NSA [C(SPE)/C(NSA)] of NSA-PSf-1 in the reaction solution for a reaction time of 3 hr. Data are expressed as the means±S.D. of four independent measurements.

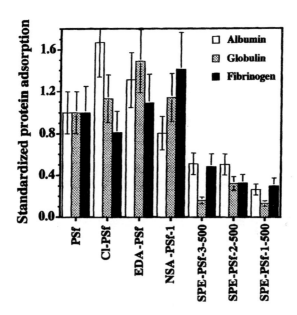

Figure 2. Adsorption of single proteins on PSf and some surface-modified PSf membranes from BSA, γ-globulin or fibrinogen solutions containing 0.15 mol/L NaCl at pH 7.4. Data are expressed as the means±S.D. of four independent measurements.

Figure 3 shows the amount of total plasma proteins on PSf and surface-modified PSf membranes from platelet-poor plasma. Clearly, the SPE-PSf membranes showed lower protein adsorption than the PSf, Cl-PSf, EDA-PSf and NSA-PSf membranes.

The specific plasma protein adsorption on the membranes was also investigated using ELISA assay. Figure 4 shows the standardized amount of adsorbed proteins (albumin, globulin, fibrinogen, and fibronectin) on PSf and surface-modified PSf membranes from platelet-poor plasma. The adsorption of specific plasma proteins on the membranes from PPP (i.e., Figure 4) showed a different amount measured from a single protein solution (Figure 2). This is because the specific protein adsorption on the membranes is influenced by other proteins co-existing in the plasma solution, and co-operative binding of each protein on the membranes from PPP.

SPE-PSf membranes showed significantly lower adsorption of albumin from PPP, while slightly reduced adsorption of globulin, fibrinogen and fibronectin on SPE-PSF-1-500 from PPP was observed relative to that on PSf, Cl-PSf, EDA-PSf and NSA-PSf membranes.

In summary, the SPE-PSf membranes showed lower protein adsorption of each of the proteins in PPP evaluated from single protein solution (Figure 2) and from PPP (Figures 3 and 4).

Platelet Adhesion on Surface-modified PSf Membranes

Platelet adhesion on the membranes from human plasma is an important test for the evaluation of the blood compatibility of the membranes. Platelet adhesion on PSf and surface-modified PSf membranes from platelet-rich plasma was investigated. A typical scanning electron micrograph of platelets adhering to PSf, EDA-PSf, NSA-PSf-1, and SPE-PSf-1-500 membranes is shown in Figure 5. Platelets were rarely observed on SPE-PSf-1-500 membranes, while numerous platelets were observed on PSf and EDA-PSf membranes. The activated platelets on EDA-PSf membranes showed spread and deformed shapes.

Figure 6 shows the number of platelets adhering to PSf, SPE-PSf, and other surface-modified PSf membranes using platelet-rich plasma. The SPE-PSf-1-500 (immobilized amount of SPE = 27 μmol/cm^2), SPE-PSf-1-1000 (immobilized amount of SPE = 35 μmol/cm^2), SPE-PSf-1-2000 (immobilized amount of SPE = 45 μmol/cm^2), SPE-PSf-1-3000 (immobilized amount of SPE = 130 μmol/cm^2), SPE-PSf-2-500, and SPE-PSf-3-500 membranes showed a reduced number of platelets adhering to the surface compared to PSf, Cl-PSf, EDA-PSf, and NSA-PSf membranes. More than 27 μmol/cm^2 of SPE should be immobilized on the SPE-PSf membranes in order to show the characteristic surfaces for extensively suppressed adhesion of platelets, although SPE-PSf-2-500 and SPE-PSf-3-500 showed suppressed platelet adhesion and have only 0.6 - 1.3 μmol/cm^2 of SPE. The different synthesis conditions of SPE on the SPE-

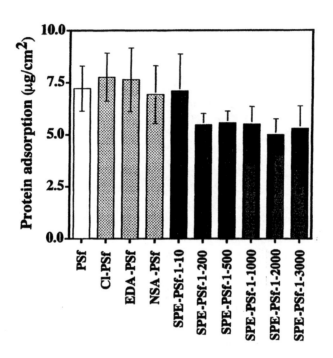

Figure 3. Adsorption of total plasma proteins on PSf and some surface-modified PSf membranes from PPP at pH 7.4. Data are expressed as the means±S.D. of four independent measurements.

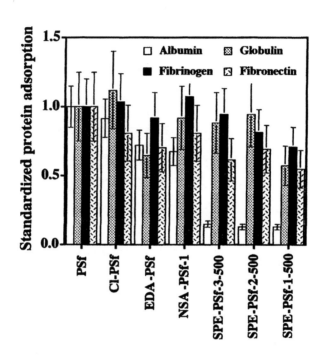

Figure 4. Standardized amount of adsorbed proteins (albumin, globulin, fibrinogen and fibronectin) on PSf and surface-modified PSf membranes from PPP evaluated from ELISA assay. Data are expressed as the means±S.D. of five independent measurements.

(a) PSf

(b) EDA-PSf

(c) NSA-PSf-1

(d) SPE-PSf-1-500

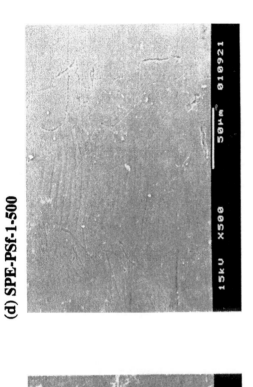

Figure 5. Scanning electron micrograph of platelets adhering to (a) PSf, (b) EDA-PSf, (c) NSA-PSf-1 and (d) SPE-PSf-1-500 membranes.

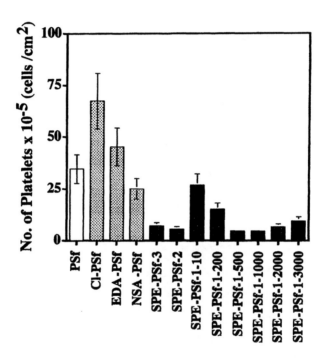

Figure 6. The number of adhering platelets on PSf, SPE-PSf, and other surface-modified PSf membranes from platelet rich-plasma estimated by SEM. Data are expressed as the means±S.D. of 16 measurements.

PSf membranes might be contributing to the homogeneous or heterogeneous distribution of the length and location of SPE on the surface of SPE-PSf membranes. It is suggested that the hydrophilic surface of SPE-PSf membranes caused the suppression of platelet adhesion on SPE-PSf membranes.

The suppression of platelet adhesion on membranes is generally believed to be due to the reduction in protein adsorption, especially the suppression of fibrinogen adsorption (2,18). The fibrinogen adsorption from plasma onto membranes is known to regulate the adhesion of platelets because the fibrinogen causes the glycoprotein, GP IIb-IIIa, of the platelet membranes to bind (2,19). SPE-PSf membranes showed a slightly lower amount of fibrinogen adsorption compared to PSf membranes based on the evaluation of the ELISA assay of PPP (i.e., Figure 4). The platelets are less adsorbed on SPE-PSf membranes than on PSf and other surface-modified membranes.

Conclusions

Polysulfone membranes covalently conjugated with zwitterionic sulfoalkylbetaine on the surface were prepared. The immobilized amount of SPE on the SPE-PSf membranes could be controlled by the concentration of SPE monomer in the reaction solution. The SPE-PSf membranes were significantly more hydrophilic than PSf and other surface-modified PSf membranes. This results from the long hydrophilic side chain of SPE which determines the hydrophilic nature of the modified PSf membranes. The SPE-PSf membranes had a lower protein adsorption from plasma solution than PSf, Cl-PSf, EDA-PSf, and NSA-PSf membranes. This can be attributed to the hydrophilic surface of the SPE-PSf membranes (5-8). The SPE-PSf membranes showed a lower number of platelets adhering on the surface than the PSf, Cl-PSf, EDA-PSf, and NSA-PSf membranes. It is suggested that the hydrophilic surface of the SPE-PSf membranes caused suppression of platelet adhesion.

References

1. Tanaka, K.; Kobayashi, T. Japanese Patent (Tokkaihei) 4-300636, 1992.
2. Ishihara, K.; Fukumoto, K.; Iwasaki, Y.; Nakabayashi, N. *Biomaterials* **1999**, *20*, 1553-1559.
3. Ulbricht, M.; Belfort, G. *J. Membrane Sci.* **1996**, *111*, 193-215.
4. Pieracci, J.; Crivello, J. V.; Belfort, G. *J. Membrane Sci.* **1999**, *156*, 223-240.
5. Viklund, C.; Irgum, K. *Macromolecules* **2000**, *33*, 2539-2544.
6. Higuchi, A.; Iwata, N.; Tsubaki, M.; Nakagawa, T. *J. Appl. Polym. Sci.* **1988**, *36*, 1753-1767.

382

7. Higuchi, A.; Iwata, N.; Nakagawa, T. *J. Appl. Polym. Sci.* **1990**, *40*, 709-717.
8. Higuchi, A.; Nakagawa, T. *J. Appl. Polym. Sci.* **1990**, *41*, 1973-1979.
9. Higuchi, A.; Koga, H.; Nakagawa, T. *J. Appl. Polym. Sci.* **1992**, *46*, 449-457.
10. Nabe, A.; Staude, E.; Belfort, G. *J. Membrane Sci.* **1996**, *133*, 57-72.
11. Higuchi, A.; Shirano, K.; Harashima, M. Yoon, B.-O.; Hara, M.; Hattori, M.; Imamura, K. *Biomaterials*, **2002**, *23*, 2659-2666.
12. Miyata, T.; Asami, N.; Uragami, T. *Nature*, **1999**, *399*, 766-769.
13. Shoemaker, S. G.; Hoffman, A. S.; Priest, J. H. *Appl. Biochem. Biotech.*, **1987**, *15*, 11-24.
14. Adalsteinsson, O.; Lamotte, A.; Baddour, R. F.; Colton, C. K.; Pollak, A.; Whitesides, G. M. *J. Mol. Cat.* **1979**, *6*, 199-225.
15. Marchant, K. K.; Veenstra, A. A.; Marchant, R. E. *J. Biomed. Mater. Res.* **1996**, *30*, 209-220.
16. Bahulekar, R.; Tamura, N.; Ito, S.; Kodama, M. *Biomaterials* **1999**, *20*, 357-362.
17. Kawakami, H.; Nagaoka, S.; Kubota, S. *ASAIO J.* **1996**, *42*, M871-M876.
18. Park, K.; Mao, F.W.; Park, H. *J. Biomed. Mater. Res.* **1991**, *25*, 407-420.
19. Phillips, D. R.; Charo, I.F.; Parise, L. V.; Fitzgerald, *Blood* **1988**, *71*, 831-843.

Chapter 26

Selective Protein Adsorption and Cell Attachment to Rubbed Fluorinated Polyimides

H. Kawakami, K. Ashiba, and Y. Okuyama

Department of Applied Chemistry, Tokyo Metropolitan University, Hachioji, Tokyo 192–0397, Japan

High-pressure mechanical rubbing was used to nanopattern a fluorinated polyimide membrane surface. Plasma protein adsorption and cell attachment to the rubbed membranes were characterized using a micro-bicinchoninic acid protein assay and phase-contrast microscopy. The morphologies of rat skin fibroblast cells attached to the rubbed membrane were three-dimensional multicellular spheroids. The amount of hydroxyproline generated on the membrane significantly increased due to rubbing. The amount of albumin adsorption on the rubbed polyimide membrane was slightly lower than that on an unrubbed membrane.

Aromatic polyimides are a class of high-performance polymers that are highly thermally stable and have high glass transition temperatures and relatively low dielectric constants. Various polyimides have become increasingly important in a variety of technological applications, such as semiconductor devices, high-temperature adhesives, and high-performance composite materials. We have reported that polyimides containing fluorinated groups are promising materials for medical devices (1-5). For example, polyimide hollow fibers showed not only high gas exchange (O_2 transfer and CO_2 removal) but also suppression of platelet adhesion, neutrophil adhesion, and complement activation, suggesting the possibility of a novel membrane oxygenator with the advantage of both increased gas exchange and excellent biocompatibility (1-4).

© 2004 American Chemical Society

Micropatterning and nanotechnology are becoming increasingly popular for the development of improved biomaterials and medical devices. In particular, surface micro- or nano-fabrication techniques have been discovered and developed to create unique materials for regulating protein or cell functions (6). Several techniques have been investigated for creating micropatterning on surfaces, including the use of conventional photoresist lithography, photochemistry, and self-assembled monolayers. However, currently, the patterning for biomaterials and medical devices has been limited to micrometer or tenths of micrometer size features.

In this paper, we report a simple method for creating nanopatterning on surfaces of biomaterials and medical devices. Mechanically rubbed polyimide membranes are widely used as alignment materials in liquid crystal displays (7). We have used high pressure rubbing to nanopattern a fluorinated polyimide membrane surface for biomaterials applications. Plasma protein adsorption and cell attachment to the rubbed polyimide membranes were evaluated using a micro-bicinchoninic acid protein assay and phase-contrast microscopy.

Experimental

Materials

2, 2'-Bis(3,4-dicarboxyphenyl)hexafluoropropane dianhydride (6FDA) was purchased from the Japanese Hoechst Co., (Tokyo, Japan) and purified by sublimation. 2,2'-Bis(4-aminophenyl)hexafluoropropane (6FAP) was purchased from Central Glass Co., (Saitama, Japan) and recrystallized twice in ethanol solution.

The fluorinated polyimide, 6FDA-6FAP, was synthesized by preparing its polyamic acid precursor and then chemically imidizing the polyamic acid as reported in the literature (8). The chemical structure of 6FDA-6FAP is presented in Scheme 1. Polyimide films were prepared by solvent casting from tetrahydrofuran solution onto a glass plate, followed by curing at 150 °C. The cast films were optically clear and were approximately 50 μm thick. Polystyrene (PSt) films were used as control materials.

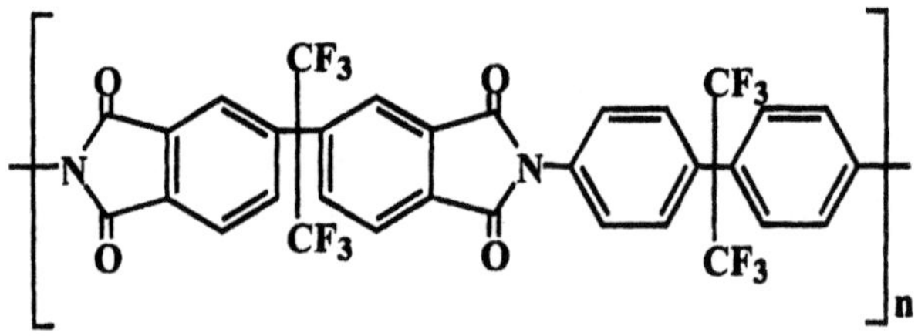

Scheme 1. Structure of 6FDA-6FAP.

Rubbed Polyimide Membrane

Nanopatterning was performed by mechanically rubbing the surface of the polyimide membranes using a constant velocity rotating cylinder covered with a rubbing cloth. The cylinder velocity was 7.5 mm/s, and the pressure added to the cylinder was 6.0×10^4 Pa.

Protein Adsorption

Bovine serum albumin (BSA) was purchased from Nacalai Tesque, Inc., (Tokyo, Japan). BSA was used without further purification. PBS (pH=7.4) was purchased from the Yatoron Co. (Tokyo, Japan). The polyimide membranes were rinsed with deionized water and ethanol prior to use. The amount of BSA adsorbed to the membrane surface from 1.0 mg/ml BSA solution was evaluated by a micro-bicinchoninic acid protein assay (Pierce, Inc., Micro BCA™ Protein Assay) at 36.5 °C in PBS as reported in the literature (4). The BSA adsorbed on the membrane was removed by rinsing with 1% sodium dodecyl sulphate. Then, the amount of BSA in the rinse solution was determined by Micro BCA™ Protein Assay at 562 nm using a spectrophotometer (Ubest-55, JASCO, Tokyo, Japan).

Cell Culture

Rat skin fibroblast (FR) cells were cultured in flasks containing EMEM and alpha-MEM supplemented with 10% FBS at 37°C in a 5% CO_2 incubator. The cells were seeded at 1×10^4 cells/well in 96-well plates and incubated for 24 h at 37°C. The cells were observed by phase-contrast microscopy (Nikon, ECLPISE TE-300, Tokyo, Japan) and their function was determined based on the collagen production as measured by a specific assay for hydroxyproline.

Analysis of Hydroxyproline

The assay was performed as reported in the literature (9). The absorption was measured using a spectrophotometer (Ubest-55, JASCO, Tokyo, Japan) operating at 560 nm.

Surface Analysis

Contact angle measurements for water were performed on dry membranes using a contact angle measurement device (Kyowa Co. Elma GI, Tokyo, Japan).

AFM Analysis

The surface morphology of the membranes was visualized using an atomic force microscope (AFM: Seiko SPI3700, Tokyo, Japan) operating in air at room temperature (*10*). The cantilevers (Seiko SN-AF01), with a spring constant of 0.021 N/m, were microfabricated from silicon nitride. The surface was continuously imaged in feedback mode with a scan area of 500 nm x 500 nm and a constant scan speed of 2 Hz. The surface roughness profile was analyzed using a parameter, R_a, which is the arithmetic average of departures of the roughness profile from the mean line.

The force-distance curves were obtained by moving the cantilever vertically at a speed of approximately 70 nm/s over a maximum travel height of 700 nm. The tip radius was 20 nm.

Results and Discussion

Characteristics of Rubbed Polyimide Membranes

The fluorinated polyimide, 6FDA-6FAP, had a molecular weight of 9.2 x 10^5 and a polydispersity index of 1.2. Therefore, a high molecular weight polyimide with a narrow molecular weight distribution was prepared by chemical imidization. Based on ^{1}H-NMR spectroscopy, the degree of imidization of 6FDA-6FAP was complete. The polyimide membranes were prepared by a solvent-casting method and were cured at 150°C for 15 hours. The amount of residual solvent in the membrane was measured using thermogravimetric analysis (Seiko, TG/FDTA300, Tokyo Japan). These measurements indicated that the curing step completely removed the solvent.

Figure 1 presents AFM images of the top surfaces of two 6FDA-6FAP membranes. The surface area imaged in these pictures is 500 x 500 nm. The surface morphology on the rubbed membrane is nanopatterned. The width between peaks was approximately 100 nm and the height of each peak was 2-3 nm. In contrast, the surface of the unrubbed polyimide membrane was essentially smooth.

To elucidate the surface properties of the rubbed 6FDA-6FAP membrane, we measured the contact angle of water and the adhesion forces on the membrane surface. As shown in Table I, there was no difference in the contact angle between the rubbed and unrubbed membranes. The level of adhesion between the cantilever tip and the polyimide surface was determined from force-distance experiments on the top and bottom surfaces of the membranes. The silicon nitride tip used in this study was hydrophilic in nature. Interestingly, the adhesion forces on the top and bottom surfaces of the rubbed membrane were different; they exhibited values of 8.5±0.3 and 4.4±0.3 nN, respectively. One explanation for this finding may be the increased hydrophilic nature of the top surface of the rubbed membrane and the increased hydrophobic nature of the bottom surface.

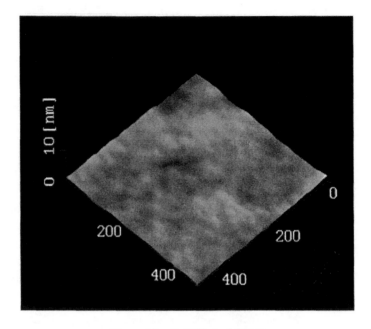

Unrubbed 6FDA-6 FAP

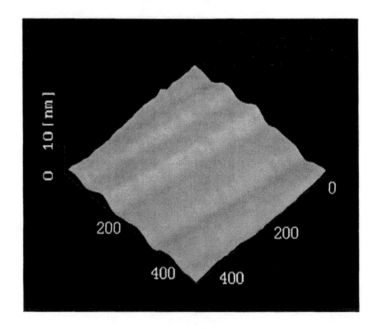

Rubbed 6FDA-6FAP

Figure 1. AFM images of top surface of 6FDA-6FAP membranes.

388

**Table I. Water contact angle and adhesion force
on 6FDA- 6FAP surfaces**

Polyimide	$\theta(°)$	Adhesion force (nN)	
		top	bottom
Rubbed 6FDA-6FAP	85	8.5±0.3	4.4±0.3
Unrubbed 6FDA-6FAP	84	7.9±0.1	7.9±0.1

Cell Attachment to Rubbed Polyimide Membranes

Figure 2 shows phase contrast micrographs of FR cells on 6FDA-6FAP membranes. Interestingly, the morphologies of FR cells attached to the rubbed 6FDA-6FAP membrane were three-dimensional multicellular spheroids. On the other hand, the cells on the unrubbed 6FDA-6FAP membrane were a two-dimensional monolayer. Recently, study of spheroids of normal cells has attracted much attention, since spheroids show functional similarities to tissues and organs, unlike the conventional two-dimensional monolayer culture of cells (*11*). Tissues and organs *in vivo* are made from cells of many types, which are systematically and functionally combined together. Therefore, the spheroid may be an important approaches for regenerating tissue or organs. The nanopatterning formed on the rubbed 6FDA-6FAP membrane induces the attachment of highly aggregated cells and generation of three-dimensional multicellular spheroids.

Conventional methods of generating spheroid morphologies include the spontaneous aggregation of cells on a non-adhesive substrate such as as agarose or polyhydroxyethylmethacrylate-coated plastics or tbe spontaneous detachment of cells from a semi-adhesive surface such as proteoglycan or positively charged polystyrene. Both methods significantly limit the cell species from which spheroids can be made. In contrast, the rubbing method described in this article is an extremely simple method for forming multicellular spheroids, and it should be applicable to a wide variety of cell species.

In cell culture systems, efficient recovery of cells from the culturing substrate is very important to maintain the cell function. In general, confluent cells in substrates were treated with trypsin and detached from the substrate. However, trypsin treatment destroys two-dimensional monolayer culture of cells formed on the substrate and damages cell membranes by hydrolyzing various membrane-associated proteins. In this study, spheroids formed on the rubbed 6FDA-6FAP membrane were easily detached from the substrate and were recovered when treated with EDTA. The detachment process of the spheroids from the rubbed 6FDA-6FAP membrane is illustrated in Figure 3.

Figure 4 shows the cell function of FR. The function was determined based on collagen production measured by a specific assay for hydroxyproline. The amount of hydroxyproline generated on the rubbed 6FDA-6FAP membrane was approximately 10-fold higher than that generated on the unrubbed membrane.

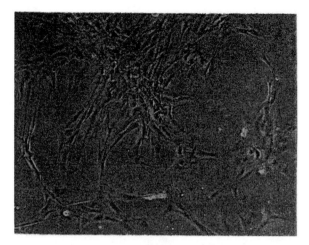

Unrubbed 6FDA-6FAP

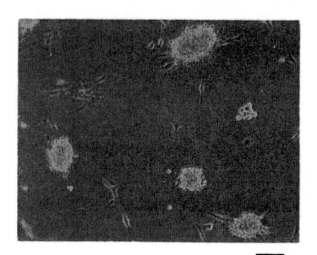

5 μm

Rubbed 6FDA-6FAP

Figure 2.Phase contrast microscopy images of FR cells cultured on 6FDA-6FAP surfaces.

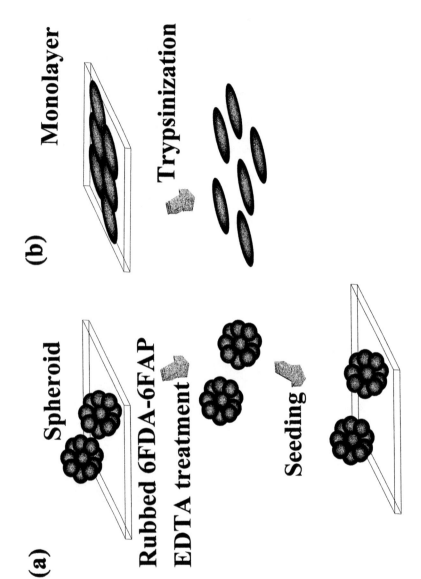

Figure 3. Detachment process of cells from polymer membrane. (a) rubbed 6FDA-6FAP surface; (b) tissue culture polystyrene surface.

Therefore, the interactions between cells, due to the formation of spheroids, were strong enough to facilitate the differentiated functions of FR cells.

Protein Adsorption

Figure 5 presents the amount of BSA adsorption on 6FDA-6FAP membranes. The BSA adsorption measurements were determined using a micro-bicinchoninic acid protein (BCA) assay. We have already reported that fluorinated polyimides show excellent biocompatibility and suppress plasma protein adsorption (*1-5*). Actually, in this study, 6FDA-6FAP also suppressed BSA adsorption relative to PSt; the amount of BSA adsorbed onto 6FDA-6FAP was 2-fold smaller than that on PSt. The amount of BSA adsorption on the rubbed 6FDA-6FAP membrane decreased slightly relative to that adsorbed on the unrubbed membrane. In general, more protein is adsorbed on a hydrophobic surface than on a hydrophilic surface. We speculate that the nano-ordered hydrophilic and hydrophobic patterns formed at the top and bottom surfaces of the rubbed 6FDA-6FAP membrane may facilitate BSA adsorption. However, these arguments require validation by evaluating the direct interaction between the rubbed polyimide membrane and plasma protein.

Conclusions

The purpose of this study was to demonstrate a simple method for creating nanopatterning on biomaterial and medical device surfaces. We prepared nanopatterned fluorinated polyimide surfaces using a conventional, high-pressure mechanical rubbing method. The plasma protein adsorption and cell attachment to the rubbed polyimide membranes were evaluated *in vitro* using a micro-bicinchoninic acid protein assay and phase-contrast microscopy. Interestingly, morphologies of FR cells attached to the rubbed 6FDA-6FAP membrane were three-dimensional multicellular spheroids, while the cells on the unrubbed membrane were two-dimensional monolayers. In addition, the amount of hydroxyproline generated on the rubbed 6FDA-6FAP membrane was about 10-fold higher than that generated on the unrubbed membrane. Therefore, the interactions between cells, due to the formation of spheroids, were strong enough to facilitate the differentiated functions of FR cells. On the other hand, the amount of BSA adsorption on the rubbed 6FDA-6FAP membrane decreased slightly relative to that on the unrubbed membrane.

In previous papers, we reported that fluorinated polyimides were promising biomaterials with excellent biocompatibility. The results obtained in this study indicate that the rubbed 6FDA-6FAP membrane may be a novel nanopatterned biomaterial for regulating protein or cell functions.

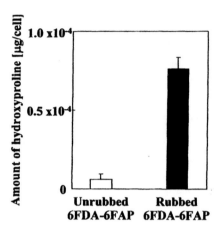

Figure 4. Amount of hydroxyproline produced from FR cultured on 6FDA-6FAP surfaces.

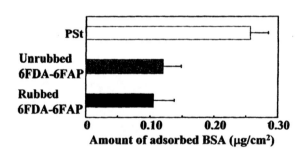

Figure 5. Amount of adsorbed BSA on polymer surfaces. BSA: 1.0 mg/ml; temperature: 36.5 °C; Mean±S.D. (n=3).

Acknowledgments

This work was partially supported by a grant from the Japan Society for the Promotion of Science.

References

1. Kawakami, H.; Nagaoka, S. *ASAIO J.* **1995**, *41*, M379-83.
2. Kawakami, H.; Nagaoka, S. *ASAIO J.* **1996, 42**, M871-876.
3. Kawakami, H.; Mori, Y.; Takagi, J.; Nagaoka, S.; Kanamori, T.; Shinbo, T.; Kubota, S. *ASAIO J.* **1997**, *43*, M490-494.
4. Niwa, M.; Kawakami, H.; Kanno, M.; Nagaoka, S.; Kanamori, T.; Shinbo, T.; Kubota, S. *J. Biomater. Sci. Polym. Ed.* **2001**, *12*, 533-542.
5. Kanno, M.; Kawakami, H.; Nagaoka, S.; Kubota, S. *J. Biomed. Mater. Res.* **2002**, *60*, 53-60.
6. Blawas, A. S.; Reichert, W. M. *Biomaterials* **1998**, *19*, 595-609.
7. Samant, M. G.; Stohr, J.; Brown, H. R.; Russell, T. P.; Sands, J. M.; Kumar, S. K. *Macromolecules* **1996**, *29*, 8334-8342.
8. Kawakami, H.; Anzai, J.; Nagaoka, S. *J. Appl. Polym. Sci.* **1995**, *57*, 789-795.
9. Acil, Y.; Terheyden, H.; Dunsche, A.; Fleiner, B.; Jepsen, S, *J. Biomed. Mater. Res.* **2000**, *51*, 703-710.
10. Kawakami, H.; Takahashi, T.; Nagaoka, S.; Nakayama, Y. *Polym. Adv. Technol.* **2001**, *12*, 244-252.
11. Kunz-Schughart, L. A.; Heyder, P.; Schroeder, J.; Knuechel, R. *Exp. Cell Res.* **2001**, *266*, 74-86.

Chapter 27

Separation of Endocrine Disruptors from Aqueous Solutions by Pervaporation: Relationship between the Separation Factor and Solute Physical Parameters

Boo Ok Yoon, Takao Asano, Kenta Nakaegawa, Mai Ishige, Mariko Hara, and Akon Higuchi

Department of Applied Chemistry, Seikei University, Tokyo 180–8633, Japan

The separation of endocrine disruptors from aqueous solution by pervaporation is reported. Hydrophobic polydimethylsiloxane membranes were used at elevated feed temperatures. Endocrine disruptors such as dibenzo-p-dioxin, diethylphthalate, co-planar PCB, etc. were concentrated 30 – 1300 times in the permeate. The endocrine disruptor separation factor, α, increased with increasing vapor pressure. An empirical linear relationship between log α and log p_{vap} was observed (because vapor pressure is directly related to the driving force of permeation through the membrane), while no correlation was observed between α and endocrine disruptor molecular weight. Theoretically, α was directly related to the product of p_{vap} and log K_{ow}. A relatively good correlation between log α and p_{vap} x log K_{ow} was found experimentally.

© 2004 American Chemical Society

Polychlorinated biphenyls (PCBs), dioxins, and certain phthalates exhibit estrogenic activity (1,2) and adversely affect the development and reproduction of humans and animals (3-5). These compounds are called endocrine disruptors, and their presence in the environment is a major concern. In this work, we examined the feasibility of separating endocrine disruptors from extremely dilute aqueous solutions by pervaporation using hydrophobic polydimethylsiloxane (PDMS) membranes.

Pervaporation has been successfully used in the dehydration of alcohols (6,7), as well as in the removal of dissolved volatile organic compounds from water (8-31). Böddeker et al. investigated the removal of low volatility aromatic hydrocarbons [i.e., p-t-butylphenol (MW 150.2 Da), thymol (MW 150.2 Da), o-dichlorobenzene (MW 147.0 Da), and vanillin (MW 152.2 Da)] from dilute aqueous solutions through polyether-block-polyamide membranes by pervaporation (8,9). In their study, very high separation factors were observed. For example, thymol and o-dichlorobenzene had separation factors over water of 380 and 1000, respectively (8).

Hoshi et al. also reported the separation of 1,1,2-trichloroethane (MW 133.4 Da) and tetrachloroethylene (MW 167.9 Da) from dilute aqueous solutions using composite membranes of various crosslinked acrylate copolymers by pervaporation. The separation factors for 1,1,2-trichloroethane and tetrachloroethylene over water were 700 and 1000, respectively (12).

In most of these previous studies (8,10-28), the organic compounds removed from water by pervaporation were volatile organic solvents such as benzene, toluene, chloroform, trichloroethylene, and tetrachloroethylene with molecular weights of less than 170 Da. Most endocrine disruptors have higher molecular weights [e.g., 1,2-dibromo-3-chloropropane, DBCP, (MW 236.4 Da)], a very low vapor pressure [0.58 Torr (77.3 Pa) for DBCP], and are nearly insoluble in water (see Table I).

Despite these factors, in an earlier study we reported the pervaporative removal of endocrine disruptors from extremely dilute aqueous solutions using hydrophobic polydimethylsiloxane membranes (29). DBCP could be separated efficiently from very dilute aqueous solutions by pervaporation through PDMS membranes provided that the vacuum line between the pervaporation cell and cold trap on the permeate side was heated to 150°C. Furthermore, the separation factor of endocrine disruptors over water was found to depend significantly on membrane thickness due to concentration polarization (29). PDMS membranes having a thickness of more than 300 mm resulted in high separation factors (29).

The aim of this study is to examine the feasibility of improving the separation of endocrine disruptors from water by controlling the temperature of the feed solution and by maintaining elevated temperatures at the permeate side in pervaporation. Separation of several endocrine disruptors, such as dibenzo-p-dioxin, diethylphthalate (DEP) and co-planar PCB (3,3',4,4'-tetrachlorobiphenyl, TCB), from aqueous solutions has been investigated by pervaporation. The relationship between endocrine disruptor separation factor and endocrine disruptor physical properties (i.e., saturated vapor pressure and hydrophobicity (log K_{ow}, octanol-water partition coefficient)) is discussed.

Table I. Endocrine disruptor properties

Endocrine disruptor	MW (Dalton)	Use	Vapor Pressure (Torr)*	Water solubility (mg/l)*	Log K_{ow}*
1 Dibenzo-p-dioxin	184.2		4.125×10^{-4}	1	4.37
2 Biphenyl	154.2	insecticide	0.0089	7.5	4.01
3 1,2-Dibromo-3-chloropropane (DBCP)	236.3	insecticide	0.58	1230	2.96
4 2-sec-Butylphenyl-methylcarbamate (BPMC)	207.3	pesticide	1.425×10^{-4}	420	2.78
5 2,2-Dimethyl-1,3-benzodioxol-4-yl methylcarbamate (Bendiocarb)	223.2	pesticide	5×10^{-6}	260	1.70
6 n-Butylbenzene	134.2	liquid crystal	1.064	11.8	4.38
7 Diethylphthalate (DEP)	222.2	plasticizer	2.1×10^{-3}	1080	2.42
8 Dibutylphthalate (DBP)	278.3	plasticizer	2.01×10^{-5}	13	4.50
9 3,3',4,4'-Tetrachloro-biphenyl (TCB)	292.0	insecticide	1.64×10^{-5}	5.69×10^{-4}	6.63

NOTE: * refers to references (30-39).

Experimental

Materials

Flat-sheet polydimethylsiloxane membranes (SR-30 and SR-50, Tigers Polymer Co., Ltd.) with a thickness of 300 mm and 500 mm, respectively, were used. Prior to use, the PDMS membranes were sequentially rinsed with 2-propanol for 2 days, a 1:1 mixture of acetone and *n*-hexane for 2 days, and dried under vacuum at 100°C for 24 h. Pre-treatment of the membranes was performed to remove unreacted monomers and impurities from the membranes.

Model endocrine disruptors were purchased from Wako Pure Chemicals Industries Ltd., and are listed together with their physical properties (*30-39*) in Table I. Other chemicals were of reagent grade and were used without further purification. Ultra pure water was used throughout the experiments.

Pervaporation

Permeation experiments were carried out using a standard pervaporation (PV) apparatus (*12,13,29*). Figure 1 shows a schematic representation of the pervaporation apparatus used in this study. The effective membrane area was 15.2 cm^2. The magnetic stir bar in the PV cell was operated at 300 rpm. The downstream (permeate) pressure was kept below 7 KPa in all experiments. The permeate solution was collected in a liquid nitrogen cold trap at regular intervals (1 h) and then analyzed using either a gas chromatograph (GC-14B, Shimadzu Co.) equipped with a flame ionization detector (FID) or a gas chromatograph-mass spectrometer (GCMS-QP5050A, Shimadzu Co.). The pervaporation separation factor (α) was calculated according to eq 1

$$\alpha = (X_p/Y_p) / (X_f/Y_f) \qquad (1)$$

where X_f and Y_f are the mole fractions of endocrine disruptor (X) and water (Y) in the feed solution, and X_p and Y_p are the mole fractions of these components in the permeate.

In this study, PV experiments were performed as a function of feed solution (T_{feed}) temperature. The PV cell was insulated using a flexible heating sheet and the feed solution temperature was regulated at the desired value using a temperature controller. The membrane temperature was regulated at the same temperature as the feed solution using a flat sheet heater located under the PV cell (*29*). The vacuum line between the PV cell and the cold trap on the permeate side was insulated with flexible heating tape and the temperature of the vacuum line ($T_{permeate}$) was regulated at the desired temperature (i.e., 150°C in

398

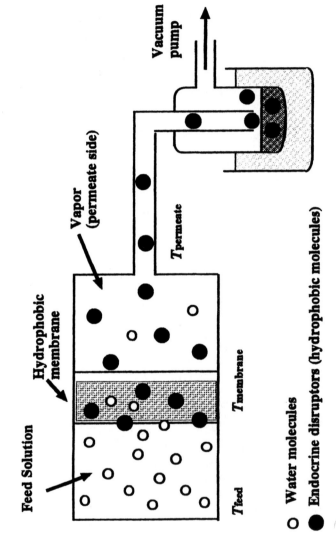

Figure 1. Schematic representation of the pervaporation apparatus.

Vacuum pump

Vapor (permeate side)

Hydrophobic membrane

Feed Solution

$T_{permeate}$

$T_{membrane}$

T_{feed}

○ Water molecules

● Endocrine disruptors (hydrophobic molecules)

this study) (see Figure 1) (29). Heating the permeate tubing was required to prevent condensation of the endocrine disruptors in this line.

The membrane flux was determined by measuring the weight of liquid collected in the cold trap.

Solubility

To measure the endocrine disruptor solubility in the membranes, isotropic PDMS membranes of known weight, thickness (500 mm), and area (15.2 cm^2) were immersed for 48 h at 25°C in an endocrine disruptor aqueous solution (volume, V, (25 cm^3), concentration, C_s (ppm)). After removal of the PDMS membranes from the solution, the membrane surface was wiped dry and the weight of the membrane (W_w) measured. Thereafter, the membrane was dried for 24 h at 80°C and the weight of the dry membrane (W_d) was determined.

The concentration (in ppm) of endocrine disruptors in the solution after immersion of the membrane (C_b) was measured and used to determine the solubility selectivity. The concentration (in ppm) of endocrine disruptors in the membrane, C_m, was calculated from:

$$C_m = 25(C_s - C_b)/(W_w - W_d) \qquad (2)$$

The solubility selectivity, α_s was obtained from

$$\alpha_s = [C_m (1-10^{-6}C_b)] / [C_b (1-10^{-6}C_m)] \qquad (3)$$

GC/GC-MS Measurements

Aliquots (5ml) of endocrine disruptor solutions, except for biphenyl, 3,3',4,4'-tetrachlorobiphenyl (co-planar PCB, TCB) and dibenzo-p-dioxin, were analyzed using a gas chromatograph (GC-14 B, Shimadzu Co.) equipped with an FID and a Thermon-3000 column (1.6 x 3.2 mm, Shincarbon A 60/80 Shinwakakou Co., Ltd.). The carrier gas was nitrogen, and it was kept constant at 40 kPa. The injector and detector temperatures were 230°C. The column temperature was programmed as follows: 140°C for 6 min, heating 20°C/ min up to 220°C, and maintained at 220°C for 3 min.

Aliquots (1 ml) of dibenzo-p-dioxin, biphenyl and TCB solutions were analyzed from splitless injection using a GC-MS equipped with an autosampler (AOC-20) and a DB-1 capillary column (inner diameter 0.25 mm, length 30 m,

J&W Scientific Co.). The injection and detection temperatures were set at 250°C. The column temperature was programmed as follows: dibenzo-p-dioxin; 150°C for 1 min, heating 10°C/min up to 250°C, and kept at 250°C for 10 min; biphenyl; 100°C for 2 min, heating 20°C/min up to 250°C, and kept at 250°C for 5 min; and TCB; 70°C for 2 min, heating at 8°C/ min up to 300°C, and kept at 300°C for 15 min.

Results and Discussion

Separation Factors and Feed Solution Temperature

Pervaporation experiments using PDMS membranes (300 μm thickness) were performed using a 1.0 liter feed solution at $T_{permeate}$ = 150°C. The dependence of membrane flux on feed solution temperature in pervaporation of (a) pure water, (b) 10 ppm of 1,2-dibromo-3-chloropropane (DBCP) solution and (c) 10 ppm of 2-sec-butylphenylmethyl-carbamate (BPMC) solution was examined. The results are presented in Figure 2. The flux of all components increased more or less exponentially with temperature. Because the concentration of the endocrine disruptors in the feed solution was very low (10 ppm), the presence of these components in the feed did not significantly change the water flux.

Figure 3 shows the effect of feed temperature on endocrine disruptor separation factors. The optimal feed solution temperature was about 60°C for a 10 ppm DBCP solution. At this temperature, the separation factor reached a maximum value of about 2000. This may be because the vapor pressure of water increases dramatically as temperature increases, especially above 60°C. This contributes to an increase in water flux through the membrane as a result of the increase in driving force for the water transport. High water flux will, in turn, contribute to a decrease in the separation factor above 60°C for pervaporation when using organic permselective membranes in this application.

On the other hand, water preferentially permeated through the PDMS membranes at T_{feed} ≤60°C, and the separation factor increased with increasing T_{feed} at T_{feed} > 60°C in the pervaporation of 10 ppm BPMC solution. This is because BPMC has a lower vapor pressure than DBCP [i.e., 1.4 x 10^{-4} Torr (1.9 x 10^{-2} Pa), see Table 1], so BMPC effectively needs the higher temperatures (*i.e.,* T_{feed} > 60°C) to have a sufficiently high vapor pressure to permit the preferential permeation of BPMC over water in the PDMS membranes. This result indicates that non-volatile organic compounds having very low vapor pressure, such as BPMC, can be successfully separated by elevating T_{feed} (e.g., 70-90°C) in pervaporation.

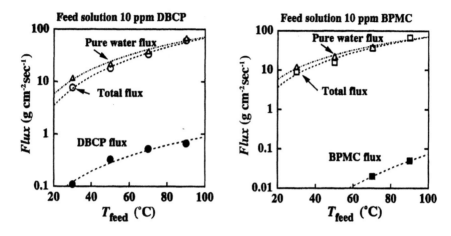

Figure 2. Dependence of the experimentally obtained water and endocrine disruptor flux on feed solution temperature. The flux of pure water through the same membrane (PDMS membrane with a 300 mm thickness, $T_{permeate}$ = 150°C) is shown for comparison.

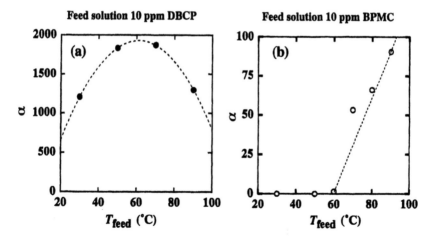

Figure 3. Effect of feed solution temperature on the separation factor of 10 ppm DBCP and BPMC solutions in PDMS membranes (300 mm thick) at $T_{permeate}$ = 150°C.

Relationship Between Separation Factors of Diffusion and Sorption

The separation factor of pervaporation, α, can be divided into two independent factors, the sorption separation factor (α_S) and the diffusion separation factor (α_D) (13):

$$\alpha = \alpha_S \cdot \alpha_D \qquad (4)$$

Table II summarizes α values obtained from pervaporation measurements at $T_{feed} = 90°C$ and $T_{permeate} = 150°C$, α_S obtained from sorption measurements, and α_D estimated from α/α_S using 1 ppm biphenyl, 1ppm dioxin, 10 ppm BPMC, 10 ppm bendiocarb and 10 ppm DBCP solutions. α_D of the endocrine disruptors exhibited a lower value than α and α_S. For biphenyl, dioxin, BPMC, bendiocarb and DBCP, α_D was less than 1, so water diffuses much faster than these endocrine disruptors in PDMS membranes. The low diffusion separation factor results from the large size of the endocrine disruptor molecules relative to that of water. Our results clearly indicate that the separation of endocrine disruptors from water by pervaporation is the result of the large solubility difference between endocrine disruptors and water in PDMS. Therefore, the separation of hydrophobic endocrine disruptors by pervaporation through PDMS membranes is mainly governed by sorption selectivity.

Relationship Between Separation Factor and Molecular Weight

The separation of certain endocrine disruptors in Table I (i.e., dibenzo-*p*-dioxin, diethylphthalate (DEP) and co-planar PCB (3,3',4,4'-tetrachlorobiphenyl, TCB)) from aqueous solutions using pervaporation was studied. The pervaporation experiments were performed using PDMS membranes having a thickness of 300 μm at $T_{feed} = 90°C$ and $T_{permeate} = 150°C$. Figure 4 shows the relationship between the separation factor of various endocrine disruptors (and some other organic chemicals) and their molecular weight using PDMS membranes from the present study and from the literature (10,15,20).

From the present study, the endocrine disruptor separation factors do not correlate well with molecular weights. However, as expected, hydrophobic endocrine disruptors (e.g., DBCP, biphenyl, and *n*-butylbenzene) exhibited in higher separation factors than hydrophilic organic chemicals (e.g., ethanol, phenol) due to the higher affinity and, therefore, higher solubility of the more hydrophobic organic compounds in the hydrophobic PDMS membranes.

**Table II. Solubility, diffusion and pervaporation selectivity
for various endocrine disruptors**

Endocrine Disruptors	MW	$\alpha_S{}^*$	α_D	α^{**}
Biphenyl	154.2	2278	0.27	621
Dioxin	184.2	3231	0.01	46
BPMC	207.3	1521	0.06	91
Bendiocarb	223.2	670	0.04	30
DBCP	236.3	2719	0.48	1295

NOTE: $\alpha = \alpha_S \alpha_D$, * at 25°C, ** at $T_{feed} = 90$°C and $T_{interface} = 150$°C

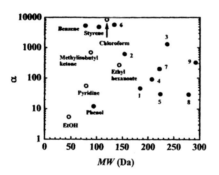

*Figure 4. The relationship between the separation factor of various endocrine
disruptors (numbers 1 to 9 correspond to those listed in Table I), other organic
chemicals measured in this study (●), and other compounds taken from the
literature (10,15,20) in PDMS membranes.*

Separation Factor of Endocrine Disruptors and Their Physical Properties

The relationship of the separation factor and the saturation vapor pressure of the endocrine disruptors (p_{vap}) is shown in Figure 5. A good correlation is obtained, and the separation factor increases with increasing vapor pressure. This correlation is observed because vapor pressure is directly related to the driving force in permeation. The separation factor of endocrine disruptors over water is high for those endocrine disruptors having a high vapor pressure. Accordingly, it is possible to estimate the separation factor of other endocrine disruptors in pervaporation through PDMS membranes once their vapor pressure is known. Estimated separation factors of DDT and 2,3,7,8-tetrachlorodioxin (2,3,7,8-TCDD) (10.0 and 1.9, respectively) are shown in Figure 5.

The linear relationship between log α and log p_{vap} ($r = 0.867$), shown in Figure 5 is regarded as an empirical relationship. The theoretical relationship between α and physical parameters such as log p_{vap} and log K_{ow} (the octanol-water partition coefficient) are developed as described below.

Flux can be expressed by Fick's first law based on the solution-diffusion theory (40).

$$J = - D \cdot dc/dx = - D \cdot S \cdot dp/dx \qquad (5)$$

where D is the diffusion coefficient, S is the solubility, c is the concentration of solute (water or endocrine disruptor) in the membrane, x is the space co-ordinate for the diffusion direction, and p is the solute vapor pressure in the membrane.

The separation factor by pervaporation is described by

$$\alpha = [J(ED)/J(H_2O)] / [X_f/Y_f] \qquad (6)$$

where $J(ED)$ and $J(H_2O)$ are the flux of the endocrine disruptor and water, respectively. Because the concentration of endocrine disruptors in the feed solution is dilute (less than 10 ppm), eq 6 reduces to eq 7:

$$\alpha = [J(ED)/J(H_2O)] / X_f \qquad (7)$$

The driving force of solute permeation through the membrane by pervaporation is described by Raoult's law (41) with the assumption that the vapor pressure of the solute on the permeate side is zero due to the vacuum on the permeate side:

$$dp/dx = - X_f \cdot p_{vap}(ED)/L \quad \text{for endocrine disruptor permeation} \qquad (8)$$

and $\qquad dp/dx = - Y_f \cdot p_{vap}(H_2O)/L \; \text{for } H_2O \text{ permeation} \qquad (9)$

where L is the membrane thickness, $p_{vap}(ED)$ and $p_{vap}(H_2O)$ are the saturation vapor pressures of the endocrine disruptor and water, respectively. Because $X_f \ll 1$, then $Y_f \cong 1$ and eq 9 reduces to eq 10:

$$dp/dx = - p_{vap} (H_2O)/L \quad \text{for } H_2O \text{ permeation} \tag{10}$$

The endocrine disruptor solubility in a PDMS membrane from an aqueous dilute solution of endocrine disruptor should be related to the partition coefficient between octanol and water (log K_{ow}) as indicated in eq 11. The solubility of an endocrine disruptor [$S(ED)$] in a hydrophobic PDMS membrane increases with increasing hydrophobicity of the endocrine disruptor. The hydrophobicity of the solute (i.e., endocrine disruptor) increases as the value of log K_{ow} increases (42). Therefore, $S(ED)$ is assumed to be linearly related to log K_{ow} of the endocrine disruptors in this study (42):

$$S(ED) = \gamma \cdot \log K_{ow} \tag{11}$$

where γ is a constant.

Combining eqs 7, 8, 10 and 11, the separation factor is obtained as eq 12:

$$
\begin{aligned}
\alpha &= [D(ED) \cdot S(ED) \cdot p_{vap}(ED)] / [D(H_2O) \cdot S(H_2O) \cdot p_{vap} (H_2O)] \\
&= \beta \cdot D(ED) \cdot \log K_{ow} \cdot p_{vap} (ED)
\end{aligned} \tag{12}
$$

where $\beta = \gamma / [D(H_2O) \cdot S(H_2O) \cdot p_{vap} (H_2O)]$, $D(ED)$ is the diffusion coefficient of an endocrine disruptor in the membrane, $D(H_2O)$ is the diffusion coefficient of water in the membrane, and $S(H_2O)$ is the solubility of water in the membrane.

Because β is a constant, the separation factor may be written as follows:

$$\alpha \propto D(ED) \cdot \log K_{ow} \cdot p_{vap} (ED) \tag{13}$$

Assuming that endocrine disruptor diffusion coefficients in PDMS are approximately the same for all of the endocrine disruptors used in this study, because their molecular weights are similar (134 – 292 Da), eq 14 is obtained:

$$\alpha \propto \log K_{ow} \cdot p_{vap} (ED) \tag{14}$$

Figure 6 presents the relationship between the separation factor of endocrine disruptors and log $K_{ow} \cdot p_{vap} (ED)$ based on eq 14. A relatively good correlation ($r = 0.883$) was obtained.

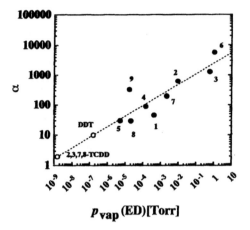

Figure 5. Relationship between experimental separation factor (●) and the saturation vapor pressure of the endocrine disruptors (p_{vap}) in the feed solution. Pervaporation experiments were performed using PDMS membranes (300 mm thick) at $T_{feed} = 90°C$. The numbers in the figure correspond to those listed in Table I. Estimated separation factors (○) of DDT and 2,3,7,8-tetrachlorodioxin are also given.

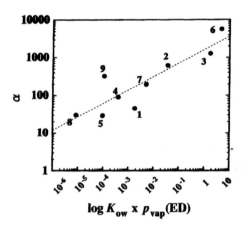

Figure 6. Relationship between the separation factor of endocrine disruptors (numbers 1 to 9 correspond to those listed in Table I) and $\log K_{ow} \cdot p_{vap}(ED)$. The dotted line is the curve fit of eq 14.

Conclusions

Pervaporation-based separation of endocrine disruptors from an aqueous solution was performed using hydrophobic PDMS membranes. Most endocrine disruptors have a very low vapor pressure [e.g., 1.4×10^{-4} Torr (1.9×10^{-2} Pa) for BPMC and 1.6×10^{-5} Torr (2.1×10^{-3}Pa) for co-planar PCB], and are nearly insoluble in water. Nevertheless, relatively high pervaporation separation factors are obtained. To improve the separation of endocrine disruptors, pervaporation experiments were performed at elevated temperatures of $T_{permeate}$ and relatively thick PDMS membranes (i.e., 300 mm) were used. The permeation flux and separation factor were observed to depend on the feed solution temperature for pervaporation of 10 ppm DBCP and BPMC solutions. In particular, non-volatile organic compounds having a very low vapor pressure, such as BPMC, can be successfully separated by elevating T_{feed} (i.e., 70-90 °C). Separation of endocrine disruptors from water by pervaporation was due to the large solubility difference between the endocrine disruptors and water in the PDMS membrane. Thus separation of hydrophobic endocrine disruptors by pervaporation through PDMS membranes is mainly governed by sorption selectivity.

Separation of several endocrine disruptors listed in Table I, such as dibenzo-p-dioxin, diethylphthalate (DEP) and co-planar PCB (TCB), was also performed by pervaporation. The relationship between the separation factor of endocrine disruptors and their molecular weight, saturation vapor pressure and hydrophobicity ($\log K_{ow}$) were evaluated. The endocrine disruptor/water separation factor did not correlate well with molecular weight. However, a linear relationship between $\log \alpha$ and $\log p_{vap}$ was found. The separation factor of endocrine disruptors increased with increasing vapor pressure because the vapor pressure is directly related to the driving force in permeation.

It was theoretically shown that the separation factor of endocrine disruptors was directly related to the product of p_{vap} and $\log K_{ow}$. A relatively good correlation between $\log \alpha$ and the product of p_{vap} and $\log K_{ow}$ was observed.

Acknowledgment

This research was partially supported by the Salt Science Foundation.

List of Symbols

α	separation factor of pervaporation
α_s	solubility selectivity
c	concentration of solute in the membrane (g/cm^3)
C_b	concentration of endocrine disruptors in the solution

	after immersion of the membrane (ppm)
C_s	initial concentration of endocrine disruptor in aqueous solution (ppm)
C_m	concentration of endocrine disruptors in the membrane (ppm)
D	diffusion coefficient (cm^2/sec)
J	flux (g $cm^{-2}sec^{-1}$)
L	membrane thickness (cm)
$\log K_{ow}$	octanol-water partition coefficient
p_{vap}	saturated vapor pressure (Pa)
S	solubility coefficient (g/(cm^3 Pa))
T_{feed}	temperature of the feed solution (°C)
$T_{permeate}$	temperature of the vacuum line (°C)
W_w	weight of the wet membrane (g)
W_d	weight of the dry membrane (g)
X	space co-ordinate for the diffusion direction
X_f	mole fraction of endocrine disruptor in the feed solution
X_p	mole fraction of the components in the permeate
Y_f	mole fraction of water in the feed solution
Y_p	mole fraction of the components in the permeate
β	constant
γ	constant

References

1. Korach, K.S.; Sarver, P.; Chae, K.; McLachlan, J.A.; McKinney, J.D. *Mol.Pharmacol.* **1988**, *33*, 120-126.
2. Jobling, S.; Reynolds, T.; White, R.; Parker, M.G.; Sumpter. J.P. *Environ. Health Perspect.* **1995**, *103*, 582-587.
3. Colborn, T.; Vom Saal, F.S.; Soto, A.M. *Environ. Health Perspect* **1993**, *101*, 378-384.
4. Sharpe, R.M.; Skakkebaek, N.F. *Lancet* **1993**, *341*, 1392-1395.
5. Sharpe, R.M. *Nature,* **1995**, *375*, 538-539.
6. Yoshikawa, M.; Ochiai, S.; Tanigaki, M.; Eguchi, W.J. *Appl. Polym. Sci.* **1991**, *43*, 2021-2032.
7. Uragami, T.; Matsuda, T.; Okuno, H.; Miyata, T.J. *Membrane Sci.* **1994**, *88*, 243-251.
8. Böddeker, K.W.; Bengtson, G.J. *Membrane Sci.* **1990**, *53*, 143-158.
9. Lau, W.W.Y.; Finlayson, J.; Dickson, J.M.; Jiang, J.; Brook, M.A.J. *Membrane Sci.* **1997**, *134*, 209-217.
10. Olsson, J.; Trägårdh, G.J. *Membrane Sci.* **2001**, *187*, 23-37.
11. Hoshi, M.; Saitoh, T.; Yoshioka, C.; Higuchi, A.; Nakagawa, T.J. *Appl. Polym. Sci.* **1999**, *74*, 983-994.
12. Hoshi, M.; Kobayashi, M.; Saitoh, T.; Higuchi, A.; Nakagawa, T.J. *Appl. Polym. Sci.* **1998**, *69*, 1483-1494.

13. Wu, P.; Field, R. W.; England, R.; Brisdon, B. J. J. *Membrane Sci.* **2001**, *190*, 147-157.
14. Karlsson, H. O. E.; Trägårdh, G. J. *Membrane Sci.* **1993**, *76*, 121-146.
15. Pereira, C. C.; Habert, A. C.; Nobrega, R.; Borges, C. P. J. *Membrane Sci.* **1988**, *138*, 227-235.
16. Oliveira, T. A. C.; Scarpello, J. T.; Livingston, A. G. J. *Membrane Sci.* **2002**, *195*, 75-88.
18. Ji, W.; Hilaly, A.; Sikdar, S. K.; Hwang, S. T. J. *Membrane Sci.* **1994**, *97*, 109-125.
19. Wijmans, J. G.; Athayde, A. L.; Daniels, R.; Ly, J. H., Kamaruddin, H. D.; Pinnau, I. J. *Membrane Sci.* **1996**, *109*, 135-146.
20. Bennett, M.; Brisdon, B. J.; England, R.; Field, R. W. J. *Membrane Sci.* **1997**, *137*, 63-88.
21. Mishima, S.; Nakagawa, T. J. *Appl. Polym. Sci.* **1999**, *73*, 1835-1844.
22. Uragami, T.; Yamada, H.; Miyata, T. J. *Membrane Sci.* **2001**, *187*, 255-269.
23. Aoki, T.; Yamagiwa, K.; Yoshino, E.; Oikawa, E. *Polymer* **1993**, *34*, 1538.
24. Raghunath, B.; Hwang, S. T. J. *Membrane Sci.* **1992**, *65*, 147-161.
25. Baker, R. W.; Wijmans, J. G.; Athayde, A. L.; Daniels, R.; Ly, J. H.; Le, M. J. *Membrane Sci.* **1997**, *137*, 159-172.
26. Schnabel, S.; Moulin, P.; Nguyen, Q. T.; Roizard D.; Aptel, P. J. *Membrane Sci.* **1998**, *142*, 129-141.
27. Smart, J.; Schucker, R. C.; Lloyd, D. R. J. *Membrane Sci.* **1998**, *143*, 137-157.
28. Yamaguchi, T.; Tominaga, A.; Nakao, S.; Kimura, S. *AIChE J.* **1996**, *42*, 892-895.
29. Higuchi, A.; Yoon, B. O.; Asano, T.; Nakaegawa, K.; Miki, S.; Hara, M.; He, Z.; Pinnau, I. J. *Membrane Sci.* **2002**, *198*, 311-320.
30. Herbert, B. J.; Dorsey, J. G. *Anal. Chem.* **1995**, *67*, 744-749.
31. Shiu, W. Y.; Doucette, W.; Gobas, F. A. P. C.; Andren, A.; Mackay, D. *Environ. Sci. Technol.* **1988**, *22*, 651-.658.
32. Yalkowsky, S. H.; Dannenfelser, R. M. *Aquasol Database of Aqueous Solubility*; Version 5; University of Arizona-Tucson: 1992.
33. Hansch, C.; Leo, A.; Hoekman, D. *Exploring QSAR – Hydrophobic, Electronic, and Steric Constants;* American Chemical Society:Washington DC, 1995; p 101.
34. Burkhard, L. P.; Armstrong, D. E.; Andren, A. W. J. *Chem. Eng. Data* **1984**, *29*, 248-250.
35. Doucette, W. J.; Andren, A. W. *Environ. Sci. Technol.* **1987**, *21*, 821-824.
36. Munnecke, D. E.; Van Guhdy, S. D. *Ann. Rev. Phytopathol* **1979**, *17*, 405-429.
37. Howard, P. H.; Banerjee, S.; Robillard, K. H. *Environ. Tox and Chem* **1985**, *4*, 653-661.
38. Donovan, S. F. J. *Chromatogr. A* **1996**, *749*, 123-129.
39. Hinckley, D. A.; Bidleman, T. F.; Foreman, W. T. *J. Chem. Eng. Data* **1990**, *35*, 232-237.

40. Crank, J.; Park, G. S. *Diffusion in Polymers* Academic Press Inc: New York, NY,1968, pp. 165-217.
41. Atkins, P. W. *Physical Chemistry*, 5th ed.; Oxford University Press: Oxford, 1994, pp. 226-274.
42. Yoon, B. O.; Koyanagi, S.; Asano, T.; Hara, M.; Higuchi, A. *J. Membrane Sci*: in press.

Chapter 28

Improvement of Selectivities of Microphase-Separated Membranes for the Removal of Volatile Organic Compounds

Tadashi Uragami, Hiroshi Yamada, Terumi Meotoiwa, and Takashi Miyata

Unit of Chemistry, Faculty of Engineering and High Technology Research Center, Kansai University, Suita, Osaka 564–8680, Japan

This paper describes the removal of benzene from an aqueous solution by pervaporation using poly(methyl methacrylate)-graft-polydimethylsiloxane (PMMA-*g*-PDMS) membranes, surface-modified with poly(perfluoro alkyl acrylate-graft-polydimethylsiloxane (PFA-*g*-PDMS/PMMA-*g*-PDMS) and tert-butylcalix [4] arene (CA/PMMA-*g*-PDMS). Membranes based on PFA-*g*-PDMS/PMMA-*g*-PDMS and CA/PMMA-*g*-PDMS showed high benzene selectivity using an aqueous feed solution containing 0.05 wt% benzene. Both the permeability and the benzene selectivity of these membranes were enhanced by increasing the PFA-*g*-PDMS and CA content because the membrane surface became more hydrophobic. Furthermore, the affinity of the CA/PMMA-*g*-PDMS membranes for benzene increased by introducing CA to the membranes. Contact angle measurements and X-ray photoelectron spectroscopy revealed that the addition of PFA-*g*-PDMS produced a hydrophobic surface at the membrane air-side due to surface localization. Transmission electron microscope observations revealed that the CA/PMMA-*g*-PDMS membranes had a microphase-separated structure consisting of a PMMA phase and a PDMS phase containing CA.

© 2004 American Chemical Society

Introduction

Recently, research has been focused on the removal of volatile organic compounds (VOCs) from wastewater or groundwater. In addition, treatment of tap water is very important as it contains various organic contaminants. A variety of separation techniques to remove VOCs from water, such as carbon adsorption, air stripping and steam stripping are available. However, the removal of VOCs from water by membrane separation can offer great advantages in potential energy cost savings. Separation of organic liquid mixtures through a variety of polymer membranes by pervaporation has therefore been the subject of many studies (1-5).

In a previous paper (6), we reported that microphase separation of graft copolymer membranes based on poly(methyl methacrylate) (PMMA) and polydimethylsiloxane (PDMS) can significantly influence their permeability and selectivity for the removal of VOCs from an aqueous solution. Furthermore, we suggested that a continuous PDMS phase in the microphase-separated structure plays an important role in the selective removal of VOCs.

Generally, surface properties of multicomponent polymers are quite different from their bulk properties, because of surface localization of a specific component. For example, a fluorine-species in a multicomponent polymer is preferentially concentrated at its surface to minimize the surface free energy. Therefore, a fluorine-containing polymer that is spontaneously localized at the polymer surface might enable a simple surface modification for pervaporation membranes. Our previous studies revealed that adding fluorine-containing polymers to ethanol-selective membranes enhances their selectivity for the separation of aqueous ethanol solutions due to their very hydrophobic surface (7,8). Nakagawa et al. improved a PDMS membrane by graft polymerization of 1H,1H,9H-hexadecafluorononyl methacrylate, which resulted in an increase of the selectivity for chlorinated hydrocarbons (2).

Calixarene is a cyclic oligomer and is composed of phenol units linked to alkylidene groups. It has a cavity for incorporation of organic compounds, which is similar to cyclodextrins or crown ethers. In particular, tert-butylcalix [4] arene (CA) can selectively sorb benzene and its derivatives.

In this study, we prepared hydrophobic, surface-modified membranes by adding a fluorine-containing graft copolymer to a microphase-separated membrane. The goal of this work was to (i) develop a high-performance pervaporation membrane for removing VOCs from water and (ii) introducing CA as a transport carrier for VOCs into multicomponent polymer membranes to improve their selectivity. We also evaluated the relationship between the benzene selectivity and the membrane structure.

Experimental

Materials

The PDMS macromonomer, which has 25 units of the pendant PDMS, was supplied by Toray Dow Corning Silicone Co., Ltd.; 1H,1H,2H,2H-heptadecafluorodecyl acrylate (perfluoroalkylacrylate: PFA) (Clariant, Japan) was used as received; tert-Butylcalix [4] arene (CA) was obtained from Aldrich Chem. Co., Ltd. The co-monomer, methyl methacrylate (MMA), was purified by distillation under reduced pressure in a nitrogen atmosphere. The initiator, 2,2'-azobisisobutyronitrile (AIBN), was recrystallized from benzene solution.

In this study, PMMA-g-PDMS with a DMS content of 74 mol% was used as a matrix polymer because this membrane showed the highest permeability and selectivity for a dilute aqueous solution of benzene in PMMA-g-PDMS membranes with various DMS contents (6).

A graft copolymer consisting of PFA and PDMS (PFA-g-PDMS) was obtained by copolymerization of a PDMS macromonomer with PFA. In this study, PFA-g-PDMS with a PFA content of 10 mol% was employed as a polymer surface modifier.

Membrane Preparation

PMMA-g-PDMS was dissolved in benzene at 25°C at a concentration of 4 wt%. Furthermore, PFA-g-PDMS or CA was added to the casting solutions. The PMMA-g-PDMS membranes containing a small amount of PFA-g-PDMS (PFA-g-PDMS/PMMA-g-PDMS membranes) or CA (CA/PMMA-g-PDMS membranes) were prepared by pouring the casting solutions onto stainless-steel plates and allowing the solvent to evaporate completely at 25°C. The resulting membranes were transparent and their thickness was about 150-160 μm.

Permeation Measurements

Pervaporation experiments were carried out under the following conditions: temperature: 40°C; permeate pressure, 1×10^{-2} Torr. The effective membrane area was 13.8 cm^2. An aqueous solution of 0.05 wt% benzene was used as the feed solution. The feed solution was circulated between the pervaporation cell and the feed tank to maintain a constant concentration in the pervaporation cell. Afterwards, the permeate was dissolved in a known amount of ethanol for mixing with water and benzene and the composition was determined using a gas chromatograph (Shimadzu GC-14A) equipped with a flame ionization detector (FID) and a capillary column (Shimadzu Co., Ltd.; Porapak Q) heated to 180°C. The permeation rates were determined from the weight of the permeate collected in a cold trap, permeation time, and effective membrane area.

Results and Discussion

Effect of Surface Modification on Membrane Characteristics

This section describes the relationship between surface modification of the PMMA-*g*-PDMS membranes and their pervaporation characteristics. To investigate the effect of the asymmetric structure of the PFA-g-PDMS/PMMA-*g*-PDMS membranes on their permeability and selectivity, an aqueous solution of 0.05 wt% benzene was permeated from the air-side surface (permeation mode I) and the stainless-steel-side surface (permeation mode II). In permeation modes I and II, the PFA-*g*-PDMS/PMMA-*g*-PDMS membranes were placed in a pervaporation cell with the air-side surface and the stainless-steel-side surface facing the feed side, respectively. Figure 1 shows the effect of the PFA-*g*-PDMS content on the normalized permeation rate and the benzene concentration in the permeate passing through the PFA-*g*-PDMS/PMMA-*g*-PDMS membranes. In Figure 1, the normalized permeation rate is given as the product of the permeation rate and the membrane thickness. There was no difference in the pervaporation characteristics of the PMMA-*g*-PDMS membranes operated in permeation modes I and II. However, the PFA-*g*-PDMS/PMMA-*g*-PDMS membranes showed substantial differences in the normalized permeation rate and the benzene concentration in the permeate for permeation modes I and II. This indicates that the asymmetrically modified surface of the membranes strongly influences the permeability and selectivity.

In permeation mode II, the normalized permeation rate through the PFA-*g*-PDMS/PMMA-*g*-PDMS membranes showed a maximum at a PFA-*g*-PDMS content of 0.5 wt%, and the benzene concentration in the permeate tended to decrease with increasing PFA-*g*-PDMS content. On the other hand, both the normalized permeation rate and the benzene concentration in the permeate in permeation mode I increased dramatically following the addition of PFA-*g*-PDMS up to 1.2 wt%. However, adding more than 1.2 wt% PFA-*g*-PDMS resulted only in a slight decrease in both the normalized permeation rate and the benzene concentration in the permeate. Generally, most modifications of pervaporation membranes cannot enhance their selectivity without lowering their permeability. Therefore, it is interesting to note that adding a small amount of PFA-*g*-PDMS enhanced both the permeability and benzene selectivity of the PMMA-*g*-PDMS membrane in permeation mode I. This might be due to the fact that the small amount of PFA-*g*-PDMS can make the air-side surface of the PFA-*g*-PDMS/PMMA-*g*-PDMS very hydrophobic without changing the inner microphase-separated structure. When the PFA-*g*-PDMS content was increased to 1.2 wt%, the behavior of the benzene concentration in the permeate corresponded to that of the surface free energy of the PFA-*g*-PDMS/PMMA-*g*-PDMS membranes. This fact suggests that the benzene selectivity of the PFA-*g*-PDMS/PMMA-*g*-PDMS membranes was related to their surface characteristics. Thus, the asymmetric surface-modification by a fluorine-containing polymer

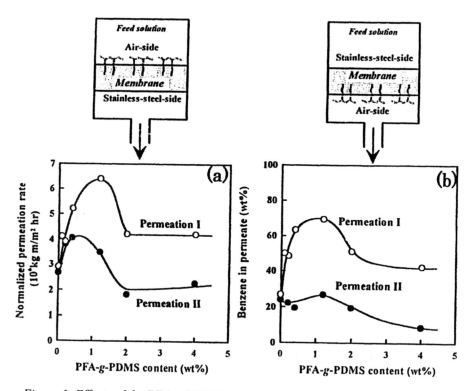

Figure 1. Effects of the PFA-g-PDMS content on the (a) normalized permeation rate and (b) benzene concentration of the permeate through PFA-g-PDMS/PMMA-g-PDMS membranes during pervaporation of an aqueous feed solution containing 0.05 wt% benzene at 40°C. (○) Mode I: permeation from the air-side surface (●) Mode II: permeation from the stainless-steel-side surface. (Reproduced from Macromolecules **2001**, 34, 8026–8033. Copyright 2001 American Chemical Society.)

additive is a very effective method for enhancing both membrane permeability and selectivity.

Surface Characterization

Figure 2 shows the effect of the PFA-*g*-PDMS content on the contact angle of water and the surface free energy of the PFA-*g*-PDMS/PMMA-*g*-PDMS membranes. The surface free energy was determined from the contact angles of water and methylene iodide on the air-side and stainless-steel-side surfaces of the membranes. The contact angles of water on the air-side and stainless-steel-side surfaces of the PFA-*g*-PDMS/PMMA-*g*-PDMS membranes increased dramatically following the addition of PFA-*g*-PDMS. This result suggests that adding PFA-*g*-PDMS to the PMMA-*g*-PDMS membrane made its surface very hydrophobic. With increasing PFA-*g*-PDMS content, the surface free energy of the air-side and stainless-steel-side surfaces of the PFA-*g*- PDMS/PMMA-*g*-PDMS membranes decreased sharply. This is due to surface localization of the very hydrophobic PFA component to minimize the surface free energy. At a PFA-*g*-PDMS content over 0.2 wt%, however, the surface free energy of the air-side surface was constant at about 11 erg/cm^2. The constant surface free energy indicates that the amount of PFA-*g*-PDMS at the air-side surface was saturated. In addition, the surface free energy of the air-side surface was lower than that of the stainless-steel-side surface. This indicates that the hydrophobic PFA-*g*-PDMS was preferentially more concentrated at the air-side surface than at the stainless-steel-side surface.

The atomic ratios of fluorine to carbon (F/C) at the air-side and stainless-steel-side surfaces of the PFA-*g*-PDMS/PMMA-*g*-PDMS membranes are shown in Figure 3. Total F/C was calculated from the PFA-*g*-PDMS content in the PFA-*g*-PDMS/PMMA-*g*-PDMS membranes. Adding PFA-*g*-PDMS up to 0.2 wt% resulted in a dramatic increase in F/C at both surfaces. The surface F/C in the membranes with a PFA-*g*-PDMS content of more than 0.2 wt% was constant in spite of increasing the PFA-*g*-PDMS content. These results suggest that PFA-*g*-PDMS was preferentially concentrated at the surface and was saturated at the surfaces of the membranes with a PFA-*g*-PDMS content of more than 0.2 wt%. In addition, the F/C ratio of the air-side surface was much larger than that of the stainless-steel-side surface. It can be concluded that PFA-*g*-PDMS is preferentially concentrated at the air-side surface than at the stainless-steel-side surface. Thus, the larger contact angle of water on the air-side surface can be attributed to preferential surface localization of PFA-*g*-PDMS. Consequently, the contact angle and XPS measurements revealed that adding a small amount of PFA-*g*-PDMS makes the air-side surface of the PMMA-*g*-PDMS membranes very hydrophobic. This is a simple surface-modification to obtain a very hydrophobic membrane surface. It is worth noting that the PFA-*g*-PDMS/PMMA-*g*-PDMS membranes have an asymmetric structure, because the air-side of the membranes has a more hydrophobic surface than the stainless-

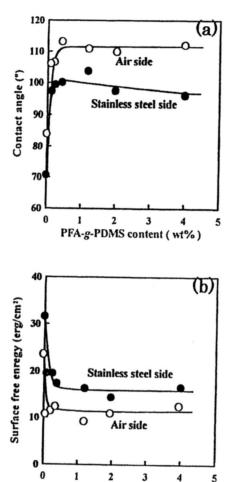

Figure 2. Effects of the PFA-g-PDMS content on the (a) contact angle of water (b) surface free energy of the PFA-g-PDMS/PMMA-g-PDMS membranes. (○) air-side surface and (●) stainless-steel surface.
(Reproduced from Macromolecules *2001, 34, 8026–8033. Copyright 2001 American Chemical Society.)*

steel-side. Therefore, the permeation and separation properties, shown in Figure 1, are significantly influenced by such an asymmetric structure.

Microphase Separation

Transmission electron micrographs of cross sections of the PFA-*g*-PDMS/PMMA-*g*-PDMS membranes with various PFA-*g*-PDMS contents are shown in Figure 4. The PDMS component was stained with RuO_4. These micrographs demonstrate clearly that all PFA-*g*-PDMS/PMMA-*g*-PDMS membranes showed distinct microphase separation. As reported in previous papers (*6,9*), the PMMA-*g*-PDMS membranes with a high DMS content exhibited distinct microphase separation consisting of a continuous PDMS phase and a continuous PMMA phase. The PMMA-*g*-PDMS membranes with a continuous PDMS phase enabled the membrane to preferentially permeate benzene from water (*6*). Figure 4 demonstrates that the PFA-*g*-PDMS/PMMA-*g*-PDMS membranes with a PFA-*g*-PDMS content of less than 1.2 wt% displayed a microphase-separated structure in which a PDMS component formed a continuous phase. This behavior is similar to microphase separation of the PMMA-*g*-PDMS membrane without PFA-*g*-PDMS. On the other hand, it is apparent from the TEM images of the membranes with more than 1.2 wt% PFA-*g*-PDMS that the addition of PFA-*g*-PDMS over 1.2 wt% made the PDMS phase discontinuous. Consequently, the morphology of microphase separation of the PFA-*g*-PDMS/PMMA-*g*-PDMS membranes with a high PFA-*g*-PDMS content is strongly dependent on the PFA-*g*-PDMS content. The effect of the PFA-*g*-PDMS additive on microphase separation of the PFA-*g*-PDMS/PMMA-*g*-PDMS membranes can be explained by the schematic model shown in Figure 5. In membranes containing a small amount of PFA-*g*-PDMS, very little PFA-*g*-PDMS exists in the microphase-separated bulk structure, because most of the PFA-*g*-PDMS is concentrated at the membrane surface. Therefore, the addition of less than 1.2 wt% PFA-*g*-PDMS (Figure 5 a) has no influence on the morphology of the PMMA-*g*-PDMS membranes. In addition, the air-side surface of the PFA-*g*-PDMS/PMMA-*g*-PDMS membranes is more hydrophobic than the stainless-steel-side surface because PFA-*g*-PDMS is more localized at the air-side surface rather than at the stainless-steel-side surface to minimize its surface and interfacial free energy. At a PFA-*g*-PDMS content of about 1.2 wt% (Figure 5 b), the amount of PFA-*g*-PDMS localized at the membrane surface is saturated and the PFA-*g*-PDMS/PMMA-*g*-PDMS membranes have the most hydrophobic surface. Membranes with a PFA-*g*-PDMS content of more than 1.2 wt% (Figure 5 c) also have a hydrophobic surface, but excess PFA-*g*-PDMS is distributed within the inner microphase-separated structure. Excess PFA-*g*-PDMS causes a morphological change in the microphase-separated structure because it may act as a surface-active agent. The PDMS component then changes from a continuous phase to a discontinuous phase to form a finely dispersed phase because of the surface-active function of excess PFA-*g*-PDMS.

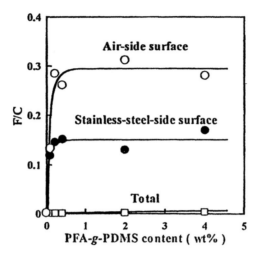

Figure 3. Effects of the PFA-g-PDMS content on the atomic ratio of fluorine to carbon (F/C) of the air-side surface (○), stainless-steel-side surface (●), and on total F/C (□) of the PFA-g-PDMS /PMMA-g-PDMS membranes. The surface F/C ratio was determined by XPS analysis; the total F/C ratio was calculated from the PFA-g-PDMS content.
(Reproduced from Macromolecules **2001,** 34, 8026–8033. Copyright 2001 American Chemical Society.)

PFA-g-PDMS content				
0wt%	0.1wt%	1.2wt%	2.0wt%	4.0wt%

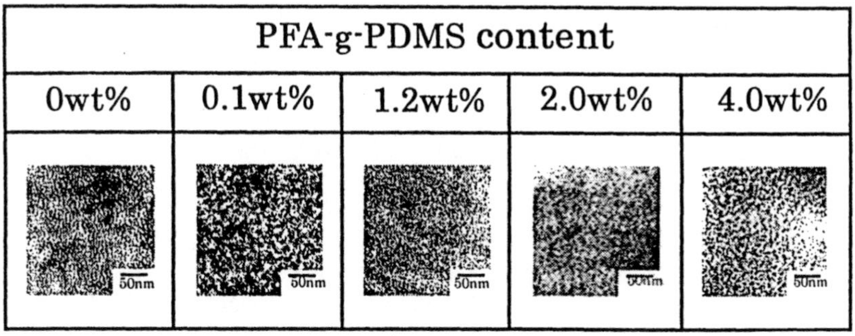

Figure 4. Transmission electron micrographs of cross sections of PFA-g-PDMS/PMMA-g-PDMS membranes with various PFA-g-PDMS contents. The dark region stained with RuO_4 represents the PDMS component.
(Reproduced from Macromolecules **2001,** 34, 8026–8033. Copyright 2001 American Chemical Society.)

Consequently, it can be concluded from surface characterizations and TEM observations that the addition of 1.2 wt% PFA-*g*-PDMS can enhance the surface hydrophobicity of the PMMA-*g*-PDMS membrane most effectively without changing its inner microphase-separated structure. It is worth noting that the PFA-*g*-PDMS/PMMA-*g*-PDMS membranes have an asymmetric structure, because their air-side surface is more hydrophobic than their stainless-steel-side surface.

Selectivity of Calixarene-Containing Membranes

Figure 6 shows the effect of the CA content on the normalized permeation rate and the benzene permeate concentration for an aqueous feed solution of 0.05 wt% benzene permeated through a CA/PMMA-*g*-PDMS membrane by pervaporation. The fact that the benzene permeate concentration was much higher than in the feed suggests that all of the CA/PMMA-*g*-PDMS membranes were benzene-selective. Both the benzene concentration in the permeate and the normalized permeation rate increased gradually with increasing CA content. This indicates that the addition of CA into the PMMA-*g*-PDMS membrane can enhance both selectivity and permeability. In general, most modifications of pervaporation membranes cannot improve their selectivity without lowering their permeability. Therefore, it is unusual that the addition of CA enables the membrane selectivity to increase without lowering the permeability.

To clarify the effects of CA on the permeation and separation characteristics of CA/PMMA-*g*-PDMS membranes, the degree of swelling of the membranes in an aqueous solution of 0.05 wt% benzene was examined. Figure 7 shows that the degree of swelling of the CA/PMMA-*g*-PDMS membranes hardly changed with an increase in the CA content. This result is in contrast to the dramatic increase in the normalized permeation rate and the benzene selectivity upon increasing the CA content, as shown in Figure 6. Therefore, the enhanced benzene selectivity created by the addition of CA cannot be explained by a change in the degree of swelling.

The benzene concentration in the CA/PMMA-*g*-PDMS membranes was examined to evaluate the affinity of benzene for the membranes. Figure 6 also includes the benzene concentration of the CA/PMMA-*g*-PDMS membranes immersed in an aqueous of 0.05 wt% benzene as a function of the CA content. The benzene concentration in the CA/PMMA-*g*-PDMS membranes increased upon increasing the CA content. This indicates that the addition of CA into the PMMA-*g*-PDMS membrane can lead to the preferential sorption of benzene into the membranes. It is well known that CA can form a complex with benzene. Thus, the enhancement of benzene concentration in the membranes may be due to specific interactions between benzene and the CA. As a result, the benzene concentration in a CA/PMMA-*g*-PDMS membrane with a CA content of 40 wt% was three times higher than that in a PMMA-*g*-PDMS membrane. The

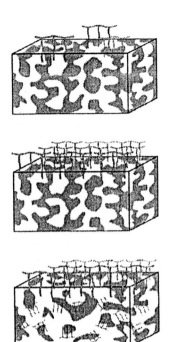

Figure 5. Schematic illustration for the surface and bulk microphase-separated structures of the PFA-g-PDMS / PMMA-g-PDMS membranes as a function of the PFA-g-PDMS content.
(Reproduced from Macromolecules **2001,** 34, 8026–8033. *Copyright 2001 American Chemical Society.)*

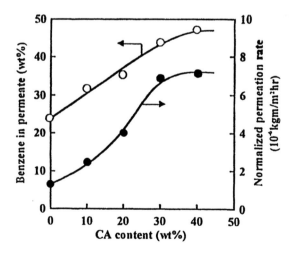

Figure 6. Effects of calixarene content on the benzene concentration in the permeate (O) and the normalized permeation rate (●) for an aqueous solution of 0.05 wt % benzene through CA/PMMA-g-PDMS membranes.
(Reproduced from Macromolecules **2001,** 34, 6806–6811. *Copyright 2001 American Chemical Society.)*

preferential sorption of benzene following the addition of CA is an important factor for the high benzene selectivity of the CA/PMMA-g-PDMS membranes.

Structure of Calixarene-Containing Membranes

In general, the permeation and separation properties of organic liquid mixtures through polymer membranes are dependent on the membrane structure. For example, an increase in membrane density often leads to a decrease of permeability due to a tighter molecular structure for permeation of the penetrant.

Figure 8 shows the effects of the CA content on the density of the CA/PMMA-g-PDMS membranes. The addition of CA into the CA/PMMA-g-PDMS membranes resulted in an increase in the membrane density. Generally, increasing the membrane density causes a decrease of the permeability due to a lower diffusivity of the penetrants in the higher density matrix. Although the membrane density increased in the CA/PMMA-g-PDMS membranes with increasing CA content, both the permeability and benzene selectivity were still enhanced. Because a benzene molecule is much larger than a water molecule, an enhancement of the benzene selectivity following an increase in membrane density cannot be attributed to the physical structure of the CA/PMMA-g-PDMS membranes.

Transmission electron micrographs of cross-sections of CA/PMMA-g-PDMS membranes at various CA contents were prepared. These micrographs showed clearly that all CA/PMMA-g-PDMS membranes exhibited distinct microphase separation. The TEM image of a PMMA-g-PDMS membrane without CA was transferred to a microcomputer to enhance the contrast of the phase separation process. It was also confirmed that the PMMA-g-PDMS membrane without CA exhibited a microphase-separated structure consisting of a continuous PDMS phase and a discontinuous PMMA phase. Similarly, because the CA/PMMA-g-PDMS membranes have a microphase-separated structure with a continuous PDMS phase, it can be assumed that benzene preferentially permeated through the PDMS phase.

To investigate the distribution of CA in the microphase-separated structure, the effects of the CA content on the Tgs of the (i) PMMA phase and (ii) PDMS phase were examined by DSC. The DSC measurements demonstrated that the CA/PMMA-g-PDMS membranes had two Tgs assigned to the PMMA phase and the PDMS phase, which were around 105°C and –100°C, respectively. Figure 9 shows the effects of the CA content on the Tgs of the PMMA and PDMS phases in dried CA/PMMA-g-PDMS membranes versus membranes immersed in an aqueous solution of 0.05 wt% benzene. Upon increasing the CA content, the Tg of the PDMS phase in both dried and wet CA/PMMA-g-PDMS membranes decreased; however, the Tg of the PMMA phase did not change. When CA was added to a PMMA homopolymer, the Tg decreased linearly with CA content similarly to the Tg of the PDMS phase in the CA/PMMA-g-PDMS membranes. The DSC measurements revealed that CA was preferentially distributed in the

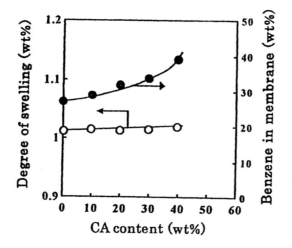

Figure 7. Effects of the calixarene content on the degree of swelling of the CA/PMMA-g-PDMS membranes immersed in an aqueous solution of 0.05 wt% benzene and the benzene concentration in the membrane at 40°C.
(Reproduced from Macromolecules *2001, 34, 6806–6811. Copyright 2001 American Chemical Society.)*

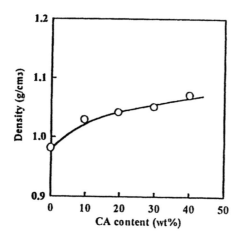

Figure 8. Relationship between the CA content and the density of the CA/PMMA-g-PDMS membranes.
(Reproduced from Macromolecules *2001, 34, 6806–6811. Copyright 2001 American Chemical Society.)*

424

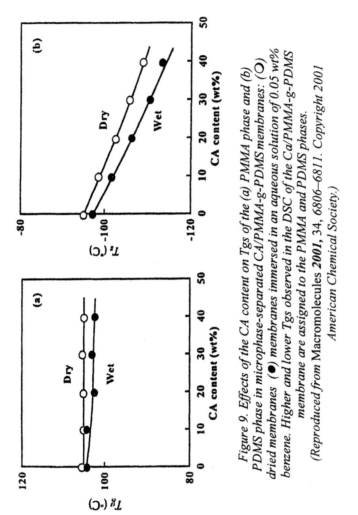

Figure 9. Effects of the CA content on Tgs of the (a) PMMA phase and (b) PDMS phase in microphase-separated CA/PMMA-g-PDMS membranes: (○) dried membranes (●) membranes immersed in an aqueous solution of 0.05 wt% benzene. Higher and lower Tgs observed in the DSC of the Ca/PMMA-g-PDMS membrane are assigned to the PMMA and PDMS phases.
(Reproduced from Macromolecules **2001**, 34, 6806–6811. Copyright 2001 American Chemical Society.)

PDMS phase rather than in the PMMA phase. In addition, immersing the CA/PMMA-g-PDMS membranes in an aqueous solution of 0.05 wt% benzene significantly decreased the Tg of the PDMS phase, whereas the Tg of the PMMA phase changed very little. This decrease in the Tg of the PDMS phase upon immersing the membranes in an aqueous benzene solution can be attributed to sorption of benzene into the PDMS phase containing the CA. Consequently, as CA was more preferentially distributed in the PDMS phase, benzene was preferentially sorbed into this phase due to strong interactions between CA and benzene. It is hypothesized that this preferential sorption of benzene into the PDMS phase containing CA is an important factor to selectively permeate benzene from water.

Acknowledgments

The authors are grateful for financial support from a Grant-in-Aid for Scientific research on Priority Area (B) "Novel Smart Membranes Containing Controlled Molecular Cavity" from the Ministry of Education, Culture, Science, Sports, and Technology, and for Scientific Research (C) from the Japan Society for Promotion of Science (JSPC).

References

1. Schnabel, S.; Moulin, P.; Nguyen, Q. T.; Roizard, D.; Aptel, P. *J. Membr. Sci.* **1998**, *142*, 129.
2. Mishima, S.; Kaneoka, H; Nakagawa, T. *J. Appl. Polym. Sci.* **1995**, *103*, 195.
3. Fang, Y.; Pham, V. A.; Matsuura, T.; Santerre, J. P.; Narbaitz, R. M. *J. Appl. Polym. Sci.* **1994**, *54*, 1937.
4. Yang, D.; Majumdar, S.; Kovenklioglu, S.; Sirkar, K. K. *J. Membr. Sci.* **1995**, *103*, 195.
5. Jian, K.; Pintauro, P. N. *J. Membr. Sci.* **1997**, *135*, 41.
6. Uragami, T.; Yamada, H.; Miyata, T. *Trans. Mater. Res., Jpn.* **1999**, *24*, 165; *J. Membr. Sci,* **2002**, *187*, 255.
7. Miyata, T.; Nakanishi, Y.; Uragami, T. *Macromolecules* **1997**, 30, 5563;
8. Uragami, T.; Doi, T.; Miyata, T. In *Membrane Formation and Modification*; Pinnau, I.; Freeman, B. D. (Eds.); ACS Symposium Series 744, American Chemical Society: Washington, DC, 2000; p.263-279.
9. Miyata, T.; Takagi, T.; Uragami, T. *Macromolecules* **1996**, *29*, 7787.
10. Uragami, T.; Meotoiwa, T.; Miyata, T. *Macromolecules* **2001**, *34*, 6806– 6811.
11. Uragami, T.; Meotoiwa, T.; Miyata, T. *Macromolecules* **2003**, *36*, 2041– 2048.

Indexes

Author Index

Asano, Takao, 394
Ashiba, K., 383
Baschetti, M. Giacinti, 55
Bélanger, Daniel, 311
Bruening, Merlin L., 269
Byrne, Catherine, 335
Carter, Roy, 335
Cornelius, Chris J., 234
Dai, Ying, 154
Doghieri, F., 55, 74
Fang, Jianhua, 253
Freeman, Benny D., 1, 106
Fried, J. R., 24
Grassia, F., 55
Guiver, Michael D., 154
Guo, Xiaoxia, 253
Hara, Mariko, 366, 394
Harrup, Mason K., 177
Hashiba, Hirokazu, 366
Hattori, Mitsuo, 366
Hayashi, Rika, 366
He, Zhenjie, 167, 218
Heuchel, M., 91
Higa, Mitsuru, 324
Higuchi, Akon, 366, 394
Hill, Anita J., 234
Hofmann, D., 91
Ishige, Mai, 394
Jho, Jae Young, 154
Kang, Yong Soo, 154
Kawakami, H., 383
Kim, Hyun-Il, 281
Kim, Sung Soo, 281
Kinoshita, T., 129

Kita, Hidetoshi, 203, 253
Laforgue, Alexis, 311
Lash, Robert P., 177
Lee, Doo Hyun, 281
Lee, Kwi Jong, 154
Lee, Young Moo, 190
Li, Bo, 24
Maeda, Hiroshi, 203
Marand, Eva, 234
Masuda, Toshio, 167
McCoy, John D., 177
Meakin, Pavla, 234
Meotoiwa, Terumi, 411
Miyata, Takashi, 411
Morisato, Atsushi, 218
Nagai, Kazukiyo, 139
Nakaegawa, Kenta, 394
Nakagawa, Tsutomu, 139, 352
Nakai, Y., 300
Nanbu, Koji, 203
Okamoto, Ken-ichi, 203, 253
Okuyama, Y., 383
Orme, Christopher J., 177
Park, Ho Bum, 190
Pinnau, Ingo, 1, 167, 218
Pintauro, Peter N., 335
Prabhakar, Rajeev S., 106
Quinzi, M., 74
Rethwisch, D. G., 74
Robertson, Gilles P., 154
Saito, Akiko, 352
Sakaguchi, Toshikazu, 167
Sarti, G. C., 55, 74
Shantarovich, V., 91

Stewart, Frederick F., 177
Takahashi, Shuichi, 352
Tan, Sophie, 311
Tanaka, Kazuhiro, 253
Tsujita, Y., 129, 300
Tsukahara, M., 129
Uragami, Tadashi, 411
van der Vegt, N. F. A., 39
Watari, Tatsuya, 253

Weinkauf, Don H., 177
Wycisk, Ryszard, 335
Yamada, Hiroshi, 411
Yamakawa, Tomoko, 324
Yamauchi, M., 300
Yampolskii, Yu., 91
Yoon, Boo Ok, 366, 394
Yoshimizu, H., 129, 300

Subject Index

A

Accessible free volume
gas diffusion coefficient and, 235–236
ortho-positronium (oPs), 235
permeability and, 247, 249
See also Polyimide/silica hybrid membranes
Acetylene-based polymers
relationship between *n*-butane permeability and *n*-butane/methane selectivity, 175*f*
See also Poly(*p-t*-butyl diphenylacetylene) (PptBDPA)
Acrylic acid. *See* Composite membranes with ultrathin skins; Surface modification by UV grafting
Alternating polyelectrolyte deposition (APD)
membrane formation, 271
See also Composite membranes with ultrathin skins
Aniline. *See* Cation-exchange membranes with sulfonate groups (CMX); Polyaniline (PANI)
Anion-exchange membranes (AEM)
electrodialysis process, 312
See also Cation-exchange membranes with sulfonate groups (CMX)
Atomic force microscopy (AFM)
carbon-silica membranes, 195, 196*f*
fluorinated polyimide membranes, 386, 387*f*
adsorption on fluorinated polyimide membranes, 391, 392*f*

Atomic ratios of fluorine to carbon, poly(perfluoro alkyl acrylate)-*g*-poly(dimethylsiloxane) /poly(methyl methacrylate)-*g*-PDMS (PFA-*g*-PDMS/PMMA-*g*-PDMS) membranes, 416, 419*f*

B

Benzene
calixarene-containing membranes, 420, 422
physical properties, 216*t*
See also Carbon membranes from phenolic resin; Microphase-separated membranes
Benzene/cyclohexane
pervaporation and vapor permeation, 214*t*
See also Carbon membranes from phenolic resin
Benzene/*n*-hexane, pervaporation and vapor permeation, 214*t*
Benzophenone
grafting initiator, 283
See also Surface modification by UV grafting
Binary polymer blends. *See* Polymer blends
Bisphenol A polysulfone
commercial use, 155
See also Polysulfones (PSF)
Blends. *See* Polymer blends
Blood. *See* Platelets; Proteins
Bovine serum albumin (BSA)
effect of grafting time on BSA adsorption to grafted membranes, 290, 291*f*

431

effect of reaction temperature on BSA adsorption to grafted poly(propylene) (PP), 293, 297*f*

effect of solvent on BSA adsorption to grafted PP, 296, 298*f*

See also Surface modification by UV grafting

n-Butane. *See* Poly(*p-t*-butyl diphenylacetylene) (PptBDPA)

n-Butane/hexane. *See* Polymer/inorganic hybrid membranes

t-Butylcalix[4]arene (CA). *See* Microphase-separated membranes

C

Calixarene
description, 412
selectivity of membranes containing, 420, 422
structure of membranes containing, 422, 425
See also Microphase-separated membranes

Carbon dioxide
comparing experimental infinite dilution solubility coefficient with predictions for CO_2 in polysulfone, 83, 85*f*
comparing experimental infinite dilution solubility coefficient with predictions for CO_2 and Ar in polycarbonate (PC), 86, 87*f*
composition of natural gas stream for removal of, 109*t*
diffusion coefficients for liquid crystalline polyester, 136, 137*t*
gas sorption and diffusion in liquid crystal polyester, 134, 136
permeation data in cellulose acetate and polyimide, 109*t*
separation, 140
solubility isotherms of CO_2 in PS–TMPC blends, 71*f*

solubility isotherms of CO_2 in pure PS and pure TMPC, 70*f*
See also Cellulose acetate membranes; Liquid crystalline polyesters; Natural gas separations; Poly(propylene oxide) (PPO); Polymer blends

Carbon membranes, performance, 191

Carbon membranes from phenolic resin
characteristics of carbonized phenolic resin, 205
chemical structure, 206*f*
comparison with zeolite and polymer membranes, 215*f*
effect of benzene feed concentration on pervaporation for benzene/cyclohexane separation, 210*f*
effect of benzene feed concentration on vapor permeation for benzene/cyclohexane separation, 211*f*
effect of benzene feed concentration on vapor permeation for benzene/cyclohexane separation through Y-type zeolite, 215*f*
effect of benzene feed concentration on vapor permeation for benzene/*n*-hexane separation, 213*f*
experimental, 204–205
gas permeances vs. molecular diameter of gases, 208*f*
gas separation, 205, 209
mass spectra of gas evolution during pyrolysis of phenolic resin, 207*f*
materials and membrane characterization, 204–205
membrane performance for pervaporation and vapor permeation of benzene/cyclohexane, 212*f*
performance, 216
pervaporation, 209, 214, 216
pervaporation and vapor permeation performances, 214*t*

physical properties of penetrants, 216
t
sorption isotherm of N_2 for
carbonized phenolic resin, 208*f*
thermograms of phenolic resin, 206*f*
Carbon-silica membranes
atomic force microscopy and surface
morphology, 195, 196*f*
class I for effect of SiO_2 content, 197
class II varying geometry of random
and block sequences, 198
diffusion coefficient, 193
energy dispersive X-ray spectrometry
(EDX), 195–196
experimental, 191–193
gas permeation experiments, 193
gas permeation properties of class II
membranes, 198–199
gas permeation properties of class I
membranes, 196–198
O_2 permeability vs. O_2/N_2 selectivity,
200, 201*f*
permeability coefficient, 193
poly(imide siloxane)s (PISs)
preparation, 191–192
precursor film preparation, 191–192
preparation, 192–193
sample designation and composition
of precursors, 192*t*
selectivity definition, 194
strategy for membranes with high
selectivity, 193–194
structural analysis, 194–196
surface composition, 195, 196*f*
transmission electron microscopy
images, 199, 200*f*
Cation-exchange membranes,
electrodialysis process, 312
Cation-exchange membranes with
sulfonate groups (CMX)
blocking efficiency of CMX–
polyaniline (PANI) membranes,
317–319
characteristics of Neosepta CMX
membrane, 313*t*
characterization of PANI layer, 315

commercial, 312
electrodialysis method, 314
emeraldine salt form of PANI, 314–
315
evidence of PANI degradation, 320–
322
evolution of chlorine components,
320
exchange capacity method, 314
exchange capacity of CMX and
CMX–PANI, 319–320
experimental, 312–314
general structure of PANI, 314
materials and chemicals, 312–313
membrane atomic composition vs.
polymerization time, 321*f*
modification of cation-exchange
membranes, 313
scanning electron microscopy (SEM)
of surface of CMX and CMX–
PANI membrane, 316*f*
surface morphology, 315, 316*f*
XPS survey spectra of CMX and
CMX–PANI membrane, 317*f*
X-ray photoelectron spectroscopy
(XPS) method, 313
Zn(II) and Cu(II) leakage for CMX
and CMX–PANI membranes vs.
immersion time in aniline, 318*f*
Zn(II) and Cu(II) leakage for CMX
and CMX–PANI membranes vs.
polymerization time, 318*f*
Cavity formation, energetics, 41–42
Cellulose acetate membranes
apparatus for measuring gas
permeability with microwave
irradiation, 303*f*
CO_2 and CH_4 permeation data in,
109*t*
CO_2 gas permeation curve during
microwave irradiation, 305*f*
dielectric dispersion method, 302
dielectric dispersion transition map,
305*f*
diffusion and solubility, 307, 308*f*
diffusion coefficient, 302, 304

effect of microwave power on average CO_2 diffusion coefficient, 308f

effect of microwave power on average CO_2 permeability coefficient, 306f

experimental, 301–304

methods, 302, 304

permeability coefficient, 302, 303f

permeation, 304, 307

samples, 301–302

Chain conformation, polysulfones, 163–164

Chain flexibility

analysis method, 29–30

silicon polymers, 31, 34

See also Silicon-based polymers

Chain packing, polysulfones, 161–162

Charge density

changes in response to temperature, 327, 329f

determination, 326

See also Temperature-responsive charged membranes

Chemical potential

low molecular weight penetrant, 62

penetrant, 67

pseudo-equilibrium conditions, 63–64

Chloromethylation, polysulfone membranes, 372–373

Cohen–Turnbull equation, free volume, 13

Combinatorial chemistry, hybrid membranes, 235

COMPASS, force field for simulation of polysilanes, 26

Composite membranes

CO_2/CH_4 selectivity, 121, 123–124

flux of penetrant through, 116–117

hydrocarbon penetrant critical volume, 124, 125f

size sieving ability, 124

See also Cation-exchange membranes with sulfonate groups (CMX); Natural gas separations

Composite membranes with ultrathin skins

alternating polyelectrolyte deposition (APD), 271

deposition of different polyelectrolytes in films, 278

deposition of hyperbranched poly(acrylic acid) (PAA), 270–271

field emission scanning electron microscopy (FESEM) of hyperbranched PAA films, 272–273

future directions, 278

gas permeabilities and selectivities of imidized poly(amic acid)/poly(allylamine hydrochloride) (PAH) membranes, 277t

gas permeabilities of imidized poly(amic acid)/PAH films, 276

gas transport studies with derivatized PAA films, 273

grafting PAA to PAA/PAH, 273–274

hyperbranched PAA skins, 271–274

idealized schematic of hyperbranched PAA film, 270f

imidization of poly(pyromellitic dianhydride-phenylenediamine) (PMDA-PDA)/PAH film, 275

membrane permeability and selectivity under different imidization and film formation conditions, 276–277

PAH content in film and selectivity, 277

polyimide structures, 276

schematic of APD, 271f

skins of imidized polyelectrolyte multilayers, 274–278

synthesis of PAA films, 271, 272

synthesizing gas separation membranes from polyelectrolyte multilayers, 275–276

Computer simulation

free volume, 97, 100, 102, 104

free volume distribution for packing models, 98f, 99f
size distribution functions, 100, 101f, 102, 103f
size distribution of free volume, 100
See also Free volume
Conductivity. *See* Proton conductivity
Conformational analysis, polysulfones, 163–164
Contact angles, water. *See* Water contact angles
Copolymers. *See* Hydrophilic membranes with hydrophobic side groups; Microphase-separated membranes
Crosslinked polymer
advantages and disadvantages of crosslinking methods, 339, 340f, 341
e-beam crosslinking for membrane preparation, 337
polyimide, 241, 244f
UV crosslinking for membrane preparation, 337
See also Ion-exchange membranes; Polyimide/silica hybrid membranes
Crystalline polymers. *See* Liquid crystalline polyesters
Crystallinity, ion-exchange membrane, 345, 346f
Cyclohexane
physical properties, 216t
See also Carbon membranes from phenolic resin

D

Degradation, polyaniline, 320–322
Degree of swelling, polyphosphazenes, 181
Density
model for binary polymer blends, 66
polyimide/silica hybrid membranes, 240, 245–246

relaxation of polymeric, with time, 62–63
Dielectric dispersion, cellulose acetate membranes, 304, 305f
Differential scanning calorimetry (DSC)
characterization of liquid crystalline polyester, 131, 132f, 133f
characterization of polysulfones, 156
polysulfones, 159
See also Liquid crystalline polyesters
Diffusion
cellulose acetate membranes, 307, 308f
Fick's law, 219–220
relationship between separation factor of, and sorption, 402, 403t
silicon-based polymers, 30–31
simulation of gas, 28–29
See also Liquid crystalline polyesters
Diffusion coefficients
accessible free volume and, 235–236
apparent, 142
carbon-silica membranes, 193
determination, 302, 304
fluoropolymer Hyflon AD 80, 119, 120f
gases in polysulfone and poly(dimethylsiloxane), 3, 4f
microwave irradiation, 307, 308f
sulfonated polyimides, 264, 266f, 267
Diffusivity
permeability dependence, 3
poly(propylene oxide) and poly(dimethylsiloxane), 144, 146f, 147, 148f
Diffusivity selectivity, gas separations, 5, 8
Diphenylsulfone
molecular modeling, 163–164
See also Polysulfones (PSF)
Dodecane, thermodynamic analysis of gas sorption, 41
Downhill transport, ions, 325

d-spacing. *See* X-ray diffraction (XRD)
Dual-mode model, glassy polymers, 56

E

Electrochromism, free volume investigation, 92*t*
Electrodialysis
limitations, 312
method, 314
process, 312
See also Cation-exchange membranes with sulfonate groups (CMX)
Electroneutrality condition, uphill transport, 331, 333
Electron spectroscopy for chemical analysis (ESCA), carbon-silica membranes, 195
Endocrine disruptors
apparatus for pervaporation, 398*f*
concern, 395
dependence of flux on feed solution temperature, 401*f*
effect of feed solution temperature on separation factor, 401*f*
flux by Fick's first law, 404
gas chromatography (GC)/GC-mass spectrometry measurements, 399–400
pervaporative removal, 395
poly(dimethylsiloxane) (PDMS) membranes, 395
properties, 396*t*
relationship between separation factor and log $K_{ow} \cdot p_{vap}$, 406*f*
relationship between separation factor and molecular weight, 402, 403*f*
relationship between separation factor and saturation vapor pressure, 406*f*

relationship between separation factors of diffusion and sorption, 402, 403*t*
separation factor of, and physical properties, 404–405
separation factors and feed solution temperature, 400
solubility, 399, 405
solubility, diffusion, and pervaporation selectivity, 403*t*
Energy dispersive X-ray spectrometry (EDX), carbon-silica membranes, 195–196
Enthalpy
solvation, 43–46
See also Thermodynamics
Entropy
effects of molecular interactions, 54
probability determining value, 52
solvation, 43–46
See also Thermodynamics
Equation of state (EoS), SAFT (Statistical Associating Fluid Theory), 80–81
Ethyl cellulose
fluoropolymer Hyflon AD 80 as coating layer for, 119, 120*f*, 121
parameter values, 121*t*
See also Natural gas separations
Ethylenediamination, polysulfone membranes, 372–373
Exchange capacity
cation-exchange composite membrane, 319–320
method, 314

F

Fick's law
diffusion, 219–220
flux, 404
Field emission scanning electron microscopy (FESEM), hyperbranched poly(acrylic acid) films, 272–273

Fillers. *See* Polymer/inorganic hybrid membranes
Flexibility. *See* Chain flexibility
Fluorinated polyimide membranes
AFM images of top surface of 2,2'-bis(3,4-dicarboxyphenyl)hexafluoropropane dianhydride with 2,2-bis(4-aminophenyl)hexafluoropropane (6FDA–6FAP) membranes, 387*f*
amount of adsorbed bovine serum albumin (BSA) on polymer surfaces, 392*f*
analysis of hydroxyproline, 385
atomic force microscopy analysis, 386
cell attachment to rubbed, 388, 391
cell culture, 385
cell function of rat skin fibroblast (FR) cells, 392*f*
characteristics of rubbed, 386
chemical structure of 6FDA–6FAP, 384
detachment process of cells from polymer membrane, 390*f*
experimental, 384–386
materials, 384
phase contrast microscopy images of FR cells on 6FDA–6FAP surfaces, 389*f*
protein adsorption, 391, 392*f*
protein adsorption method, 385
rubbed polyimide membrane, 385
surface analysis, 385
water contact angle and adhesion force on 6FDA–6FAP surfaces, 388*t*
Fluorine to carbon ratios, poly(perfluoro alkyl acrylate)-*g*-poly(dimethylsiloxane)/poly(methyl methacrylate)-*g*-PDMS (PFA-*g*-PDMS/PMMA-*g*-PDMS) membranes, 416, 419*f*
Fluoropolymers
CO$_2$/CH$_4$ selectivity, 112*t*

commercial Hyflon AD 80 as coating layer for hydrocarbon polymer, 119, 121
dependence of solubility on penetrant condensability, 110
membrane plasticization, 110
parameter values, 121*t*
solubility to hydrocarbon gases, 10
See also Natural gas separations
Flux conditions
gas transport through composite membrane, 116–117
steady state equation, 169–170
Fouling
internal membrane, 16
mechanisms, 14, 15*f*
surface, 14
See also Liquid separations
Free volume
accessible, 235–236
cavity sizes, 102
Cohen–Turnbull equation, 13
computer modeling, 102
distribution functions for poly(trimethylsilyl-*p*-styrene) (PTMSS) and poly(trimethylsilyl-1-propyne) (PTMSP), 100, 102
effect on diffusion coefficient, 13
estimation of size of free volume elements (FVE) in glassy polymers, 96*f*
fractional FVE, 140, 142, 169, 236
FVE, 92
hybrid membranes by positron annihilation lifetime spectroscopy (PALS), 251*t*
hybrids, 251*t*
inverse gas chromatography, 95, 97
model, 92
molecular modeling, 97, 100, 102, 104
PALS, 93–95
parameters of free volume size distribution in glassy polymers, 95*t*

principal view of approaches to connect free grid points in specific free volume region, 101*f*

probe methods for investigation, 92*t*

radii of spherical FVEs, 97*t*

representation of, of PTMSP packing models, 98*f*

representation of, of PTMSS packing models, 99*f*

silicon-based polymers, 31

size distribution, 100

size distribution of FVE for PTMSP vs. cavity radius of spheres, 103*f*

size distribution of FVE for PTMSS vs. cavity radius of spheres, 101*f*

strong non-equilibrium state of highly permeable glassy materials, 102, 104

structure and properties of various glassy polymers, 94*t*

Transition State Theory (TST) method, 29

Fuel cells

current-voltage curves, 349*f*

ion-exchange membrane, 348, 350

proton conducting membranes containing Kynar FLEX, 336–337

See also Ion-exchange membranes

G

Gas diffusion

silicon-based polymers, 30–31

simulation, 28–29

Gas diffusivity, poly(propylene oxide) and poly(dimethylsiloxane), 144, 146*f*, 147, 148*f*

Gases, thermodynamics of small permanent, in liquid poly(ethylene), 46–49

Gas flux, equation, 2–3

Gas permeability

imidized poly(amic acid)/poly(allylamine hydrochloride (PAH) membranes, 277*t*

poly(propylene oxide) and poly(dimethylsiloxane), 143–144, 145*f*

polypyrrole membranes, 301

See also Permeability

Gas permeation

carbon-silica class II membranes, 198–199

carbon-silica class I membranes, 196–198

experiments for carbon-silica membranes, 193

See also Carbon-silica membranes; Poly(*p-t*-butyl diphenylacetylene) (PptBDPA)

Gas selectivity

imidized poly(amic acid)/poly(allylamine hydrochloride (PAH) membranes, 277*t*

poly(propylene oxide) and poly(dimethylsiloxane), 151–152

See also Selectivity

Gas separations

n-butane permeability and *n*-butane/methane selectivity in poly(4-methyl-2-pentyne) (PMP0, 15*f*

carbon membrane from phenolic resin, 205, 209

CO_2/H_2 solubility selectivity, 11–12

Cohen–Turnbull equation, 13

diffusion coefficients of gases in polysulfone (PSF) and poly(dimethylsiloxane) (PDMS), 4*f*

effect of free volume on diffusion coefficients, 13

fluoropolymers and solubility to hydrocarbon gases, 10

gas diffusion, 3

gas flux, 2–3

gas solubility, 3

hydration problem, 10–11

ideal selectivity, 3
industrial membrane uses, 2
material science approaches to
    olefin/paraffin membranes, 9*t*
materials for membranes, 5*t*
molecular-sieving carbon
    membranes, 10
more permeable component as larger
    component in mixture, 11–14
novel membranes, 191
olefin/paraffin separation, 10
olefins and Group I-B metal cations,
    10
PDMS, 3, 4*f*, 5, 11
permeability coefficients of gases in
    PSF and PDMS, 4*f*
permeability depending on solubility
    and diffusivity, 3
plasticization, 8
PMP, 13–14
poly(1-trimethylsilyl-1-propyne)
    (PTMSP), 13
polymeric membranes, 2
polysulfone, 3, 4*f*
properties of membrane materials for
    CO_2/H_2 separations, 12*t*
selective removal of small molecules,
    5–8
separation of natural gas liquids, 13
solubility-selective separations, 11–
    12
solubility selectivity and diffusivity
    selectivity, 8–11
solution-diffusion mechanism, 2
tradeoff between H_2 permeability and
    H_2/N_2 selectivity in glassy and
    rubbery polymers, 7*f*
transport mechanism in polymeric
    membranes, 6, 8
See also Carbon membranes from
    phenolic resin; Liquid separations;
    Natural gas separations
Gas solubility
non equilibrium statistical
    associating fluid theory (NE–
    SAFT) model for pseudo-

equilibrium, in glassy polymers,
    82
poly(propylene oxide) and
    poly(dimethylsiloxane), 147, 149,
    150*f*
polymer/gas interactions, 3
solution-diffusion membranes, 40
See also Solubility
Gas sorption and diffusion
CO_2 and Xe sorption isotherms in
    annealed and melt-cooled liquid
    crystalline polyester, 135*f*
diffusion coefficients of CO_2 and Xe
    for annealed and melt-cooled
    liquid crystal polyester, 136, 137*t*
Henry's law constants of CO_2 and Xe
    for liquid crystal polyester, 134,
    136*t*
See also Liquid crystalline polyesters
Gas transport
dense polymer membranes, 140
derivatized poly(acrylic acid) films,
    273
permselectivity vs. permeability for
    CO_2/N_2 gas pair, 160*f*
permselectivity vs. permeability for
    O_2/N_2 gas pair, 160*f*
polysulfones, 159, 161
See also Polysulfones (PSF)
Gibbs energy
nonequilbrium free energy, 62
solvation, 43
See also Thermodynamics
Glass transition temperature
dynamic mechanical thermal analysis
    (DMTA) of hybrids, 247*t*
hybrids, 251*t*
substituted polysulfones, 159
Glassy polymeric solvents, infinite
    dilution solubilities, 51–52
Glassy polymers
basic assumptions and results of non
    equilibrium lattice fluid (NELF)
    approach, 76
comparing experimental infinite
    dilution solubility coefficient with

440

predictions for CO_2 and Ar in
polycarbonate (PC), 86, 87*f*

comparing experimental infinite
dilution solubility coefficient with
predictions for CO_2 in polysulfone
(PSF), 83, 85*f*

departure from thermodynamic
equilibrium, 56

dual-mode (DMS) model, 56

estimation of size of free volume
elements, 96*f*

fitting procedure for PC and PSF
with statistical associating fluid
theory (SAFT) equation of state
(EoS), 83, 84*f*

Helmholtz free energy density, 77–
78

infinite dilution solubility coefficient,
82–83

inverse gas chromatography, 95, 97

NELF model, 57–58

non equilibrium SAFT (NE–SAFT)
model for pseudo-equilibrium gas
solubility, 82

parameters of free volume size
distribution, 95*t*

polymer density as internal variable
of state, 61–63

polymer network density, 76–77

positron annihilation lifetime
spectroscopy, 93–95

pseudo-equilibrium states for glassy
phases, 77

pure component parameter for SAFT
EoS, 85*t*

SAFT EoS, 80–81

solubility calculation, 78–80

solute chemical potential, 77–78

See also Free volume; Non
equilibrium lattice fluid (NELF)
model

Graft copolymers
microphase separation, 412
See also Microphase-separated
membranes

Grafting

effect of grafting time on bovine
serum albumin (BSA) adsorption
to grafted poly(propylene) (PP)
and polysulfone (PSF)
membranes, 290, 291*f*

effect of grafting time on contact
angle of grafted PP, 291*f*

effect of grafting time on flux and
rejection of grafted PP and PSF
membranes, 289*f*

effect of UV grafting time, 284, 290

poly(acrylic acid) to multilayer
polyelectrolyte film, 273, 274

See also Composite membranes with
ultrathin skins; Surface
modification by UV grafting

Grand Canonical Monte Carlo
(GCMC), simulation of gas
solubility, 29

Group contributions
Hansen parameters and molar
volumes, 182*t*
nitrogen and phosphorus, 182*t*

H

Hansen parameters
determinations, 180–181
estimation for polyphosphazenes,
182–183
group contributions, 182*t*
plots for polyphosphazenes, 184*f*,
185*f*
types of interactions, 179

Helmholtz free energy
glassy phase, 77–78
statistical associating fluid theory
(SAFT), 80

Henry's law constants, CO_2 and Xe for
liquid crystalline polyester, 134,
136*t*

n-Hexane
physical properties, 216*t*
See also Carbon membranes from
phenolic resin

n-Hexane, liquid membrane, relationship between *n*-butane permeability and *n*-butane/methane selectivity, 175*f*

Hybrid membranes
morphology, 235
predicting flux and separation properties, 235
See also Polyimide/silica hybrid membranes; Polymer/inorganic hybrid membranes

Hydration, water-swollen polymer membranes, 10–11

Hydrocarbon polymers
membrane plasticization, 110
overcoated membranes, 114, 116
See also Natural gas separations

Hydrophilic membranes with hydrophobic side groups
behavior of water in modified membranes, 360, 362*f*
characterization of copoly(3-methacryloxypropyl tris(trimethylsiloxy)silane (SiMA)—4-vinylpyridine (4VP)) membranes, 357*t*
characterization of copoly(SiMA—*N*-vinyl-2-pyrrolidone (NVP)) membranes, 357*t*
characterization of copolymer membranes, 356, 357*t*
characterization of modified copoly(SiMA–4VIP) membranes, 363*t*
characterization of used polymer membranes, 355
comparing oxygen gas permeability coefficients in dry and wet state, 360, 361*f*
effect of quaternization ratio on water-dissolved oxygen permeability coefficient, 363*f*, 364
experimental, 353, 355–356
material and synthesis of copolymer, 353

modification of copolymer membranes, 353
monomers SiMA, 4VP, and NVP, 353, 354*f*
oxygen permeability measurement in dry and wet states, 355–356
oxygen permeation properties in dry and wet states, 356, 359–360
properties of modified copolymer membranes, 360, 364
quaternization for modified copolymer, 353, 360, 364
relationship between absorbed water and hydrophilic monomer mole fraction in copolymer membranes, 358*f*
solution-diffusion mechanism, 359–360
synthesis and structure of copolymer and modified copolymer, 353, 354*f*
transport, 359–360
water-dissolved oxygen permeability coefficients, 355–356
weight of freezing water, 356, 359

Hydrophobic side groups. *See* Hydrophilic membranes with hydrophobic side groups

Hyperbranched poly(acrylic acid). *See* Composite membranes with ultrathin skins

I

Infinite dilution solubilities, glassy polymeric solvents, 51–52

Infinite dilution solubility coefficient
calculations and comparisons with experiment, 82–83, 86
comparing experimental with predictions for CO_2 and Ar in polycarbonate (PC), 86, 87*f*
comparing experimental with predictions for CO_2 in polysulfone (PSF), 83, 85*f*

definition, 82–83
Interactions, Hansen parameters, 179
Inverse gas chromatography
  estimation of size of free volume
    elements in glassy polymers, 96f
  free volume, 95, 97
  free volume investigation, 92t
  radii of spherical free volume
    elements, 97t
  See also Free volume
Ion exchange capacity (IEC)
  ion-exchange membranes, 337–338
  sulfonated polyimides (SPIs), 257t
Ion-exchange membranes
  blending, 339, 341
  blending with sulfonated poly[bis(3-
    methylphenoxy)phosphazene]
    (SPOP), 336
  comparing crosslinking methods,
    340f
  conductivity measurements, 338
  containing sulfonated
    polyphosphazene, 336
  correlation between equilibrium
    water swelling and conductivity,
    345, 347f
  crosslinking procedures, 339, 341
  crystallinity, 345, 346f
  current-voltage curves in liquid
    methanol-air fuel cell test station,
    349f
  experimental, 337–338
  fuel cell test results, 348, 350
  ion-exchange capacity (IEC)
    measurements, 337–338
  mechanical properties, 348
  membrane preparation procedure,
    339, 349f
  membrane preparation with e-beam
    crosslinking, 337
  membrane preparation with UV
    crosslinking, 337
  miscibility of blends, 341
  morphology, 341, 344f, 345
  photographs of three membrane
    samples from SPOP in different

salt forms, 342f, 343f
  proton conducting membranes
    containing Kynar FLEX, 336–337
  relationship between IEC of SPOP
    and its water affinity, 339, 340f
  scanning electron microscopy (SEM)
    method, 338
  SEM micrographs of three
    membrane samples from SPOP in
    different salt forms, 344f
  sulfonation of polyphosphazene, 337
  swelling and proton conductivity,
    345, 348
  swelling measurements, 338
  synthesis of groups of membranes,
    341
  tensile tests, 338
  tensile tests on water-swollen blends,
    348t
  uses, 312
  WAXS diffractograms for blend and
    pure components, 346f
  wide-angle X-ray scattering (WAXS)
    method, 338
  See also Cation-exchange
    membranes with sulfonate groups
    (CMX)
Ion transport
  modes, 325
  See also Temperature-responsive
    charged membranes
Irradiation
  effect of distance on performance of
    grafted polymer membranes, 290,
    292f
  See also Surface modification by UV
    grafting

L

Layered structures
  crystalline and liquid crystalline
    states, 129–130
  layer spacing and density of liquid
    crystalline polyester, 134t

schematic of liquid crystalline polyester, 132*f*
See also Liquid crystalline polyesters
Liquid crystalline polyesters
B-C14 sample synthesis, 130
characterization by differential scanning calorimetry (DSC), 131, 132*f*
characterization by X-ray diffraction (XRD), 131, 134
CO_2 and Xe Henry's law constants in B-C14 and polycarbonate (PC), 136*t*
CO_2 sorption isotherms of melt-cooled and annealed samples of B-C14, 135*f*
composition, 129
differential scanning calorimetry (DSC) of as-cast sample of B-C14, 132*f*
diffusion coefficients of CO_2 and Xe for B-C14, 137*t*
DSC of melt-cooled B-C14, 133*f*
experimental, 130
gas sorption and diffusion properties, 134, 136
layered structures, 129–130
layer spacing and density of B-C14, 134*t*
methods, 130
schematic model of B-C14, 132*f*
Xe sorption isotherms of melt-cooled and annealed samples of B-C14, 135*f*
XRD profiles of B-C14, 133*f*
Liquid separations
fouling, 14, 15*f*, 16
fouling resistant reverse osmosis membranes, 17, 20
fouling-resistant ultrafiltration (UF) membranes, 16
internal membrane fouling, 16
membrane-based processes, 14
poly(vinylidene fluoride) (PVDF), 16, 18f
polymeric membranes, 2

reverse osmosis (RO), 17, 18*f*
scanning electron micrograph (SEM) of surface of coated membrane, 19*f*
surface fouling, 14
surface modification, 17, 20
water fluxes of commercial RO membrane and composite membrane vs. permeation time, 19*f*
water fluxes of porous and nonporous membrane modules vs. permeation time, 18*f*
See also Gas separations

M

Mechanical rubbing. *See* Fluorinated polyimide membranes
Mechanism
solution-diffusion, 307
transport in polymeric membranes, 6, 8
Medical field, polymer membranes, 353
Membranes
applications and challenges, 270
chemical structure and separation properties, 300–301
gas separations, 191
polyphosphazenes, 181
separation advantages, 204
See also Carbon membranes from phenolic resin; Carbon-silica membranes; Cellulose acetate membranes; Composite membranes with ultrathin skins; Fluorinated polyimide membranes; Gas separations; Hydrophilic membranes with hydrophobic side groups; Liquid separations; Microphase-separated membranes; Polymer/inorganic hybrid membranes; Polyphosphazenes; Surface-modified polysulfone (PSf) membranes

3-Methacryloxypropyl tris(trimethylsiloxy)silane (SiMA). *See* Hydrophilic membranes with hydrophobic side groups

Methane
permeation data in cellulose acetate and polyimide, 109*t*
solubility isotherms of, in PS–TMPC blends, 69*f*
solubility isotherms of, in pure PS and pure TMPC, 68*f*
See also Natural gas separations; Poly(*p-t*-butyl diphenylacetylene) (PptBDPA); Polyimide/silica hybrid membranes; Polymer blends

Methanol. *See* Carbon membranes from phenolic resin

Methanol/benzene, pervaporation and vapor permeation, 214*t*

Microfiltration (MF), membrane process, 14

Micropatterning
surface fabrication, 384
See also Fluorinated polyimide membranes

Microphase-separated membranes
atomic ratios of fluorine to carbon (F/C) at air-side and stainless-steel size surfaces, 419*f*
t-butylcalix[4]arene (CA), 412
calixarene, 412
degree of swelling of CA/poly(methyl methacrylate)-*g*-poly(dimethylsiloxane) (PMMA-*g*-PDMS) membranes, 423*f*
effect of poly(perfluoro alkyl acrylate)-*g*-PDMS (PFA-*g*-PDMS) content on water contact angle and surface free energy, 417*f*
effect of surface modification on membrane characteristics, 414, 416
effects of CA content on Tg of PMMA and PDMS phases, 424*f*

effects of calixarene content on benzene concentration in permeate and permeation rate, 421*f*
effects of PFA-*g*-PDMS content on permeation rate and benzene concentration of permeate, 415*f*
experimental, 413
graft copolymers, 412
materials, 413
membrane preparation, 413
microphase separation, 418, 420
permeation measurements, 413
relationship between CA content and density of CA/PMMA-*g*-PDMS membranes, 423*f*
selectivity of calixarene-containing membranes, 420, 422
structure of calixarene-containing membranes, 422, 425
surface and bulk microphase-separated structures of membranes vs. PFA-*g*-PDMS content, 421*f*
surface characterization, 416, 418
transmission electron microscopy (TEM) of cross sections of PFA-*g*-PDMS/PMMA-*g*-PDMS membranes, 419*f*

Microwave irradiation
diffusion coefficients for cellulose acetate membranes, 307, 308*f*
permeability coefficients in cellulose acetate membranes, 304, 306*f,* 307
uses, 301
See also Cellulose acetate membranes

Mixed-gas permeation
n-butane/methane selectivity of poly(*p-t*-butyl diphenylacetylene) (PptBDPA) vs. *n*-butane feed concentration, 174*f*
effect of filler content in poly(4-methyl-2-pentyne) (PMP)/silica hybrid membranes, 229–232
effect of filler type in PMP/silica hybrid membranes, 227–229

experiments for polymer/inorganic hybrid membranes, 223

methane and *n*-butane permeability of PptBDPA vs. *n*-butane feed concentration, 174*f*

PptBDPA, 172–173

See also Poly(*p-t*-butyl diphenylacetylene) (PptBDPA); Polymer/inorganic hybrid membranes

Mobility, effect of silica domains in hybrid membranes, 246–247

Models

dual-mode (DMS), 56

free volume, 92

non equilibrium lattice fluid (NELF), 57–58

non equilibrium SAFT model for pseudo-equilibrium gas solubility in glassy polymers, 82

rule of mixture, 235

SAFT (Statistical Associating Fluid Theory), 80–81

solution-diffusion model for gas transport in polymers, 113–114, 219–220

See also Free volume; Molecular modeling; Non equilibrium lattice fluid (NELF) model

Molar volumes, group contributions, 182*t*

Molecular modeling

distribution functions, 100, 102

free volume characterization, 93

free volume distribution for packing models, 97, 98*f*, 99*f*

polysulfones, 163–164

size distribution of free volume, 100, 101*f*, 103*f*

See also Free volume; Models

Molecular simulations

diffusion and solubility of silicon polymers, 30–31

gas diffusion, 28–29

gas solubility, 29

hybrid membranes, 235

See also Silicon-based polymers; Simulations

Molecular weight, relationship between separation factor and, 402, 403*f*

Monomer concentration

acrylic acid (AA) grafting to polymer membranes, 290, 293, 294*f*

See also Surface modification by UV grafting

Morphology

atomic force microscopy of carbon-silica membranes, 195, 196*f*

carbon-silica membranes by transmission electron microscopy, 199, 200*f*

ion-exchange membrane, 341, 344*f*, 345

polyimide/silica hybrid membranes, 241, 242*f*, 243*f*

See also Carbon-silica membranes; Polyimide/silica hybrid membranes

N

Nanofiltration (NF), membrane process, 14

Nanostructure

glassy polymer, 100

See also Free volume; Polymer/inorganic hybrid membranes

Nanotechnology

rubbed polyimide membrane, 385

surface fabrication, 384

See also Fluorinated polyimide membranes

Naphthalene-1,4,5,8-tetracarboxylic dianhydride (NTDA), polymerization with sulfonated diamines, 254, 255

Natural gas

composition of stream, 109*t*

separations of liquids, 13

world market, 107
See also Natural gas separations
Natural gas separations
adjustable constants, 114, 121t
analysis, 116–119
CO_2/CH_4 selectivity condition, 119, 126–127
coating selective hydrocarbon polymer with fluoropolymer layer, 112–113
commercial fluoropolymer Hyflon AD 80 as coating layer, 119, 121
composition of natural gas stream, 109t
correlation of gas solubility with critical temperature, 112t
critical properties of penetrants, 123t
diffusion coefficients of gases in poly(vinyl chloride) (PVC), 108f
effect of plasticization on diffusion coefficients in PVC, 107, 108f
ethyl cellulose/Hyflon AD 80 composite membrane interface, 122f, 123–124
flux condition, 116–117
flux of penetrant through polymer membrane, 113
gas and vapor solubility in n-heptane and fluorinated analog, 110, 111f
gas diffusion coefficients and critical volume, 113–114
hydrocarbon penetrant critical volume, 124, 125f
loss in CO_2/CH_4 selectivity, 123
mass transfer resistance of membrane layer to gas permeation, 116
membrane plasticization, 110
mixed gas CO_2 permeance and CO_2/CH_4 selectivity of polyimide membrane, 108f
model cases, 119, 121
negative effect of higher hydrocarbon sorption on CO_2/CH_4 separation, 107, 108f
overcoated hydrocarbon polymer membrane, 114, 116

parameter values for polysulfone, ethyl cellulose, and Hyflon AD 80, 121t
partial pressure condition, 117–119
penetrant solubility in polymers, 110
performance of fluoropolymer-coated HC membranes, 119, 121
permanent gas and hydrocarbon solubility in low density poly(ethylene), 111f
polysulfone/Hyflon AD 80 composite membrane interface, 122f, 123–124
problem definition, 114, 116
pure gas CO_2 and CH_4 permeation data, 109t
pure gas CO_2/CH_4 selectivity of fluoropolymers, 112t
schematic of hydrocarbon (HC) polymer membrane and composite membrane, 115f
size sieving ability, 124
solubility coefficients of hydrocarbons in hydrocarbon-based membranes, 107
solubility coefficients of polysulfone, ethyl cellulose, and Hyflon AD 80 vs. penetrant critical temperature, 120f
solubility coefficients of polysulfone, ethyl cellulose, and Hyflon AD 80 vs. penetrant critical volume, 120f
solubility dependence for fluorinated liquids and fluoropolymers, 110
solution-diffusion model for gas transport in polymers, 113
steady state flux of penetrant through composite membrane, 116
theory, 113–114
world market for natural gas, 107
NELF. See Non equilibrium lattice fluid (NELF) model
Neural networks, hybrid membranes, 235
Nitrogen, group contributions, 182t

Non equilibrium lattice fluid (NELF)
model
  binary parameters, 59–60
  binary polymer blends, 65–68
  characteristic density, 59
  characteristic pressure, 59, 60, 67
  characteristic volumes, 60
  chemical potential of low molecular
    weight penetrant, 62
  chemical potential under pseudo-
    equilibrium conditions, 63–64
  comparison with experimental data
    for polymer blends, 68–71
  lattice fluid theory by Sanchez and
    Lacombe, 58
  NELF correlations for swelling
    penetrants, 64–65
  nonequilibrium Gibbs free energy in
    glassy phase, 62
  non equilibrium states of system,
    76
  partial polymer densities, 66
  polymer densities, 62
  polymer density as internal variable
    of state, 61–63
  polymer dilation, 64–65
  polymer network density, 76–77
  pseudoequilibrium solubility, 64
  pseudo-equilibrium states for glassy
    phases, 77
  relaxation of polymeric density with
    time, 62–63
  Sanchez–Lacombe Gibbs free energy
    and mixing rules, 58–61
  sorption in glassy polymers, 57
  swelling coefficient, 64–65
  swelling coefficients, 67–68
  temperature, 59–60
  testing reliability, 57–58
  theoretical basis, 57
  thermodynamic properties of glassy
    polymeric phases, 76
  See also Polymer blends
N-vinyl-2-pyrrolidone (NVP). See
Hydrophilic membranes with
hydrophobic side groups

O

Olefin/nitrogen separation, mixed-gas
  permeation of poly(p-t-butyl
  diphenylacetylene), 173t
Olefin/paraffin membranes, gas
  separations, 8, 9t, 10
Ortho-positronium (oPs)
  accessible free volume, 235
  accessible volume and permeability,
    247, 249
  probe size comparison to penetrant
    size, 236t
  See also Polyimide/silica hybrid
    membranes
Oxygen permeability
  measurement in dry and wet states,
    355–356
  permeation properties in dry and wet
    states for copolymers, 356, 359–
    360
  water-dissolved oxygen permeability
    coefficients, 363f, 364
  See also Hydrophilic membranes
    with hydrophobic side groups

P

Packing
  effect of silica domains in hybrid
    membranes, 246–247
  polymer chain, polysulfones, 161–
    162
Paraffin/olefin membranes, gas
  separations, 8, 9t, 10
Partial pressure, condition for gas
  transport through polymer, 117–119
Permeability
  accessible free volume and, 247, 249
  calculation for mixed-gas, 170
  effect of microwave power in
    cellulose acetate membrane, 304,
    306f, 307
  equation for film, 170
  hybrids, 251t

O₂, vs. O_2/N_2 selectivity of carbon-silica membranes, 200*f*
penetrants in polyimides and hybrids, 246*t*
poly(propylene oxide) and poly(dimethylsiloxane), 143–144, 145*f*
polyphosphazenes, 34, 36*t*
polysilanes, 34, 35*t*
polysiloxanes, 34, 35*t*
See also Polyimide/silica hybrid membranes
Permeability coefficients
carbon-silica membranes, 193
determination in presence of microwave irradiation, 302, 307
gases in polysulfone and poly(dimethylsiloxane), 3, 4*f*
sulfonated polyimides, 264, 266*f*, 267
Permeation. *See* Microphase-separated membranes; Poly(*p-t*-butyl diphenylacetylene) (PptBDPA)
Pervaporation
carbon membrane from phenolic resin, 209, 214, 216
experiments for polyphosphazenes, 181
polyphosphazenes, 186, 188
procedure, 397, 399
schematic of apparatus, 398*f*
separation factor, 397
uses, 395
variable temperature, through polyphosphazene membranes, 186*f*
water/2-propanol separation, 187*f*
water/methanol separation, 187*f*
See also Carbon membranes from phenolic resin; Endocrine disruptors
Phase contrast microscopy, rat skin fibroblast cells on fluorinated polyimide, 388, 389*f*
Phenolic resin
characteristics of carbonized, 205
chemical structure, 206*f*

pyrolysis mass spectrum, 207*f*
sorption isotherm of N_2, 208*f*
thermograms, 206*f*
See also Carbon membranes from phenolic resin
Phosphorus, group contributions, 182*t*
Photochromism, free volume investigation, 92*t*
Photoinitiator
benzophenone, 283
UV grafting of acrylic acid to polymer, 293, 295*f*
See also Surface modification by UV grafting
Plasma. *See* Platelets; Proteins
Plasticization
effect on diffusion coefficients in poly(vinyl chloride), 107, 108*f*
membrane, 110
membrane selectivity, 8, 9*t*
phenomenon, 8
Platelets
adhesion on surface-modified polysulfone (PSf) membranes, 375, 381
adsorption on membrane method, 372
number of adhering, on modified PSf membranes, 380*f*
scanning electron micrographs of, adhering to PSf membranes, 378*f*, 379*f*
See also Surface-modified polysulfone (PSf) membranes
Poly(acrylic acid) (PAA)
deposition of hyperbranched PAA, 270–271
future work for hyperbranched, 278
hyperbranched PAA skins, 271–274
See also Composite membranes with ultrathin skins
Polyamides, diffusivity selectivity, 5
Polyaniline (PANI)
characterization of PANI layer of composite, 315, 316*f*, 317*f*
degradation evidence, 320–322

emeraldine salt form, 314–315
general structure, 314
modification of cation-exchange
    membranes, 313
See also Cation-exchange
    membranes with sulfonate groups
    (CMX)
Poly(*p-t*-butyl diphenylacetylene)
    (PptBDPA)
experimental, 168–170
fractional free volume (FFV), 169
mixed-gas methane and *n*-butane
    permeability vs. *n*-butane feed
    concentration, 174*f*
mixed-gas *n*-butane/methane
    selectivity vs. *n*-butane feed
    concentration, 174*f*
mixed-gas permeability equation,
    170
mixed-gas permeation for
    olefin/nitrogen separation, 173*t*
mixed-gas permeation properties,
    172–173
permeability equation, 170
permeation experiments, 169–170
permeation study, 168
polymer synthesis, characterization
    and film formation, 168–169
pure-gas permeabilities and
    gas/nitrogen selectivities, 172*t*
pure-gas permeability vs. critical gas
    volume, 171*f*
pure-gas permeation properties, 170–
    171
relationship between *n*-butane
    permeability and *n*-
    butane/methane selectivity, 175*f*
repeat unit, 169*f*
selectivity equation, 170
steady-state flux, 169–170
Polycarbonate (PC)
CO_2 and Xe Henry's law constants in
    PC, 134, 136*t*
comparing experimental infinite
    dilution solubility coefficient with

predictions for CO_2 and Ar in, 86,
    87*f*
gas sorption and diffusion properties,
    130
See also Glassy polymers
Poly(dimethylsiloxane) (PDMS)
chain flexibility, 31, 34
characterization, 143
CO_2/H_2 separations, 12*t*
diffusion and solubility from
    simulation, 30–31
diffusion coefficients of gases in, 4*f*
gas diffusivity, 144, 146*f*, 147, 148*f*
gas permeability, 25, 143–144, 145*f*
gas selectivity, 151–152
gas solubility, 147, 149, 150*f*
gas solubility coefficients, 32*f*
materials for endocrine disruptor
    separation, 397
mixed-gas *n*-butane/methane
    selectivity vs. *n*-butane
    permeability, 221*f*
permeability coefficients of gases in,
    3, 4*f*
relationship between *n*-butane
    permeability and *n*-
    butane/methane selectivity, 175*f*
repeat units of PDMS, 141*f*
selectivity, 220–221
self-diffusion coefficients from
    simulation vs. gas diameter, 32*f*
separating endocrine disruptors, 395
simulated and experimental
    permeabilities, 35*t*
solubility-selective separations, 11
structures and properties, 27*t*
vectorial autocorrelation function
    (VACF), 31, 33*f*, 34
See also Endocrine disruptors;
    Microphase-separated membranes;
    Poly(propylene oxide) (PPO);
    Polysilanes; Silicon-based
    polymers
Polyelectrolyte multilayers, skins of
    imidized, 274–278

Polyesters. *See* Liquid crystalline polyesters

Poly(ethylene) (PE)
solute binding energy in PE vs. solute energy parameter, 49, 50*f*
solvation Gibbs energy, enthalpy, and entropy in, 49, 50*f*
thermodynamic analysis of gas sorption, 41
thermodynamics of small permanent gases in liquid, 46–49

Poly(ethylene oxide) (PEO)
CO_2/H_2 separations, 12*t*
permeability and selectivity, 11–12
See also Gas separations

Poly(2-hexyne) (P-2H), mixed-gas *n*-butane/methane selectivity vs. *n*-butane permeability, 221*f*

Poly(imide siloxane)s (PISs)
preparation, 191–192
preparation of $C–SiO_2$ membranes, 192–193
See also Carbon-silica membranes

Polyimides
applications, 383
CO_2 and CH_4 permeation data in, 109*t*
hollow fibers, 383
micropatterning and nanotechnology, 384
mixed gas CO_2 permeance and CO_2/CH_4 selectivity, 107, 108*f*
sulfonated, 253–254
See also Fluorinated polyimide membranes; Sulfonated polyimides (SPIs)

Polyimide/silica hybrid membranes
correlation between ortho-positronium (oPs) free volume element size and oxygen permeability, 247, 248*f*
decrease in permeability of 3-aminopropyltriethoxysilane (APTEOS) crosslinked polyimide vs. penetrant critical volume, 241, 244*f*
density, 240
density data, 245–246
differential scanning calorimetry (DSC) results for pure tetraethoxysilane (TEOS) and TEOS-polyimide hybrid, 249, 250*f*
effect of silica domains on packing and mobility of polymer chains, 246–247
experimental, 237, 240–241
gas permeability, 237, 240
increase in methane permeability of hybrid materials vs. APTEOS crosslinked polyimide, 241, 244*f*
logarithm of permeability coefficients of gases vs. reciprocal oPs accessible volume for polyimides and hybrids, 248*f*
materials, 237
morphology of silica domains by transmission electron microscopy (TEM), 241, 242*f*, 243*f*
oPs probe evaluating accessible free volume, 247
PALS (positron annihilation lifetime spectroscopy) and Tg results by dynamic mechanical thermal analysis (DMTA), 247*t*
permeability, PALS free volume, Tg, and ideal selectivity for hybrids, 249, 251*t*
permeability and accessible free volume, 247, 249
permeability data for penetrants in polyimides and hybrids, 246*t*
polymer density, 245–246
positron annihilation (PALS), 240
procedure for film formation, 241, 242*f*
synthesis scheme, 238*f*, 239*f*
tetraethoxysilane (TEOS) alkoxide and permeability, 241
thermal analysis, 240–241
transmission electron microscopy (TEM), 240, 245*t*

Poly(*N*-isopropylacrylamide)
(polyNIPAAm)
lower critical solution temperature
(LCST), 325, 328*f*
See also Temperature-responsive
charged membranes
Polymer blends
comparing model with experimental
data, 68–71
compatibility, 341
dilution of polymers, 339
solubility isotherms of CO_2 in
polystyrene–
tetramethylpolycarbonate (PS–
TMPC) blends, 71*f*
solubility isotherms of CO_2 in pure
PS and pure TMPC, 70*f*
solubility isotherms of methane in
PS–TMPC blends, 69*f*
solubility isotherms of methane in
pure PS and pure TMPC, 68*f*
See also Ion-exchange membranes;
Non equilibrium lattice fluid
(NELF) model
Polymer chain packing, polysulfones,
161–162
Polymer density
internal variable of state, 61–63
model for binary polymer blends,
66
polyimide/silica hybrid membranes,
240, 245–246
Polymer dilation, non equilibrium
lattice fluid (NELF) correlations,
64–65
Polymer/gas interactions, gas
solubility, 3
Polymeric solvents, gas and vapor
solubility, 40
Polymer/inorganic hybrid membranes
background, 219–221
diffusivity selectivity, 220–221
disadvantage of poly(1-
trimethylsilyl-1-propyne)
(PTMSP), 219

effect of filler content on mixed-gas
permeation properties of poly(4-
methyl-2-pentyne) (PMP)/silica,
229–232
effect of filler on pure-gas
permeation properties of
PMP/silica, 224, 226
effect of filler type on mixed-gas
permeation properties of
PMP/silica, 227–229
experimental, 222–223
Fick's law of diffusion, 219–220
gas permeability, 220
ideal selectivity, 220
mechanism and applicability, 230
membrane selectivity, 220
mixed-gas *n*-butane/methane
selectivity ratio vs. *n*-butane
permeability ratio of filled/unfilled
membrane, 231*f*
mixed-gas *n*-butane/methane
selectivity vs. *n*-butane
permeability, 221*f*
mixed-gas *n*-butane/methane
selectivity vs. *n*-butane
permeability for various polymers
and filled PMP, 232*f*
mixed-gas permeation
measurements, 223
mixed-gas permeation properties of
PMP and PMP/silica, 228*f*,
229*f*
mixed-gas permeation properties of
PMP and PMP/silica vs. filler
content, 230*f*
permeability of membrane, 224
physical properties of filler types,
223*t*
polymer film preparation, 222–223
polymer synthesis, 222
pure-gas nitrogen permeability of
filled PMP films vs. nano-sized
filler content, 225*f*
pure-gas permeability of filled PMP
films vs. filler content, 225*f*

pure-gas permeation measurements, 223
pure-gas selectivity of filled PMP films vs. silica filler content, 226, 227*f*
schematic of PMP and silica-filled PMP, 226*f*
solubility selectivity, 220–221
See also Polyimide/silica hybrid membranes
Polymer membranes
gas and liquid separations, 2
gas and vapor permeation, 219–221
medical field, 353
See also Membranes
Polymer network density
glassy phase, 76–77
internal state variable for system, 77
Polymer–penetrant interactions, solution-diffusion membranes, 40–41, 52
Poly(methyl methacrylate) (PMMA). *See* Microphase-separated membranes
Poly(4-methyl-2-pentyne) (PMP)
n-butane permeability and *n*-butane/methane selectivity, 15*f*
diffusion coefficients and permeability, 13–14
mixed-gas *n*-butane/methane selectivity vs. *n*-butane permeability, 221*f*
pure- and mixed-gas permeation, 168
selectivity and permeability, 219
synthesis, 222
See also Gas separations; Polymer/inorganic hybrid membranes
Polyorganophosphazenes
hybrid polymers, 336
See also Ion-exchange membranes
Poly(perfluoro alkyl acrylate)-*g*-poly(dimethylsiloxane) (PFA-*g*-PDMS). *See* Microphase-separated membranes
Polyphosphazenes

background, 178–179
chemical description of HPP1 and HPP2, 180
degree of swelling, 181
estimation of Hansen parameters, 182–183
experimental, 179–181
future work, 188–189
general structure for HPP, 178*f*
group contribution parameters for nitrogen and phosphorus, 182*t*
Hansen parameter determinations, 180–181
Hansen parameters and molar volumes from group contributions, 182*t*
Hansen parameters characterizing interactions, 179
Hansen solubility parameter plot for HPP1, 185*f*
Hansen solubility parameter plots, 184*f*
membrane formation, 181
methods and materials, 179–180
2-(2-methoxyethoxy)ethanol (MEE) groups, 178–179
permeability, 34
pervaporation experiments, 181, 186, 188
polymer synthesis of MEEP and poly[bis(4-methoxyphenoxy)phosphazene] (PMEOPP), 180
solubility characteristics of HPP1 and HPP2, 185
solvent classifications, 183
sorption behavior, 183, 185
sorption experiments, 180–181
structure and properties, 36*t*
structures, 178*f*
sulfonation, 337
variable temperature pervaporation through membranes from HPP1 and HPP2, 186*f*
water/methanol separation using HPP1, 187*f*

water/2-propanol separation using HPP1, 187*f*

See also Ion-exchange membranes

Poly(propylene) (PP)

attenuated total reflectance (ATR) spectra of acrylic acid (AA)-grafted PP, 285*f*

comparison of flux, dextran rejection, and protein adsorption before and after AA grafting, 298*t*

effect of grafting time on bovine serum albumin (BSA) adsorption to, 290, 291*f*

effect of grafting time on contact angle of AA-grafted PP, 290, 291*f*

effect of monomer concentration, 290, 293, 294*f*

effect of photoinitiator concentration, 293, 295*f*

effect of reaction temperature, 293, 295*f*, 297*f*

effect of solvent, 296

effect of solvent on BSA adsorption, 296, 298*f*

effect of solvent on performance, 296, 297*f*

effect of UV grafting time, 284, 289*f*, 290

effect of UV irradiation distance, 290, 292*f*

materials, 283

membrane characterization, 283–284

surface modification, 282–283

UV grafting reaction, 283

X-ray photoelectron spectroscopy (XPS) spectra of AA-grafted PP, 287*f*

See also Surface modification by UV grafting

Poly(propylene oxide) (PPO)

adjustable constants, 147

apparent diffusion coefficient, 142

apparent solubility coefficient, 142

characterization, 140, 142, 143

diffusivity of gases in silane-crosslinked PPO and

poly(dimethylsiloxane) (PDMS) vs. critical volume, 144, 146*f*

diffusivity selectivity of gases relative to N_2 in PPO and PDMS, 151*t*

experimental, 140, 142

fractional free volume (FFV), 140, 142

gas diffusivity, 144, 147

gas permeability, 143–144

gas permeation measurements, 142

gas selectivity, 151–152

gas solubility, 147, 149

ideal selectivity, 142

overall ideal selectivity of gases relative to N_2 in PPO and PDMS, 151*t*

parameter for polymer–penetrant interactions and polymer free volume, 149

permeability of gases in silane-crosslinked, and PMDS vs. critical volume, 145*f*

preparation of silane-crosslinked PPO films, 140, 141

properties of silane-crosslinked PPO and poly(dimethylsiloxane) films, 143*t*

ratio of diffusivity in PDMS to PPO vs. critical volume, 147, 148*f*

ratio of gas solubility in PDMS to PPO vs. critical temperature, 149, 150*f*

ratio of permeability in PDMS to PPO vs. critical volume, 144, 145*f*

relationship between gas diffusivity and critical volume, 144, 147

repeat units of PPO, 141*f*

solubility of gases in silane-crosslinked PPO and PDMS vs. critical temperature, 150*f*

solubility selectivity of gases relative to N_2 in PPO and PDMS, 152*t*

structures, 141*f*

uses, 140

Polypyrrole membranes, gas permeability, 301

Polysilanes
amorphous cell construction and characterization, 26, 28
atomistic simulations, 25–26
COMPASS force field, 26
computational methods, 26, 28
neutral charge groups, 27*t*
properties and potential uses, 25
structures and properties, 27*t*
See also Silicon-based polymers

Polysiloxanes
atomistic simulations for gas diffusivity and solubility, 25–26
See also Poly(dimethylsiloxane) (PDMS); Silicon-based polymers

Polystyrene (PS)
solubility isotherms of CO_2 in PS–tetramethylpolycarbonate (TMPC) blends, 71*f*
solubility isotherms of CO_2 in pure PS, 70*f*
solubility isotherms of methane in PS–TMPC blends, 69*f*
solubility isotherms of methane in pure PS, 68*f*
See also Polymer blends

Polysulfones (PSF)
attenuated total reflectance spectra of acrylic acid (AA)-grafted PSF, 286*f*
bisphenol A polysulfone, 155
chain conformation, 163–164
characterization methods, 155–156
CO_2/N_2 gas pair, 160*f*, 161
comparing experimental infinite dilution solubility coefficient with predictions for CO_2 in, 83, 85*f*
comparison of flux, dextran rejection, and protein adsorption before and after AA grafting, 298*t*
conformational analysis, 163–164
correlation between $P_{(CO2)}$ and d-spacing for various, 163*f*

cross sectional images of AA-grafted PSF, 288*f*
differential scanning calorimetry (DSC), 156
diffusion coefficients of gases in, 3, 4*f*
diffusivity selectivity, 5
effect of grafting time on bovine serum albumin (BSA) adsorption, 290, 291*f*
effect of trimethylsilyl (TMS) substitution level on d-spacing, 162*f*
effect of UV grafting time, 284, 289*f*, 290
effect of UV irradiation distance, 290, 292*f*
experimental, 155–156
fluoropolymer Hyflon AD 80 as coating layer for, 119, 120*f*, 121
gas transport, 159, 161
glass transition temperature, 159
materials, 283
membrane characterization, 283–284
molecular modeling, 163–164
parameter values, 121*t*
permeability coefficients of gases in, 3, 4*f*
permselectivity vs. permeability for CO_2/N_2 gas pair, 160*f*
permselectivity vs. permeability for O_2/N_2 gas pair, 160*f*
physical property data, 159*t*
polymer chain packing, 161–162
preparation of TMS derivatives by bromination-lithiation, 157
preparation of TMS derivatives by direct lithiation, 158
specific volume and TMS groups, 162
steady-state pressure rate, 156
syntheses of tetramethylbisphenol-A polysulfone (TMPSf) and tetramethylbiphenol polysulfone (TMPPSf), 155, 156, 158

ultraviolet (UV) grafting of AA to, 282–283, 296
UV grafting reaction, 283
X-ray diffraction method, 156
See also Glassy polymers; Natural gas separations; Poly(propylene) (PP); Surface modification by UV grafting; Surface-modified polysulfone (PSf) membranes
Polytrifluoropropylmethylsiloxane (PTFPrMS)
permeability, 34
structure and properties, 36*t*
Poly(1-trimethylsilyl-1-propyne) (PTMSP)
mixed-gas *n*-butane/methane selectivity vs. *n*-butane permeability, 221*f*
pure- and mixed-gas permeation, 168
selectivity and permeability, 219
solubility-selective, 13
See also Free volume; Gas separations
Poly(trimethylsilyl-*p*-styrene) (PTMSS). *See* Free volume
Poly(vinyl alcohol) (PVA)
CO_2/H_2 separations, 12*t*
lower critical solution temperature (LCST), 325, 328*f*
thermodynamic analysis of gas sorption, 41
See also Temperature-responsive charged membranes
Poly(vinyl chloride) (PVC)
effect of plasticization on diffusion coefficients, 107, 108*f*
thermodynamic analysis of gas sorption, 41
Poly(vinylidene fluoride) (PVDF), liquid separations, 16, 18*f*
Positron annihilation lifetime spectroscopy (PALS)
free volume, 235
free volume investigation, 92*t*
free volume of glassy polymers, 93–95

free volume of hybrids, 251*t*
measurements in hybrid membranes, 240
parameters of free volume size distribution, 95*t*
results for hybrid polyimides, 247*t*
See also Free volume; Polyimide/silica hybrid membranes
Pressure
partial, condition for gas transport through polymer, 117–119
model for binary polymer blends, 59, 60, 67
Proteins
adsorption of single proteins on modified polysulfone (PSf) membranes, 374*f*
adsorption of total plasma proteins on modified PSf, 376*f*
assay, 371
ELISA assay, 375, 377*f*
standardized adsorbed proteins on PSf, 377*f*
surface-modified PSf membranes, 373, 374*f*, 375
See also Surface-modified polysulfone (PSf) membranes
Proton conductivity
ion-exchange membrane, 345, 347*f*, 348
measurement, 256, 338
relative humidity dependence, 260*f*
sulfonated polyimides, 257*t*, 258, 261
temperature dependence, 262*f*
See also Sulfonated polyimides (SPIs)
Pseudoequilibrium solubility, polymer blends, 64
Pure-gas permeation
effect of filler for poly(4-methyl-2-pentyne) (PMP)/silica hybrid membranes, 224–226
experiments for polymer/inorganic hybrid membranes, 223

poly(*p-t*-butyl diphenylacetylene)
(PptBDPA), 170–171
See also Poly(*p-t*-butyl
diphenylacetylene) (PptBDPA);
Polymer/inorganic hybrid
membranes

Q

Quaternization
characterization of modified
membrane by, 363*t*
effect of ratio on water-dissolved
oxygen permeability coefficients,
363*f*, 364
procedure for modification of
copolymer membranes, 353, 354*f*
properties of modified copolymer
membranes, 360, 364
See also Hydrophilic membranes
with hydrophobic side groups

R

Reorganization energies
polymer, 47–48
solvent, 40, 53
Reverse osmosis (RO)
membrane process, 14
membranes for liquid separations,
17, 18*f*
surface modification of membranes,
17, 19*f*, 20
See also Liquid separations
Rubbery phases, thermodynamic
equilibrium, 56
Rubbing, mechanical. *See* Fluorinated
polyimide membranes

S

Scanning electron microscopy (SEM)
field emission scanning electron

microscopy (FESEM) of
hyperbranched PAA films, 272–
273
ion-exchange membrane, 341, 344*f*,
345
platelets adhering to polysulfone
membranes, 378*f*, 379*f*
surface morphology of composite
membrane, 315, 316*f*
surface of coated membrane, 19*f*
Selectivity
calculation, 170
calixarene-containing membranes,
420, 422
definition, 194
ideal, 3, 142, 220
ideal, for hybrids, 251*t*
O_2 permeability vs. O_2/N_2, of carbon-
silica membranes, 200*f*
poly(propylene oxide) and
poly(dimethylsiloxane), 151–152
strategy for membranes with high,
193–194
See also Natural gas separations;
Poly(*p-t*-butyl diphenylacetylene)
(PptBDPA)
Self-diffusion coefficient, equation,
28–29
Separation factors
endocrine disruptors and physical
properties, 404–405
feed solution temperature and, 400,
401*f*
pervaporation, 397
relationship between, and molecular
weight, 402, 403*f*
relationship between, of diffusion
and sorption, 402, 403*t*
See also Endocrine disruptors
Separations
advantages of membranes, 204
carbon dioxide, 140
mixed-gas permeation of poly(*p-t*-
butyl diphenylacetylene) for
olefin/nitrogen separation, 173*t*
solubility control, 178

water/2-propanol through
polyphosphazene membrane, 187*f*
water/methanol through
polyphosphazene membrane, 187*f*
See also Carbon membranes from
phenolic resin; Carbon-silica
membranes; Gas separations;
Liquid separations; Natural gas
separations; Polyphosphazenes
Silica. *See* Carbon-silica membranes;
Polyimide/silica hybrid membranes;
Polymer/inorganic hybrid
membranes
Silicon-based polymers
amorphous cell construction and
characterization, 26, 28
chain flexibility, 31, 34
chain flexibility analysis, 29–30
comparing simulated and
experimental permeabilities, 35*t*
COMPASS force field for
simulation, 26
computational methods, 26, 28
diffusion and solubility, 30–31
free volume, 29, 31
gas solubility coefficients vs.
Lennard–Jones potential well
parameter, 31, 32*f*
methyl group vectorial
autocorrelation functions
(VACFs), 33*f*
neutral charge groups, 27*t*
non-methyl group VACFs, 35*f*
permeability, 34
self-diffusion coefficient, 28
self-diffusion coefficient vs. effective
gas diameter from simulation, 32*f*
simulation of gas diffusion, 28–29
simulation of gas solubility, 29
solubility coefficient, 29
structures and properties, 27*t*
structures and properties of inorganic
polymers, 36*t*
VACF of main chain, 33*f*
Simulations

amorphous cell construction and
characterization, 26, 28
atomistic, of polysilanes, 25
COMPASS force field for, of
polysilanes, 26
diffusion and solubility of silicon
polymers, 30–31
gas diffusion, 28–29
gas solubility, 29
neutral charge groups for silicon
polymers, 27*t*
polysiloxanes, 25–26
See also Silicon-based polymers
Size distribution
free volume of glassy polymers, 100
poly(trimethylsilyl-1-propyne)
(PTMSP) packing models, 98*f*
poly(trimethylsilyl-*p*-styrene)
(PTMSS) packing models, 99*f*
See also Free volume
Solubility
calculation for solute in glassy
polymer, 78–80
cellulose acetate membranes, 307,
308*f*
control, 178
endocrine disruptors, 399
infinite dilution in glassy polymeric
solvents, 51–52
permeability dependence, 3
poly(propylene oxide) and
poly(dimethylsiloxane), 147, 149,
150*f*
pseudoequilibrium, 64
silicon-based polymers, 30–31
simulation of gas, 29
swelling penetrants, 65
See also Hansen parameters; Natural
gas separations
Solubility coefficients
apparent, 142
fluoropolymer Hyflon AD 80, 119,
120*f*
sulfonated polyimides, 264, 266*f*,
267

Solubility isotherms, glassy polymers, 56

Solubility selectivity
CO_2/H_2, 11–12
endocrine disruptors, 399
gas separations, 8

Solute partitioning, entropy, 54

Solution-diffusion mechanism, water transport through membrane, 359–360

Solution-diffusion membranes
energetics affecting gas solubility, 41
energetics of cavity formation, 41–42
features determining entropy change, 45
free enthalpy change, 41
gas and vapor solubility studies, 40
increasing temperature in He, H_2, and Ne, 48–49
infinite dilution solubilities in glassy polymeric solvents, 51–52
penetrant binding energy, 40
penetrant binding energy to polymeric fluid, 47–48
polymer–penetrant interactions, 40, 52
polymer reorganization energies, 47–48
solubility, enthalpy, and entropy, 46
solute binding energy in PE vs. solute energy parameter, 49, 50f
solvation enthalpies and entropies, 43–46
solvation enthalpy equation, 44
solvation entropy equation, 44
solvation Gibbs energy, 43
solvation Gibbs energy, enthalpy and entropy in PE, 49, 50f
solvation thermodynamics quantities in poly(ethylene) (PE), 47t, 48t
solvent and solute contributions to thermodynamics, 40
solvent energy change and gas solubility, 45–46
solvent reorganization energy, 40, 53
theory, 42–46

thermodynamics analysis of gas sorption, 41
thermodynamics of small permanent gases in liquid poly(ethylene), 46–49

Solution-diffusion model, gas transport in polymers, 113–114, 219–220

Solvation
enthalpies and entropies, 43–46
Gibbs energy, 43
Gibbs energy, enthalpy, and entropy in poly(ethylene) (PE), 49, 50f
solubility, enthalpy, and entropy, 46
thermodynamic quantities for gases in poly(ethylene), 47t, 48t

Solvent
effect on bovine serum albumin (BSA) adsorption, 296, 298f
effect on performance of grafted PP membranes, 296, 297f
infinite dilution solubilities in glassy polymeric, 51–52
UV grafting of acrylic acid to poly(propylene) (PP) membranes, 296
See also Surface modification by UV grafting

Solvent reorganization energies, solution-diffusion membranes, 40, 53

Sorption
behavior of polyphosphazenes, 183, 185
experiments for polyphosphazenes, 180–181
relationship between separation factor of diffusion and, 402, 403t

Specific volume, trimethylsilyl (TMS) groups on polysulfones, 162

Spin probe method, free volume investigation, 92t

Stability. See Water stability

Statistical Associating Fluid Theory (SAFT)

comparing experimental infinite dilution solubility coefficient with predictions for CO_2 and Ar in polycarbonate (PC), 86, 87f

comparing experimental infinite dilution solubility coefficient with predictions for CO_2 in polysulfone (PSF), 83, 85f

equation of state (EoS), 80–81

non equilibrium SAFT model for pseudo-equilibrium gas solubility in glassy polymers, 82

pure component parameter for SAFT EoS, 85t

Structural analysis, carbon-silica membranes, 194–196

Sulfoalkylbetaine group. *See* Surface-modified polysulfone (PSf) membranes

Sulfonated poly[bis(3-methylphenoxy)phosphazene] (SPOP). *See* Ion-exchange membranes

Sulfonated polyimides (SPIs)

activity dependence of permeability (P), diffusion (D), and solubility (S) coefficients of water vapor, 266f

equation for water uptake, 256

experimental, 254, 256

ion exchange capacity (IEC), water uptake, proton conductivity, and water stability, 257t

measurements, 256

polymerization of naphthalene-1,4,5,8-tetracarboxylic dianhydride (NDTA) and sulfonated diamines, 254, 255

preparation, 254, 256

preparation and properties, 256, 258

proton conductivity, 258, 260f, 261

relative humidity dependence of proton conductivity, 260f

S, P, and D of water vapor, 267t

sorption isotherms of water vapor, 258, 259f

synthesis, 255

temperature dependence of proton conductivity, 261, 262f

type I and type II sulfonated diamines, 264, 265

water stability of membrane, 261, 263–264

water vapor permeation, 264, 267

water vapor sorption, 258, 259f

Surface free energy, poly(perfluoro alkyl acrylate)-*g*-poly(dimethylsiloxane)/poly(methyl methacrylate)-*g*-PDMS (PFA-*g*-PDMS/PMMA-*g*-PDMS) membranes, 416, 417f

Surface modification

effect on poly(methyl methacrylate)-*g*-poly(dimethylsiloxane) (PMMA-*g*-PDMS) membrane characteristics, 414, 416

membranes for liquid separations, 17, 19f, 20

See also Microphase-separated membranes

Surface modification by UV grafting

ATR (attenuated total reflectance) spectra of polysulfone (PSf) membrane UV grafted with acrylic acid (AA), 286f

ATR spectra of poly(propylene) (PP) membrane UV grafted with acrylic AA, 285f

benzophenone as initiator for photo-induced reaction, 283

comparison of flux, dextran rejection, and protein adsorption before and after UV grafting to PP and PSf, 298t

cross sectional images of AA-grafted PSf membranes, 288f

effect of bovine serum albumin (BSA) absorption for PP and PSf membranes, 291f

effect of grafting time on contact angle, 291f

effect of monomer concentration, 290, 293, 294f

effect of photoinitiator concentration, 293, 295f

effect of reaction temperature, 293, 295f

effect of reaction temperature on BSA adsorption to AA-grafted PP membranes, 297f

effect of solvent, 296, 297f

effect of solvent on BSA adsorption to AA-grafted PP membranes, 298f

effect of UV grafting time, 284, 290

effect of UV grafting time on AA-grafted PP and PSf membranes, 289f

effect of UV irradiation distance, 290, 292f

experimental, 283–284

materials, 283

membrane characterization, 283–284

membrane performance, 282–283

microfiltration membranes of PP, 283

UV grafting reaction, 283

X-ray photoelectron spectroscopy (XPS) spectra of AA-grafted PP, 287f

Surface-modified polysulfone (PSf) membranes

adsorption of single proteins, 374f

adsorption of total plasma proteins, 375, 376f

characterization, 371

chemical modification of PSF, 372–373

covalent bonding of poly(sulfoalkylbetaine) (SPE), 367

dependence of polymerized number of SPE on molar ratio of SPE to N-succinimidylacrylate (NSA), 374f

ethylenediamination of chloromethylated PSF membranes, 369

experimental, 367–372

materials, 367–368

modifications, 367

NSA synthesis, 368

number of platelets adhering to, 375, 380f, 381

platelet adhesion, 375, 381

platelet adsorption on membranes, 372

preparation of NSA–PFS membranes, 368, 370, 371

protein adsorption, 373, 375

protein adsorption assay on membranes, 371

scanning electron micrographs (SEM) of platelets adhering to, 378f, 379f

standardized amount of adsorbed proteins, 375, 377f

suppression of platelet adhesion, 381

surface chemical modification of PSF membranes, 368, 371

surface-modified PSF hollow fibers, 367

water contact angles, 373

Swelling

calixarene-containing membranes, 420, 423f

degree of, polyphosphazenes, 181

ion-exchange membrane, 345, 347f, 348

Swelling coefficients, fitting parameter for solubility isotherm, 67–68

Swelling penetrants, non equilibrium lattice fluid (NELF) correlations, 64–65

T

Tangent hard sphere chain models, statistical associating fluid theory (SAFT), 80

Temperature

effect of feed solution, on separation factor, 400, 401f

endocrine disruptor separation, 395
gas solubility and critical, 112*t*
model for polymer blends, 59–60
polymer ordering, 48–49
UV grafting of acrylic acid to
poly(propylene) (PP) membranes,
293, 295*f*, 297*f*
See also Natural gas separations;
Surface modification by UV
grafting
Temperature-responsive charged
membranes
apparatus and system for permeation
experiments, 328*f*
concentration gradient of ions in
membrane, 330–331, 332*f*
control of transport modes with
temperature, 329–330
determination of effective charge
density, 326
Donnan equilibrium constant, 330–
331
effective charge density changes with
temperature, 327, 329*f*
electroneutrality condition, 331, 333
experimental, 325–327
ionic concentration changes at low-
concentration side with stepwise
temperature changes, 332*f*
ion transport modes, 325
membrane potential and effective
charge density vs. temperature,
329*f*
permeation experiments, 238*f*, 327
preparation, 325–326
preparation of sample membrane,
326
suggested mechanism for
temperature dependent charge
densities, 325, 328*f*
synthesis of temperature-responsive
polymer, 325–326
theoretical background of uphill
transport, 330–333
Tensile tests, ion-exchange
membranes, 338, 348*t*

tert-butylcalix[4]arene (CA). *See*
Microphase-separated membranes
Tetramethylbiphenol polysulfone
(TMPPSf). *See* Polysulfones (PSF)
Tetramethylbisphenol-A polysulfone
(TMPSf). *See* Polysulfones (PSF)
Tetramethylpolycarbonate (TMPC)
solubility isotherms of CO_2 in, 70*f*
solubility isotherms of CO_2 in
polystyrene (PS)–TMPC blends,
71*f*
solubility isotherms of methane in
PS–TMPC blends, 69*f*
solubility isotherms of methane in
pure TMPC, 68*f*
See also Polymer blends
Theory
gas transport in polymers, 113–114
uphill transport of ions, 330–333
Thermodynamics
energetics affecting gas solubility, 41
energetics of cavity formation, 41–42
gas sorption, 41
penetrant binding energy to
poly(ethylene) (PE) vs. solute
surface area weighted energy, 49,
50*f*
permanent gases in liquid PE, 46–49
polymer reorganization energies, 47–48
solute binding energy in PE vs.
solute energy parameter, 49, 50*f*
solvation, quantities for gases in PE,
47*t*, 48*t*
solvation enthalpies and entropies,
43–46
solvation Gibbs energy, 43
solvation Gibbs energy, enthalpy,
and entropy in PE, 49, 50*f*
solvent and solute contributions, 40
See also Solution-diffusion
membranes
Time. *See* Grafting
Transition State Theory (TST), free
volume, 29
Transmission electron microscopy
(TEM)

carbon-silica membranes, 199, 200*f*
observations for hybrid membranes, 245*t*
poly(perfluoro alkyl acrylate)-*g*-poly(dimethylsiloxane) /poly(methyl methacrylate)-*g*-PDMS (PFA-*g*-PDMS/PMMA-*g*-PDMS) membranes, 418, 419*f*
polyimide/silica hybrid membranes, 241, 242*f*, 243*f*
Transport modes
concentration gradient of ions in membrane for uphill, 330–331
control in response to temperature, 329–330
downhill and uphill, 325
electroneutrality condition, 331, 333
schematic of concentration gradients of Ca^{2+}, K^+, and Cl^- ions, 332*f*
theoretical background for uphill, 330–333
See also Temperature-responsive charged membrane
Trimethylsilyl (TMS) groups
effect on d-spacing, 162*f*
polymer chain packing in polysulfones, 161–162
preparation of TMS polysulfones by bromination-lithiation, 157
preparation of TMS polysulfones by direct lithiation, 158
specific volume and, 162
synthesis of substituted polysulfones, 156, 158
See also Polysulfones (PSF)

U

Ultrafiltration (UF)
membrane process, 14
See also Liquid separations
Ultrathin skins. *See* Composite membranes with ultrathin skins
Ultraviolet (UV) grafting. *See* Surface modification by UV grafting

Uphill transport
concentration gradient of ions, 330–331
electroneutrality condition, 331, 333
ions, 325
schematic of concentration gradients of Ca^{2+}, K^+, and Cl^-, 332*f*
theory, 330–333
See also Temperature-responsive charged membranes

V

Vapor/gas separation. *See* Polymer/inorganic hybrid membranes
Vapor permeation
benzene/cyclohexane through NaY zeolite, 214, 215*f*
carbon membrane performance for benzene/cyclohexane separation, 212*f*
effect of benzene feed concentration for benzene/cyclohexane separation, 211*f*
performance of carbon membrane from phenolic resin, 209, 214*t*
sulfonated polyimides, 264, 267
See also Carbon membranes from phenolic resin
Vapor solubility, solution-diffusion membranes, 40
Vectorial autocorrelation function (VACF)
chain flexibility analysis, 29–30
chain flexibility for silicon polymers, 31, 34
methyl group VACFs, 33*f*
non-methyl group VACFs, 35*f*
plot of VACF of main chain of silicon polymers, 33*f*
See also Silicon-based polymers
4-Vinylpyridine (4VP). *See* Hydrophilic membranes with hydrophobic side groups

N-Vinyl-2-pyrrolidone (NVP). *See*
Hydrophilic membranes with
hydrophobic side groups
Volatile organic compounds (VOCs)
removal, 412
See also Microphase-separated
membranes
Volume, gas diffusion coefficients
and critical, 113–114

W

Water
endocrine disruptor separation, 395
relationship between absorbed, and
hydrophilic monomer mole
fraction, 356, 358*f*
weight of freezing, 356, 359
See also Hydrophilic membranes
with hydrophobic side groups
Water contact angles
fluorinated polyimide membranes,
386, 388*t*
poly(perfluoro alkyl acrylate)-*g*-
poly(dimethylsiloxane)
/poly(methyl methacrylate)-*g*-
PDMS (PFA-*g*-PDMS/PMMA-*g*-
PDMS) membranes, 416, 417*f*
surface-modified polysulfone
membranes, 373
Water/ethanol
pervaporation and vapor permeation,
214*t*
See also Carbon membranes from
phenolic resin
Water/methanol separation,
polyphosphazene membrane, 187*f*
Water/2-propanol separation,
polyphosphazene membrane, 187*f*
Water stability
polyimide membranes, 261, 263–264
sulfonated polyimides (SPIs), 257*t*
Water transport, hydrophilic
membranes with hydrophobic side
groups, 359–360

Water uptake, equation, 256
Water vapor permeation, sulfonated
polyimides, 264, 267
Water vapor sorption
isotherms of water vapor, 259*f*
measurement, 256
sulfonated polyimides, 258
Wide-angle X-ray scattering (WAXS),
ion-exchange membrane, 345, 346*f*

X

[129]Xe-nuclear magnetic resonance,
free volume investigation, 92*t*
Xenon
CO_2 and Xe sorption isotherms in
annealed and melt-cooled liquid
crystalline polyester, 135*f*
diffusion coefficients for liquid
crystalline polyester, 136, 137*t*
gas sorption and diffusion in liquid
crystal polyester, 134, 136
See also Liquid crystalline polyesters
X-ray diffraction (XRD)
characterization of liquid crystalline
polyester, 131, 133*f*, 134
correlation between $P_{(CO2)}$ and d-
spacing for polysulfones, 163*f*
method for polysulfones, 156
polymer chain packing for
polysulfones, 161–162
See also Liquid crystalline
polyesters; Polysulfones (PSF)
X-ray photoelectron spectroscopy
(XPS), cation-exchange composite
membrane, 315, 317*f*

Z

Zwitterionic sulfoalkylbetaine group.
See Surface-modified polysulfone
(PSf) membranes